BIOLOGY

Fourth Edition

Sylvia S. Mader

Part One
The Cell

Part Two
Genetic Basis of Life

Part Three
Evolution and Diversity

Part Four
Plant Structure and Function

Part Five
Animal Structure and Function

Part Six
Behavior and Ecology

WCB **Wm. C. Brown Publishers**
Dubuque, Iowa • Melbourne, Australia • Oxford, England

Book Team

Editor *Kevin Kane*
Developmental Editor *Carol Mills*
Production Editors *Anne E. Scroggin/Ann Fuerste*
Designer *Mark Elliot Christianson*
Art Editor *Miriam J. Hoffman*
Photo Editor *Lori Gockel*
Permissions Editor *Karen Storlie*
Art Processor *Andréa Lopez-Meyer*

Wm. C. Brown Publishers
A Division of Wm. C. Brown Communications, Inc.

Vice President and General Manager *Beverly Kolz*
National Sales Manager *Vincent R. Di Blasi*
Assistant Vice President, Editor-in-Chief *Edward G. Jaffe*
Marketing Manager *Paul Ducham*
Advertising Manager *Amy Schmitz*
Managing Editor, Production *Colleen A. Yonda*
Manager of Visuals and Design *Faye M. Schilling*
Design Manager *Jac Tilton*
Art Manager *Janice Roerig*
Publishing Services Manager *Karen J. Slaght*
Permissions/Records Manager *Connie Allendorf*

Wm. C. Brown Communications, Inc.

Chairman Emeritus *Wm. C. Brown*
Chairman and Chief Executive Officer *Mark C. Falb*
President and Chief Operating Officer *G. Franklin Lewis*
Corporate Vice President, President of WCB Manufacturing *Roger Meyer*

Wild horses gallop across the Patagonian plains of South America's Argentine coast at the foot of the Fitzroy Mountains. In 1832, Charles Darwin came here aboard the HMS Beagle, captained by Robert Fitzroy. During his 5-year, around-the-world voyage on the Beagle, Darwin gathered data that would later support his theory of evolution.

Much has been learned since Darwin's day about the 65-million-year evolution of the horse, which began with a much smaller ancestor that had many toes and teeth with low crowns. As grasslands replaced a forestlike environment, the horse evolved into a much larger animal with fewer toes and teeth with high crowns. Today, the horse has only one toe and is highly adapted for running fast and far, allowing it to seek new sources of food and water when needed.

Cover Credits
Biology Front cover photo on complete text and background photo for all parts © Galen Rowell/Mountain Light Photography
Part 1: The Cell Front cover inset photo © Carolina Biological Supply Co.
Part 2: Genetic Basis of Life Front cover inset photo © Gunter Ziesler/Peter Arnold, Inc.
Part 3: Evolution and Diversity Front cover inset photo © Douglas Faulkner/Photo Researchers, Inc.
Part 4: Plant Structure and Function Front cover inset photo © Astrid and Hahns-Frieder Michler/ Science Photo Library/Photo Researchers, Inc.
Part 5: Animal Structure and Function Front cover inset photo © Runk/Schoenberger/Grant Heilman
Part 6: Behavior and Ecology Front cover inset photo © Carl R. Sams II/Peter Arnold, Inc.

Copy Edited by Laura Beaudoin
Photo Research by Michelle Oberhoffer

The Credits section for this book begins on page C-1 and is considered an extension of the copyright page.

Library of Congress Catalog Card Number: 91-76060

ISBN *Biology* Casebound 0-697-12383-9
ISBN *Biology* Paper binding 0-697-15096-8
ISBN *Part 1: The Cell* 0-697-15098-4
ISBN *Part 2: Genetic Basis of Life* 0-697-15099-2
ISBN *Part 3: Evolution and Diversity* 0-697-15100-X
ISBN *Part 4: Plant Structure and Function* 0-697-15101-8
ISBN *Part 5: Animal Structure and Function* 0-697-15102-6
ISBN *Part 6: Behavior and Ecology* 0-697-15103-4
ISBN *Biology* Boxed set: 0-697-15097-6

Printed in the United States of America by Wm. C. Brown Communications, Inc., 2460 Kerper Boulevard, Dubuque, IA 52001

10 9 8 7 6 5 4

PUBLISHER'S NOTE TO THE INSTRUCTOR

Binding Option	Description	ISBN
Biology, casebound	The full-length text (chapters 1–50), with hard-cover binding.	0-697-12383-9
Biology, paperbound	The full-length text, paperback covered and available at a significantly reduced price, when compared with the casebound version.	0-697-15096-8
Part 1 The Cell, paperbound	Part 1 features the first unit or chapters 1–9 of the text, covering the scientific method, evolution and unity of life, basic chemistry, cell biology, photosynthesis, and respiration. This paperback option is available at a significantly reduced price when compared with both the full-length casebound and paperbound versions.	0-697-15098-4
Part 2 Genetic Basis of Life	Part 2 features chapters 10–18 and covers cell reproduction, Mendelian and molecular genetics, gene activity, and recombinant DNA and biotechnology. This paperback is also available at a significantly reduced price when compared with the full-length versions of the text.	0-697-15099-2
Part 3 Evolution and Diversity	Part 3 features chapters 19–28 on Darwin and evolution, genetics in evolution and diversity. Paperbound, it is available for a fraction of the full-length casebound or paperbound prices.	0-697-15100-X
Part 4 Plant Structure and Function	Part 4 features chapters 29–32 on plant structure and function. Paperbound, it is also available for a fraction of the full-length casebound or paperbound prices.	0-697-15101-8
Part 5 Animal Structure and Function	Part 5 features chapters 33–44 on animal systems and reproduction and development. Paperbound, it also sells for a fraction of the full-length book price.	0-697-15102-6
Part 6 Behavior and Ecology	Part 6 features chapters 45–50 on behavior and ecology. Paperbound, it sells for a fraction of the full-length book price.	0-697-15103-4
Biology, the Boxed Set	The entire text, offered in an attractive, boxed set of all six paperbound "splits." It is available at the same price as the full-length casebound text.	0-697-15097-6

Binding Options and Recycled Paper

Biology—in all its numerous binding options (listed here)—is printed on **recycled paper stock**. All of its ancillaries, as well as all advertising pieces will be printed on recycled paper, subject to market availability.

Our goal in offering the text and its ancillary package on **recycled paper** is to take an important first step toward minimizing the environmental impact of our products. If you have any questions about recycled paper use, Biology, its package, any of its binding options, or any of our other biology texts, feel free to call us at 1-800-331-2111. Thank you.

Kevin Kane
Executive Editor
Life Sciences

BRIEF CONTENTS

CONTENTS

READINGS

THE BIOLOGY LEARNING SYSTEM

3

Basic Chemistry

Your study of this chapter will be complete when you can

1. name and describe the subatomic particles of an atom, indicating which one accounts for the occurrence of isotopes;
2. describe and discuss the energy levels (electron shells) of an atom, including the orbitals of the first 2 levels;
3. draw a simplified atomic structure of any atom with an atomic number less than 20;
4. distinguish between ionic and covalent reactions, and draw representative atomic structures for ionic and covalent molecules;
5. tell which atom has been reduced and which has been oxidized in a particular reaction;
6. describe the chemical properties of water, and explain their importance for living things;
7. define an acid and a base; describe the pH scale, and state the significance of buffers.

A hippopotamus keeps cool by remaining in water. Water has many biological functions, both without and within organisms. Cells are largely composed of this inorganic molecule, which facilitates chemical reactions and helps maintain a normal body temperature because it is slow to heat.

24

Concept Summaries

At the ends of major sections within each chapter, concept summaries briefly highlight key concepts in the section, helping students focus their study efforts on the basics.

Bold-Faced Key Terms

Important terms are boldfaced and defined on first mention.

Behavioral Objectives

Each chapter begins with a list of behavioral objectives designed to help students identify the major concepts of the chapter. Their study of the chapter is complete when they can satisfy these objectives.

Text Line Art

Graphic diagrams placed within textual passages help clarify difficult concepts and enhance learning.

HAVE YOU THANKED A GREEN PLANT TODAY? Plants dominate our environment, but most of us spend little time thinking about the various services they perform for us and for all living things. Chief among these is their ability to carry on photosynthesis, during which they use sunlight (*photo*) as a source of solar energy to produce food (*synthesis*):

Other organisms, called algae, are also photosynthetic. Algae (chap. 24) are a diverse group, but many are water dwelling, microscopic organisms related to plants.

The organic food produced by photosynthesis is not only used by plants themselves, it is also the ultimate source of food for all other living things (fig. 8.1). Also, when organisms convert carbohydrate energy into ATP energy they make use of the oxygen (O_2) given off by photosynthesis. Nearly all living things are dependent on atmospheric oxygen derived from photosynthesis.

At one time in the distant past, plant and animal matter accumulated without decomposing. This matter became the fossil fuels (e.g., coal, oil, and gas) that we burn today for energy. This source of energy, too, is the product of photosynthesis.

Photosynthesis is absolutely essential for the continuance of life because it is the source of food and oxygen for nearly all living things.

Sunlight

Photosynthesis is an energy transformation in which solar energy in the form of light is converted to chemical energy within carbohydrate molecules. Therefore, we will begin our discussion of photosynthesis with the energy source—sunlight.

Solar radiation can be described in terms of its energy content and its wavelength. The energy comes in discrete packets called **photons**. So, in other words, you can think of radiation as photons that travel in waves:

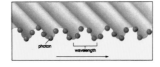

Figure 8.2*a* illustrates that solar radiation, or the **electromagnetic spectrum,** can be divided on the basis of wavelength—gamma rays have the shortest wavelength and radio waves have the longest

Figure 8.1
This squirrel is a herbivore. It feeds directly on plant material produced by a photosynthesizer. Carnivores, such as a hawk that may feed on this squirrel, are also dependent, although indirectly, on food produced by photosynthesizers.

wavelength. The energy content of photons is inversely proportional to the wavelength of the particular type of radiation; that is, short-wavelength radiation has photons of a higher energy content than long-wavelength radiation. High-energy photons, such as those of short-wavelength ultraviolet radiation, are dangerous to cells because they can break down organic molecules. Low-energy photons, such as those of infrared radiation, do not damage cells because they only increase the vibrational or rotational energy of molecules; they do not break bonds. But photosynthesis utilizes only the portion of the electromagnetic spectrum known as **visible light.** (It is called visible light because it is the part of the spectrum that allows us to see.) Photons of visible light have just the right amount of energy to promote electrons to a higher electron shell in atoms without harming cells.

We have mentioned previously that only about 42% of the solar radiation that hits the earth's atmosphere ever reaches the surface, and most of this radiation is within the visible-light range.

122

The Cell

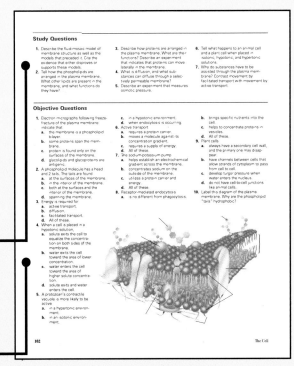

Readings

Throughout *Biology*, selected readings reinforce major concepts in the book. Most readings are written by the author, but a few are excerpted from popular magazines. A reading may provide insight into the process of science or show how a particular kind of scientific knowledge is applicable to the students' everyday lives.

Study Questions

These questions appear at the end of each chapter. They call for specific, short essay answers that challenge students' mastery of the chapter's basic concepts.

Objective Questions

Multiple-choice questions at the end of each chapter test basic recall of the chapter's key points. The answers to the objective questions are listed in appendix D.

Practice Problems

Practice problems at the end of passages in the genetics chapters help students master basic genetic quantifications. Answers to these problems are in appendix D.

Concepts and Critical Thinking

These end-of-chapter questions require students to apply the chapter's basic concepts to biological concerns. Writing the answers to these or the Study Questions fulfills any Writing-Across-the-Curriculum requirement.

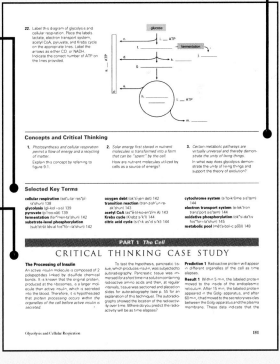

Selected Key Terms

A selected list of bold-faced, key terms from the chapter appears at the end of each chapter. Each term is accompanied by its phonetic spelling and a page number indicating where the term is introduced and defined in the chapter.

Critical Thinking Case Studies

Each part ends with a case study designed to help students think critically by participating in the process of science. At many institutions, instructors are encouraged to develop the writing skills of their students. In such cases, instructors could require students to write out their answers to the questions in each case study. Suggested answers for each of these questions appear in the Instructor's Manual for the text.

Chapter Summaries

At the end of each chapter is a summary. This listing of important points should help students more readily identify important concepts and better facilitate their learning of chapter content.

Further Readings

A list of readings at the end of each part suggests references that can be used for further study of topics covered in the chapters of that part. The items listed in this section were carefully chosen for readability and accessibility.

PREFACE

Biology is an introductory college text that covers the basic concepts and principles of general biology. While the text is clear and straightforward enough to be read by liberal arts students, it is also comprehensive and authoritative, so it can just as easily be read by science majors. The text strives to use other forms of life, in addition to humans, as examples, making frequent reference to plants and invertebrates. The text takes a hierarchical approach, proceeding from the chemistry of the cell up to the organization of the biosphere.

Science Process and Critical Thinking

Biology stresses the process of science in many ways. Chapter 2 is entirely devoted to describing the scientific method with examples. And throughout the text, experiments that have led us to our present level of knowledge are described. *New to this edition are the "How Do You Do That?" boxes, which help to emphasize technological literacy.* The boxes are meant to take the mystery out of laboratory procedures and show students that everyone is capable of utilizing the methods by which data are collected.

Understanding the concepts of biology and engaging in the process of science encourages one to think critically. *New to this edition, each chapter ends with a list of the major concepts within the chapter and asks critical thinking questions based on those concepts.* Those questions allow students to see that thinking conceptually leads to predictions and deductions concerning particular aspects of biology. Suggested answers to the end-of-chapter critical thinking questions appear in appendix D. Each major part in the text ends with a critical thinking case study. As before, the case studies were written by Dr. Robert Allen and myself. They provide students with an opportunity to use scientific methodology as they think critically. Suggested answers to the case study questions appear in the Instructor's Manual for the text.

Writing Across the Curriculum

Students need a chance to practice their writing skills and at the same time test their ability to fulfill the behavioral objectives that begin each chapter. *This is provided by the "Writing Across the Curriculum Study Questions" at the close of the chapter,* which is based on an educational approach to improving the skills of students. Instructors who have their students write out the answers to the study questions and critical thinking questions will have answered the challenge of this new approach.

The objective questions prepare students to take examinations that contain recall-based questions. These, too, have been improved, and instructors will appreciate the addition of questions that require students to complete and/or label diagrams.

Organization of the Text

Each chapter in *Biology* introduces particular concepts and describes biological experiments that have contributed to our present level of knowledge. As in the third edition, the chapters are of a suitable length to be read in one sitting. *A new chapter on animal organization and homeostasis has been added to part 5.* This chapter includes a discussion of animal tissues. The text has the following parts, which have been revised as discussed.

Introduction

Chapter 1, which concerns the characteristics of life and introduces important biological concepts, was completely rewritten to increase student interest. The chapter shows that the unity of life is due to descent from a common ancestor and that diversity is due to adaptations to particular ways of life in an ecosystem.

Chapter 2 thoroughly explains scientific methods and gives examples of both experimental and observational biological research. The biological understanding of the word *theory* is also fully explained.

Part 1 The Cell

Part 1 concerns cell structure and function including energy metabolism. *Membrane protein diversity is more thoroughly discussed and illustrated in this edition.* The first of the 3 chapters devoted to energetics gives an overview. Instructors can use just this one chapter or proceed on to cellular respiration and/or photosynthesis.

Part 2 Genetic Basis of Life

There is a strong historical emphasis in this part, but practical aspects are not neglected. Students are given an opportunity to test their ability to do problems as they proceed. New advances in the field of genetics appear daily, and this part has been thoroughly updated to reflect our changing knowledge. *The very latest that is known about human genetic disorders is included. The biotechnology chapter includes a discussion of new techniques in gene therapy in humans and the human genome project.* The presentation is at an appropriate level for the beginning student.

Part 3 Evolution and Diversity

The evolution chapters were completely reorganized and rewritten with the valuable help of many instructors and scholars who specialize in this area. The geological time scale was completely updated. Chapter 22, concerning the origin of life, was also rewritten and now includes several hypotheses on this topic. The diversity chapters were revised and the most up-to-date classification system is used. Instructors will appreciate a new expanded table that compares the major features of the plant divisions.

Part 4 Plant Structure and Function

Four chapters are devoted to flowering plant anatomy and physiology. The first chapter provides a foundation for the others on nutrition and transport, reproduction, and growth and development.

Part 5 Animal Structure and Function

This part begins with a new chapter on animal organization and homeostasis. There is a discussion of animal tissues and organ systems before homeostasis is discussed in some detail. *The comparative section that begins each of the animal physiology chapters has been rewritten for clarity and improvement. A new extended section on AIDS is included in the reproduction chapter.*

Part 6 Behavior and Ecology

A new animal behavior chapter takes a more modern approach to this topic. The ecology chapters have a logical sequence from populations to communities to the biosphere. Environmental concerns are emphasized in the last 2 chapters. Some instructors may wish to begin the year's work with this part, which is certainly a workable alternative.

Aids to the Reader

Biology was written so that students can enjoy, appreciate, and come to understand the concepts of biology and the scientific process. The following text features are especially designed to assist student learning.

Text Introduction

The introduction section (chapters 1 and 2) lays a foundation upon which the rest of the text depends. The first chapter reviews the characteristics of life and presents the fundamental concepts of biology. The second chapter explains the scientific process in detail and gives examples of studies done by biologists. Various other experiments are described throughout the text.

Study Objectives

Each chapter begins with a list of objectives designed to guide students as they study the chapter. Their study of the chapter is complete when they can satisfy these objectives, therefore, they help students prepare for an examination.

Boxed Readings

Throughout *Biology,* selected boxed readings reinforce major concepts in the book. Most readings are of general interest and heighten student involvement by expanding on a topic discussed in the chapter.

"How Do You Do That?" Boxes

New to this edition are the "How Do You Do That?" boxes, which describe laboratory methods. They are designed to take the mystery out of the manner in which biological data are collected.

Drawings, Photographs, and Tables

The drawings, photographs, and tables in *Biology* have been designed to help students learn basic biological concepts as well as the specific content of the chapters, and are consistent with multicultural educational goals. Often it is easier to understand a given process by studying a drawing, especially when it is carefully coordinated with the text, as is the case here. The photographs were selected not only to please the eye but also to emphasize specific points in the text. The tables summarize and list important information, making it readily available for efficient study.

Chapter Summaries

The summary is a numbered series of statements that follow the organization of the chapter. The summary helps students identify the concepts of the chapter and facilitates their learning of chapter content.

Chapter Questions

Three kinds of questions—study questions, objective questions, and critical thinking questions—appear at the close of each chapter. They allow students to test their ability to fulfill the study objectives. *Writing across the curriculum* recognizes that students need an opportunity to practice writing in all courses. The study questions review the chapter, and their sequence follows that of the chapter. When students write out the answers to the study questions, they are writing while studying biology. The critical thinking questions are based on biological concepts that pertain to the chapter. They show that knowledge of a biological concept allows one to reason about some particular aspect of biology. Writing out the answers to the critical thinking questions also fulfills any writing across the curriculum requirements. The objective questions allow students to test their ability to answer recall-based, multiple choice questions. The objective questions have been expanded in this edition to include questions that require the completing or labeling of diagrams. Answers to the objective questions and critical thinking questions appear in appendix D.

Selected Key Terms

Each chapter ends with a selected key term list. Key terms are boldfaced in the chapter, defined in context, and also appear in the glossary. Especially significant key terms appear in the selected key term list. Each term is accompanied by its phonetic spelling and is referenced to the page on which it is introduced and defined.

Further Readings

The list of readings at the end of each part suggests references that can be used for further study of the topics covered in the chapters of that part. The items listed in this section were carefully chosen for readability and accessibility.

Critical Thinking Case Studies

Each part ends with a case study designed to help students think critically by participating in the process of science. At many institutions, instructors are encouraged to develop the writing skills of their students. In such cases, instructors could require students to write out their answers to the questions in each case study. Suggested answers for each of these questions appear in the Instructor's Manual.

Glossary

The glossary contains the terms that are boldfaced in the text. Many of the glossary entries are accompanied by a phonetic spelling and a definition, and each is referenced to the page on which it was first introduced and defined in the text.

ADDITIONAL AIDS

Instructor's Manual/Test Item File

Revised by Jean Helgeson, the Instructor's Manual/Test Item File is designed to assist instructors as they plan and prepare for classes using *Biology*. For each chapter in the text, the Instructor's Manual provides a chapter outline, key terms, an extended lecture outline, enrichment ideas, and a listing of selected films. The Test Item File contains approximately 40 multiple choice, true/false, and critical thinking essay questions per text chapter.

Suggested answers for the critical thinking case studies that appear at the end of each part in the text are placed at the end of the corresponding parts in the Instructor's Manual.

In addition, the Instructor's Manual includes an answer key for the Critical Thinking Case Study Workbook by Robert Allen, along with listings of the transparencies and micrograph slides available to instructors using *Biology*, fourth edition.

Student Study Guide

The Student Study Guide that accompanies the text was revised by Jay Templin. For each text chapter, there is a corresponding study guide chapter that includes a list of study objectives, a chapter review, vocabulary terms that are page-referenced to the text, learning activities with text page references that correlate to the study objectives and require students to apply textual information, critical thinking questions, and chapter test questions with an answer key providing the student immediate feedback.

Micrograph Slides

A boxed set of 99 slides includes all photomicrographs and electron micrographs printed in the text.

Laboratory Manual

The Laboratory Manual that accompanies *Biology* was thoroughly revised by Kenneth Kilborn and myself. Its 34 exercises provide enough variety to meet the needs of a broad spectrum of class designs. Student aids include a list of learning objectives at the beginning of each exercise and numerous full-color illustrations throughout. Each exercise includes an introduction, and ample space is provided for students to record their observations as the lab proceeds. A laboratory review, consisting of a series of questions, ends each exercise. Answers to the review questions are provided in an appendix.

Customized Laboratory Manual

The Laboratory Manual's 34 exercises are available as individual "lab separates," so instructors can custom-tailor the manual to their particular course needs. The separates, which are published in one color, can be individually selected at a greatly reduced price and will be collated and bound by WCB on request.

Laboratory Resource Guide

Helpful and thorough information regarding each lab preparation can be found in the Laboratory Resource Guide. Completely revised by Kenneth Kilborn, the guide is designed to help instructors make the laboratory experience a more meaningful one for the student. For handy reference, a list of suppliers is printed on the inside front cover. The Resource Guide is now divided into 3 parts. They are Laboratory Preparation and Instructions; Laboratory Exercises and Expected Results; and Answers to the Laboratory Review Questions.

Transparencies and Lecture Enrichment Kit

A full set of transparency acetates also accompanies the text. These feature key illustrations from the text in both 2 and full color. They are accompanied by a Lecture Enrichment Kit, which is a set of lecture notes featuring additional high-interest information about the pictured process or concept and not presented in the text.

Critical Thinking Case Study Workbook

Written by Robert Allen, this ancillary includes 30 additional critical thinking case studies of the type found in the text. Like the text case studies, they are designed to immerse students in the "process of science" and challenge them to solve problems in the same way biologists do. The case studies here are divided into 3 levels of difficulty (introductory, intermediate, and advanced) to afford instructors greater choice and flexibility. An answer key is printed in the Instructor's Manual.

Extended Lecture Outline Software

This instructor software features extensive outlines of each text chapter with a brief synopsis of each subtopic to assist in lecture preparation. Written in ASCII files for maximum utility, it is available in IBM, Apple, or Mac formats. It is free to all adopters, upon request.

Biology Art Masters

A set of 150 art masters consisting of one-color line art with labels can be used for additional transparencies or can be copied and used for student hand-outs.

The 7 packages include the following titles: Cell Biology, Genetics, Diversity, Plant Biology, Animal Biology, Ecology and Behavior, and Evolution.

Adopters of *Biology* can order one or all 7 packages for free.

WCB TestPak with Enhanced QuizPak and GradePak

WCB TestPak, a computerized testing service, provides instructions with either a mail-in/call-in testing program or the complete test item file on diskette for use with the IBM PC, Apple, or Macintosh computer. **WCB** TestPak requires no programming experience.

WCB QuizPak, a part of TestPak, provides students with true/false, multiple choice, and matching questions for each chapter in the text. Using this portion of the program will help students to prepare for examinations. Also included with the QuizPak is an on-line testing option to allow professors to prepare tests for students to take on the computer. The computer will automatically grade the test and update a gradebook file.

WCB GradePak, also a part of TestPak, is a computerized grade management system for instructors. This program tracks student performance on examinations and assignments. It will compute each student's percentage and corresponding letter grade, as well as the class average. Printouts can be made utilizing both text and graphics.

Other Titles of Related Interest from Wm. C. Brown Publishing

You Can Make a Difference
by Judith Getis

This short, inexpensive supplement offers students practical guidelines for recycling, conserving energy, disposing of hazardous wastes, and other pollution controls. It can be shrink wrapped with the text, at minimal additional cost (ISBN 0-697-13923-9)

How to Study Science
by Fred Drewes, Suffolk County Community College

This excellent new workbook offers students helpful suggestions for meeting the considerable challenges of a college science course. It offers tips on how to take notes; how to get the most out of laboratories; as well as how to overcome science anxiety. The book's unique design helps students develop critical thinking skills, while facilitating careful note-taking. (ISBN 0-697-14474-7)

The Life Science Lexicon
by William N. Marchuk, Red Deer College

This portable, inexpensive reference helps introductory-level students quickly master the vocabulary of the life sciences. Not a dictionary, it carefully explains the rules of word construction and derivation, in addition to giving complete definitions of all important terms. (ISBN 0-697-12133-X)

Biology Study Cards
by Kent Van De Graaff, R. Ward Rhees, and Christopher H. Creek, Brigham Young University

This boxed set of 300, 2-sided study cards provides a quick yet thorough visual synopsis of all key biological terms and concepts in the general biology curriculum. Each card features a masterful illustration, pronunciation guide, definition, and description in context. (ISBN 0-697-03069-5)

The Gundy-Weber Knowledge Map of the Human Body
by G. Craig Gundy, Weber State University

This 13-disk, Mac-Hypercard program is for use by instructors and students alike. It features carefully prepared computer graphics, animations, labeling exercises, self-tests, and practice questions to help students examine the systems of the human body. Contact your local Wm. C. Brown representative or call 1-800-351-7671.

The Knowledge Map Diagrams
1. Introduction, Tissues, Integument System (0-697-13255-2)
2. Viruses, Bacteria, Eukaryotic Cells (0-697-13257-9)
3. Skeletal System (0-697-13258-7)
4. Muscle System (0-697-13259-5)
5. Nervous System (0-697-13260-9)
6. Special Senses (0-697-13261-7)
7. Endocrine System (0-697-13262-5)
8. Blood and the Lymphatic System (0-697-13263-3)
9. Cardiovascular System (0-697-13264-1)
10. Respiratory System (0-697-13265-X)
11. Digestive System (0-697-13266-8)
12. Urinary System (0-697-13267-6)
13. Reproductive System (0-697-13268-4)

Demo—(0-697-13256-0)
Complete Package—(0-697-13269-2)

GenPak: A Computer Assisted Guide to Genetics
by Tully Turney, Hampden-Sydney College

This Mac-Hypercard program features numerous interactive/ tutorial (problem-solving) exercises in Mendelian, molecular, and population genetics at the introductory level. (ISBN 0-697-13760-0)

Acknowledgments

Many persons have contributed to the success of previous editions and helped me make this edition our very best effort. My editor, Kevin Kane, directed the efforts of all. Carol Mills, my developmental editor, served as a liaison between the editor, me, and many other people. I especially want to mention Susan Murray and Anne Packard of Plymouth State College, who carefully read the galleys and made many helpful suggestions.

The production team at Wm. C. Brown was ever-faithful to their duties. Anne Scroggin and Ann Fuerste were the production editors; Miriam Hoffman the art editor; Lori Gockel the photo researcher; Michelle Oberhoffer the photo researcher; and Mark Christianson the designer. My thanks to each of them for a job well done!

The Contributors

With this edition, as with the last edition, there were several contributors to various sections of the book. They critically reviewed my revised chapters and told me how they could be improved. With their help, it was easier to make sure the content was sufficient, accurate, and up to date. The contributors were:

W. Dennis Clark (Ph.D. University of Texas), professor of botany at Arizona State University. Dr. Clark's expertise in the areas of chemistry, evolution, and physiology of plants was extremely helpful in the revision of the plant and energy chapters.

Thomas C. Emmel (Ph.D. Stanford University), professor of zoology at the University of Florida. He assisted invaluably by providing updated information in ecology, where he has published numerous articles and 2 books.

Robert M. Kitchin (Ph.D. University of California-Berkeley), professor of genetics at the University of Wyoming. He reviewed the genetics chapters and outlined new information and events taking place in this rapidly changing and expanding field.

The Ancillary Contributors

The ancillary package that accompanies *Biology* also involved the efforts of many instructors currently teaching biology.

Critical Thinking Case Study Book: Robert Allen (Ph.D. University of California-Los Angeles) is vice-president for instruction at Victor Valley College, whose special interests include research on the development of critical thinking skills in science. He is a frequent presenter on the subject for the National Science Foundation Chatauqua series.

Instructor's Manual-Test Item File: Jean Helgeson, (M. A. The University of Texas Southwestern Medical Center) is professor and coordinator of biology and health science at Collin County Community College.

Student Study Guide: Jay Templin (Ed.D. Temple University) is an instructor at Widener University who has authored numerous student and instructor's ancillaries for several college biology textbooks. He is also a co-author of the *College Board Achievement Test* in biology.

Laboratory Manual and Laboratory Resource Guide: Kenneth Kilborn (M. A. San Jose State University) has been teaching general biology and general botany in addition to 20 years of teaching in the biology laboratory at Shasta College.

QuizPak and The Lecture Enrichment Kit: Jane Aloi (Ph.D. University of California, Davis) is an ecologist teaching courses in introductory biology, zoology, and human anatomy at Saddleback College.

Reviewers

Many instructors of introductory biology courses around the country reviewed portions of the manuscript. Others assisted by taking the time to fill out survey forms that helped us make important decisions about content. With many thanks, we list their names here.

First Edition

A. Lester Allen *Brigham Young University*
William E. Barstow *University of Georgia*
Lester Bazinet *Community College of Philadelphia*
Eugene C. Bovee *University of Kansas*
Larry C. Brown *Virginia State University*
L. Herbert Bruneau *Oklahoma State University*
Carol B. Crafts *Providence College*
John D. Cunningham *Keene State College*
Dean G. Dillery *Albion College*
H. W. Elmore *Marshall University*
David J. Fox *University of Tennessee*
Larry N. Gleason *Western Kentucky University*
E. Bruce Holmes *Western Illinois University*
Genevieve D. Johnson *University of Iowa*
Malcolm Jollie *Northern Illinois University*
Karen A. Koos *Rio Hondo College*
William H. Leonard *University of Nebraska—Lincoln*
A. David Scarfe *Texas A & M University*
Carl A. Scheel *Central Michigan University*
Donald R. Scoby *North Dakota State University*
John L. Zimmerman *Kansas State University*

Second Edition

David Ashley *Missouri Western State College*
Jack Bennett *Northern Illinois University*
Oscar Carlson *University of Wisconsin-Stout*
Arthur Cohen *Massachusetts Bay Community College*
Rebecca McBride DeLiddo *Suffolk University*
Gary Donnermeyer *St. John's University (Minnesota)*
D. C. Freeman *Wayne State University*
Sally Frost *University of Kansas*
Maura Gage *Palomar College*
Betsy Gulotta *Nassau Community College*
W. M. Hess *Brigham Young University*
Richard J. Hoffmann *Iowa State University*
Trudy McKee
Brian Myres *Cypress College*
John M. Pleasants *Iowa State University*
Jay Templin *Widener University*

Third Edition

Wayne P. Armstrong *Palomar College*
Mark S. Bergland *University of Wisconsin-River Falls*
Richard Blazier *Parkland College*

William F. Burke *University of Hawaii*
Donald L. Collins *Orange Coast College*
Ellen C. Cover *Manatee Community College*
John W. Crane *Washington State University*
Calvin A. Davenport *California State University-Fullerton*
Robert Ebert *Palomar College*
Darrell R. Falk *Pt. Loma Nazarene College*
Jerran T. Flinders *Brigham Young University*
Sally Frost *University of Kansas*
Elizabeth Gulotta *Nassau Community College*
Madeline M. Hall *Cleveland State University*
James G. Harris *Utah Valley Community College*
Kenneth S. Kilborn *Shasta College*
Donald R. Kirk *Shasta College*
Jon R. Maki *Eastern Kentucky University*
Ric Matthews *Miramar College*
Joyce B. Maxwell *California State University-Northridge*
Leroy McClenaghan *San Diego State University*
Leroy E. Olson *Southwestern College*
Barbara Yohai Pleasants *Iowa State University*
David M. Prescott *University of Colorado*
Robert R. Rinehart *San Diego State University*
Mary Beth Saffo *University of California—Santa Cruz*
Walter H. Sakai *Santa Monica College*
Frederick W. Spiegel *University of Arkansas-Fayetteville*
Gerald Summers *University of Missouri-Columbia*
Marshall Sundberg *Louisiana St. University*
Kathy S. Thompson *Louisiana State University*
Anna J. Wilson *Oklahoma City Community College*
Timothy S. Wood *Wright State University*

Fourth Edition
Michael S. Gaines *University of Kansas*
Kerry S. Kilburn *West Virginia State College*
W. Sylvester Allred *Northern Arizona University*
Helen L. Grierson *Morehead State University*
Deborah K. Meinke *Oklahoma State University*
Robert Snetsinger *Queens University*
Gail Kingrey *Pueblo Community College*
Barbara Yohai Pleasants *Iowa State University*
Eugene Nester *University of Washington*

Survey Respondents
Sister Julia Van Demack *Silver Lake College*
Ken Beatty *College of the Siskiyous*
John W. Metcalfe *Potsdam College at the State University of New York*
Brenda K. Johnson *Western Michigan University*
Tom Dale *Kirtland Community College*
Marlene Kayne *Trenton State College*
Peter M. Grant *Southwestern Oklahoma State University*
Janet Carter *Delaware Technical & Community College*
Frank F. Escobar *Holyoke Community College*
John A. Chisler *Glenville State College*

Nancy Goodyear *Bainbridge College*
John F. Lyon *University of Lowell*
Rob Snetsinger *Queens University*
Fred Stevens *Schreiner College*
Paula H. Dedmon *Gaston College*
Bruce G. Stewart *Murray State College*
Leo R. Finkenbinder *Southern Nazarene University*
Mary P. Greer *Macomb Community College*
Kenneth M. Allen *Schoolcraft College*
Robert Cerwonica *SUNY-Potsdam*
Darcy Williams *Cecil Community College*
Dave McShaffrey *Marietta College*
Elizabeth L. Nebel *Macomb Community College*
Jal S. Parakh *Western Washington University*
Dwight Kamback *Northampton Community College*
David L. Haas *Fayetteville State University*
A. Quinton White *Jacksonville University*
Paul J. Hummer, Jr. *Hood College*
Dee Forrest *Western Wyoming College*
Linden C. Haynes *Hinds Community College*
John De Banzie *Northeastern State University*
Anne T. Packard *Plymouth St. College*
L. H. Buff, Jr. *Spartanburg Methodist College*
Dennis Vrba *North Iowa Area Community College*
Donald L. Collins *Orange Coast College*
Larrie E. Stone *Dana College*
Brian K. Mitchell *Longview College*
Dennis M. Forsythe *The Citadel*
Bonnie S. Wood *University of Maine at Presque Isle*
Eunice R. Knouso *Spartanburg Methodist College*
Larry C. Brown *Virginia State University*
Clyde M. Senger *Western Washington University*
James V. Makowski *Messiah College*
James R. Coggins *University Wisconsin-Milwaukee*
Roland Vieira *Green River Community College*
Donna Barleen *Bethany College*
S. M. Cabrini Angelli *Regis College*
W. Brooke Yeager, III *Penn State-Wilkes-Barre*
Diana M. Colon *Northwest Technical College*
Carolyn K. Jones *Vincennes University*
Rita M. O'Clair *University of Alaska Southeast*
Fred M. Busroe *Morehead State University*
Edward R. Garrison *Navajo Community College*
Joseph V. Faryniarz *Maitatuck Community College*
K. Dale Smoak *Piedmont Technical College*
Donald A. Wheeler *Edinboro University of Pennsylvania*
W. Lee Williams *Alamance Community College*
Eugene A. Oshima *Central Missouri State University*
Richard C. Renner *Laredo Junior College*
Peggy Rae Doris *Henderson State University*
Charles H. Owens *Virginia Highlands Community College*
Jeanette Oliver *Flathead Valley Community College*
E. L. Beard *Loyola University*
Monica Macklin *Northeastern State University*

PART 5

Animal Structure and Function

Land snails have organ systems that carry on functions necessary to life. This part studies how the organ systems maintain homeostasis, the relative constancy of the internal environment. The circulatory system carries nutrients from the digestive system and oxygen from the respiratory system to the cells. Then it carries metabolic wastes from the cells to the excretory system. The antennae of the snail send messages to the nervous system which coordinates the functions of all the other systems.

33

Animal Organization and Homeostasis

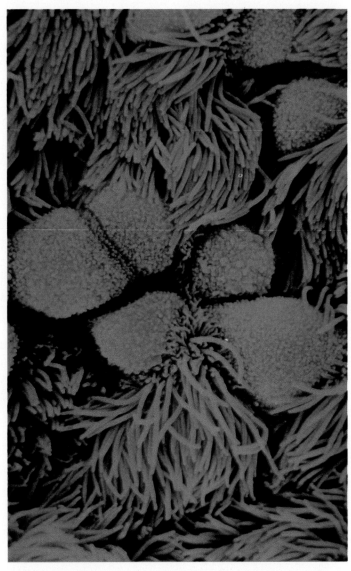

An animal's body is highly organized. This is a scanning electron micrograph of the tissue that lines the trachea, an organ that is part of the respiratory system. Some of the cells have cilia, which sweep impurities away from the lungs and toward the throat, where they can be swallowed. Magnification, X3,000.

Your study of this chapter will be complete when you can

1. name the 4 types of tissues in animals, and give a general function for each;
2. describe the structure and function of 3 types of epithelium along with their possible modifications;
3. describe the structure and function of the various types of connective tissue;
4. describe the structure and function of 3 types of muscular tissue;
5. describe the structure and function of 2 types of cells found in nervous tissue;
6. describe the structure of human skin, and tell how this illustrates the composition of an organ;
7. name the various organ systems in humans, associate each with a particular life process, and tell the specific functions of each system;
8. name the body cavities in a human, and tell which organ systems are in each cavity;
9. define homeostasis and tell how control by negative feedback works;
10. describe how body temperature is controlled in humans, and explain why body temperature fluctuates above and below a mean.

nimals, like plants, have levels of organization (fig. 33.1). In this chapter, we begin at the level of cells. The same type of cells form a tissue, and different types of tissues make up organs. Several organs are found within an organ system, and organ systems make up the organism.

Note that the structure and function of an organ system are dependent upon the structure and function of the organ, tissue, and cell type contained therein. For example, the structure and the function of the skeletal muscle system are the same as that of the skeletal muscles, the muscular tissue, and the muscle cells.

Types of Tissues

There are 4 major types of **tissue** in animals: *epithelial tissue* covers body surfaces and lines body cavities; *connective tissue* binds and supports body parts; *muscular tissue* causes body parts to move; and *nervous tissue* responds to stimuli and transmits impulses from one body part to another.

The embryological development of an animal begins with a single cell that divides to produce 3 fundamental layers, called the germ layers. The different tissues we have just mentioned arise from these germ layers:

Embryonic Germ Layer	Vertebrate Adult Structures
Ectoderm (outer layer)	Epidermis of skin; epithelial lining of mouth and rectum; nervous system
Mesoderm (middle layer)	Notochord; skeleton; muscular system; dermis of skin; circulatory system; excretory system; reproductive system—including most epithelial linings; outer layers of respiratory and digestive systems
Endoderm (inner layer)	Epithelial lining of digestive tract and respiratory tract; associated glands of these systems; epithelial lining of the urinary bladder

Epithelial Tissue

Epithelial tissue, also called epithelium, forms a continuous layer, or sheet, over body surfaces and most of the inner cavities. There are 3 types of epithelial tissue. *Squamous epithelium* is composed of flat cells; *cuboidal epithelium* contains cube-shaped cells; and in *columnar epithelium,* the cells resemble pillars or columns because they have an oblong shape. Figure 33.2 describes the structure and function of epithelium in vertebrates. Any epithelium can be simple or stratified. Simple means that the tissue has a single layer of cells, and *stratified* means that the tissue has layers piled one on top of the other. One type of epithelium is pseudostratified— it appears to be layered, but actually true layers do not exist because each cell touches a baseline.

Epithelial tissues have various functions. Epithelial cells can have hairlike extensions called cilia, which bend and move materials in a particular direction. In the human body, the ciliated

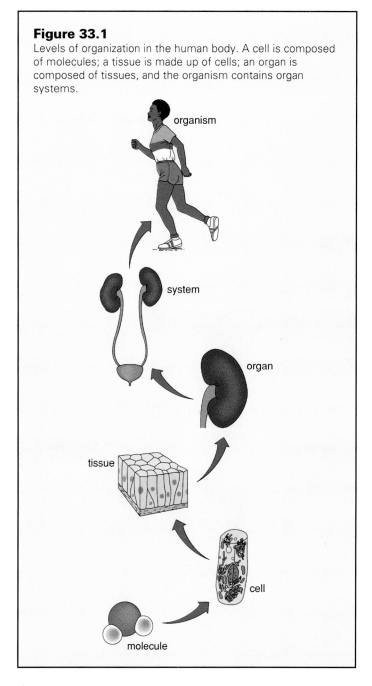

Figure 33.1

Levels of organization in the human body. A cell is composed of molecules; a tissue is made up of cells; an organ is composed of tissues, and the organism contains organ systems.

epithelium lining the respiratory tract sweeps impurities toward the throat so they do not enter the lungs. An epithelium sometimes secretes a product and is described as glandular. A gland can be a single epithelial cell, such as the mucus-secreting globlet cells in the lining of the human intestine, or a gland can contain numerous cells. *Exocrine glands* secrete their product into ducts, and *endocrine glands* secrete their product directly into the bloodstream.

Epithelium forms the skin of many animals. The skin of earthworms and snails is glandular and produces mucus that lubricates the body, helping to ease movement through a dry environment. In roundworms, annelids, and arthropods, an outer

Figure 33.2

Type of epithelial tissues in vertebrates. Epithelial tissues are classified according to shape of cell, whether they are simple or stratified, and whether they have cilia. Epithelial tissues have functions associated with protection, absorption, and secretion.

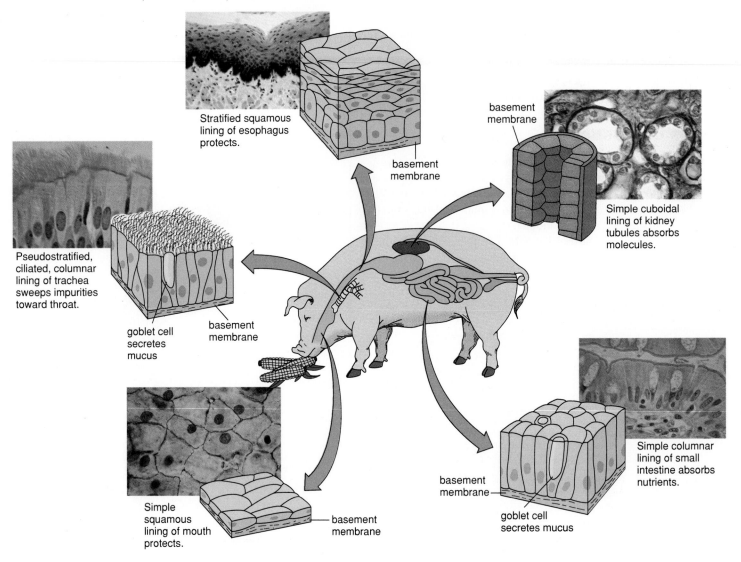

Stratified squamous lining of esophagus protects.

basement membrane

basement membrane

Simple cuboidal lining of kidney tubules absorbs molecules.

Pseudostratified, ciliated, columnar lining of trachea sweeps impurities toward throat.

goblet cell secretes mucus

basement membrane

Simple squamous lining of mouth protects.

basement membrane

basement membrane

goblet cell secretes mucus

Simple columnar lining of small intestine absorbs nutrients.

---

nonliving and protective cuticle is produced by epithelium. In terrestrial vertebrates, skin cells contain keratin, a substance that makes the skin protective against the possible loss of water.

Epithelial tissue cells are packed tightly and joined to one another in one of 3 ways: spot desmosomes, tight junctions, and gap junctions, (see fig. 6.16). Epithelium is often attached to a basement membrane, which is glycoprotein reinforced by fibers supplied by an underlying connective tissue.

Epithelial tissue is classified according to the shape of the cell. There can be one or many layers of cells, and the epithelium can be ciliated and/or secretory.

Connective Tissue

Connective tissue binds structures together, provides support and protection, fills spaces, stores fat, and forms blood cells. It provides the source cells for muscular and skeletal cells in animals that can regenerate lost parts.

Connective tissue cells are separated widely by a *matrix,* a noncellular material found between cells.

Loose Connective Tissue and Fibrous Connective Tissue

The cells of loose and fibrous connective tissues, called **fibroblasts,** are located some distance from one another and are separated by a jellylike matrix that contains white collagen fibers and yellow elastic fibers. Collagen fibers provide flexibility and strength; elastic fibers provide elasticity.

Figure 33.3

Loose connective tissue. *a.* Loose connective tissue has plenty of space between components. This type of tissue supports epithelial tissue and is found within organs. It also surrounds organs and binds them to one another. Magnification, X250. *b.* Adipose tissue looks like white ghosts because the fat has been washed out during preparation of the tissue. Magnification, X250.

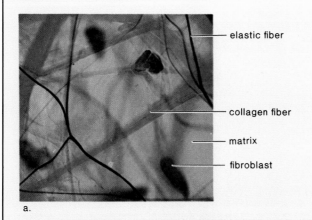

elastic fiber

collagen fiber

matrix

fibroblast

a.

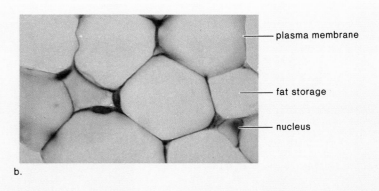

plasma membrane

fat storage

nucleus

b.

Loose connective tissue lies beneath epithelium in the skin and also beneath most internal vertebrate organs (fig. 33.3*a*). Its presence in lungs, arteries, and the urinary bladder allows these organs to expand. It forms a protective covering encasing many internal organs, such as muscles, blood vessels, and nerves.

Adipose tissue is a type of loose connective tissue in which the fibroblasts enlarge and store fat and in which the intercellular matrix is reduced (fig. 33.3*b*). In vertebrates, it is found particularly beneath the skin, around the kidneys, and on the surface of the heart. It insulates the body and provides padding.

Fibrous connective tissue contains many collagenous fibers that are packed closely together. This type of tissue has more specific functions in vertebrates than does loose connective tissue. For example, fibrous connective tissue is found in **tendons,** which connect muscles to bones, and in **ligaments,** which connect bones to other bones at joints.

Loose connective tissue and fibrous connective tissue contain fibroblasts separated by a matrix, which contains collagen and elastic fibers.

Cartilage and Bone

Cartilage and bone are rigid connective tissues in which structural proteins (cartilage) or calcium salts (bone) are deposited in the intercellular matrix (fig. 33.4).

In **cartilage,** the cells lie in small chambers called lacunae (lacuna, sing.), separated by a matrix that is strong yet flexible. There are various types of cartilage, which are classified according to type of collagen and elastic fiber found in the matrix. In some vertebrates, notably sharks and rays, the entire skeleton is made of cartilage. In humans, the fetal skeleton is cartilage, but it is later replaced by bone. Cartilage is retained at the ends of long bones, at the end of the nose, in the framework of the ear, in the walls of respiratory ducts, and within intervertebral disks.

In **bone,** the matrix of calcium salts is deposited around protein fibers. The minerals give bone rigidity, and the protein fibers provide elasticity and strength, much as steel rods do in reinforced concrete.

In compact bone, bone cells (osteocytes) are located in lacunae that are arranged in concentric circles around tiny tubes called Haversian canals. Nerve fibers and blood vessels are in these canals. The latter bring the nutrients that allow bone to renew itself. The nutrients can reach all of the cells because there are minute canals (canaliculi) containing thin processes of the osteocytes that connect them with one another and with the Haversian canals.

The ends of a long bone contain spongy bone, which has an entirely different structure. Spongy bone contains numerous bony bars and plates separated by irregular spaces. Although lighter than compact bone, spongy bone still is designed for strength. Just as braces are used for support in buildings, the solid portions of spongy bone follow lines of stress.

Cartilage and bone are support tissues. Cartilage is more flexible than bone because the matrix is rich in protein and not calcium salts, like that of bone.

Blood

Blood is a connective tissue in which the cells are separated by a liquid called plasma. In vertebrates, blood cells are of 2 types: red blood cells (erythrocytes), which carry oxygen, and white blood cells (leukocytes), which aid in fighting infection (fig. 33.5). Also present in plasma are platelets, which are important to the initiation of blood clotting. Platelets are not complete cells; rather, they are fragments of giant cells found in the bone marrow.

Figure 33.4

Anatomy of a long bone. *a.* A long bone is encased by fibrous connective tissue except where it is covered by cartilage at the ends. The central shaft is composed of compact bone, but the ends are spongy bone, which can contain red bone marrow. A central medullary cavity contains yellow bone marrow. *b.* Photomicrograph of cartilage. Magnification, X250. *c.* Micrograph of compact bone. Magnification, X320.

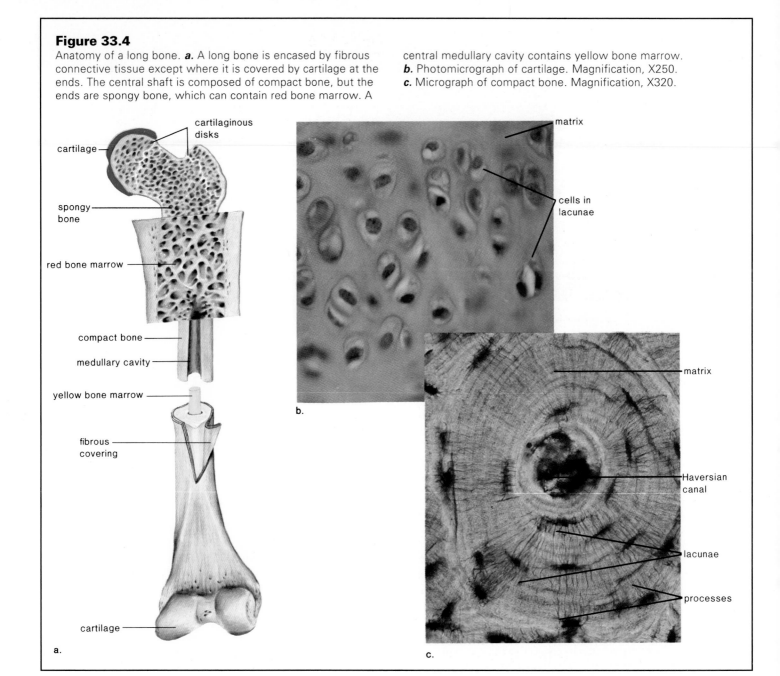

Blood is unlike other types of connective tissue in that the intercellular matrix (i.e., plasma) is not made by the cells. Plasma is a mixture of different types of molecules that enter the blood at various locations.

Blood is a connective tissue in which the matrix is plasma.

Muscular Tissue

Muscular (contractile) tissue is composed of cells called *muscle fibers*. Muscle fibers contain actin filaments and myosin filaments, whose interaction accounts for the movements we associate with animals.

Some muscle fibers are striated and some are smooth (fig. 33.6). *Striated muscle fibers* have light and dark bands perpendicular to the length of the cell. These bands reflect the placement of actin filaments and myosin filaments in the cell. *Smooth muscle fibers* are spindle-shaped cells that lack striations. They form layers in which the thick middle portion of one cell is opposite the thin ends of adjacent cells. Consequently, the nuclei form an irregular pattern in smooth muscular tissue.

Both striated and smooth muscular tissue assist locomotion in many invertebrates. In vertebrates, *smooth muscle* moves materials through internal organs such as the digestive

Animal Structure and Function

Figure 33.5

Blood, a liquid tissue. Blood is classified as connective tissue because the cells are separated by a matrix—plasma. Plasma, the liquid portion of blood, usually contains several types of cells. This shows the cells in human blood (red blood cells, white blood cells, and platelets, which are actually fragments of a larger cell).

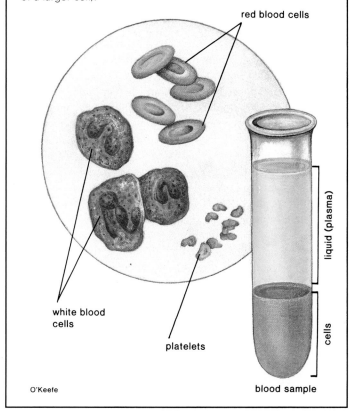

red blood cells

white blood cells

platelets

liquid (plasma)

cells

blood sample

O'Keefe

Figure 33.6

Muscular tissue. How do you distinguish an animal from a plant? One way is to detect rapid motion—only animals have contractile fibers that permit movement. ***a.*** Smooth muscle. Note the single nucleus and the lack of striations. Magnification, X250. ***b.*** Skeletal muscle is found attached to the skeleton in vertebrates. It is a type of striated muscle. Magnification, X100. ***c.*** Cardiac muscle pumps the heart. Note the branching of the fibers, the central position of the nuclei, and the presence of intercalated disks, which are folded plasma membranes between adjacent individual cells. Magnification, X160.

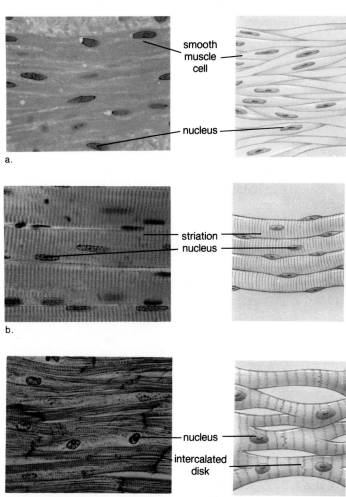

smooth muscle cell

nucleus

a.

striation
nucleus

b.

nucleus

intercalated disk

c.

and reproductive tracts and the blood vessels. Only striated muscular tissue makes up the *skeletal muscles,* which are under voluntary control and are attached to the bones of the skeleton. Skeletal muscle cells are cylindrical and quite long—they arise during development when several cells fuse, giving one multinucleated cell. The nuclei are placed at the periphery of the cell, just inside the plasma membrane. A type of striated muscular tissue, called *cardiac muscle,* is found in the heart. Cardiac muscle cells differ from skeletal muscle cells in that they have a single, centrally placed nucleus. The cells are branched and seemingly fused one with the other, and the heart appears to be composed of one large interconnecting mass of muscle cells. Actually, cardiac muscle cells are separate and individual, but they are bound end to end at intercalated disks, areas of folded plasma membrane between the cells.

All muscle tissue contains actin and myosin filaments; these form a striated pattern in skeletal and cardiac muscle but not in smooth muscle.

Nervous Tissue

Nervous tissue contains nerve cells called **neurons** (fig. 33.7). A neuron is a specialized cell that has 3 parts: (1) dendrites, which conduct impulses (send a message) to the cell body; (2) the cell body, which contains most of the cytoplasm and the nucleus of the neuron; and (3) the axon, which conducts impulses away from the cell body. Long axons and dendrites are called nerve fibers, and they are bound together by connective tissue to form nerves. Neurons are specialized to detect environmental stimuli and then to conduct signals to other neurons and to muscle or glands, thereby bringing about a coordinated response to the stimulus.

Figure 33.7
Micrograph of neuron. Conduction of the nerve impulse is dependent on neurons, each of which has the 3 parts indicated. A dendrite takes nerve impulses to the cell body, and an axon takes them away from the cell body. Magnification, X200.

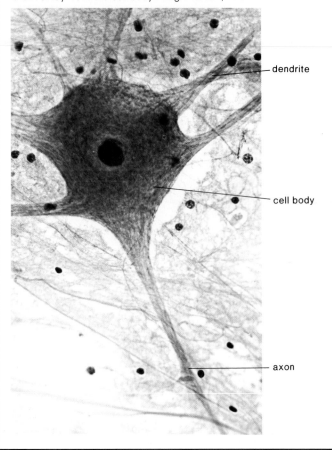

— dendrite

— cell body

— axon

In addition to neurons, nervous tissue contains neuroglial cells. These cells maintain the tissue by supporting and protecting neurons. They also provide nutrients to neurons and help to keep the tissue free of debris.

> Nerve cells, called neurons, have fibers (processes) called axons and dendrites. Long fibers are found in nerves.

Organs and Organ Systems

In most animals each **organ** consists of several types of tissues (fig. 33.1). For example, human skin contains epithelium, connective tissue, nervous tissue, and blood (fig. 33.8). Each organ is also specialized to perform specific functions. In humans, the skin covers the body, protecting underlying parts from physical trauma, microbial invasion, and water loss. The skin also helps to regulate body temperature, and because it contains receptors (sense organs), the skin helps us to be aware of our surroundings and to communicate with others.

Some authorities speak of the skin as an organ within the integumentary system. They maintain that the hair follicles, oil and sweat glands, receptors, and skin are separate organs, and these organs work together to perform various functions.

Human Skin

Skin has an outer epidermal layer (the epidermis) and an inner layer (the dermis). Beneath the dermis, there is a subcutaneous layer that binds the skin to underlying organs.

The **epidermis** is the outer, thinner layer of the skin. It is a stratified squamous epithelium whose cells are derived from the basal cells, which undergo continuous cell division. As newly formed cells are pushed to the surface, they gradually flatten and harden. Eventually, they die and are sloughed off. Hardening is caused by cellular production of a waterproof protein called keratin. Over much of the body, keratinization is minimal, but the palm of the hand and the sole of the foot have a particularly thick outer layer of dead keratinized cells arranged in spiral and concentric patterns. We call these patterns fingerprints and footprints.

Particular cells in the epidermis, called melanocytes, produce melanin, the pigment responsible for skin color in dark-skinned persons. When you sunbathe, the melanocytes become more active, producing melanin in an attempt to protect the skin from the damaging effects of the ultraviolet (UV) radiation in sunlight.

The **dermis** is a layer of fibrous connective tissue that is deeper and thicker than the epidermis. It contains elastic fibers and collagen fibers. The collagen fibers form bundles that interlace and run, for the most part, parallel to the skin surface. There are several types of structures in the dermis. A hair, except for the root, is formed of dead, hardened epidermal cells; the root is alive and resides in a hair follicle found in the dermis. Each follicle has one or more oil (sebaceous) glands that secrete sebum, an oily substance that lubricates the hair and the skin. A smooth muscle called the arrector pili muscle is attached to the hair follicle in such a way that when contracted, the muscle causes the hair to stand on end. When you are frightened or are cold, goose bumps develop due to the contraction of these muscles.

Sweat (sudoriferous) glands are quite numerous and are present in all regions of the skin. A sweat gland begins as a coiled tubule within the dermis, but then it straightens out near its opening. Some sweat glands open into hair follicles, and others open onto the surface of the skin.

Small receptors are present in the dermis. There are different receptors for touch, pressure, pain, and temperature. The fingertips contain the most touch receptors, and these add to our ability to use our fingers for delicate tasks. The dermis also contains nerve fibers and blood vessels. When blood rushes into these vessels, a person blushes, and when blood is reduced in them, a person turns blue.

The subcutaneous layer, which lies below the dermis, is composed of loose connective tissue, including adipose tissue. Adipose tissue helps to insulate the body by minimizing both heat gain and heat loss. A well-developed subcutaneous layer gives a rounded appearance to the body. Excessive development of this layer accompanies obesity.

Animal Structure and Function

Figure 33.8

Human skin anatomy. Skin contains 3 layers: epidermis, dermis, and subcutaneous.

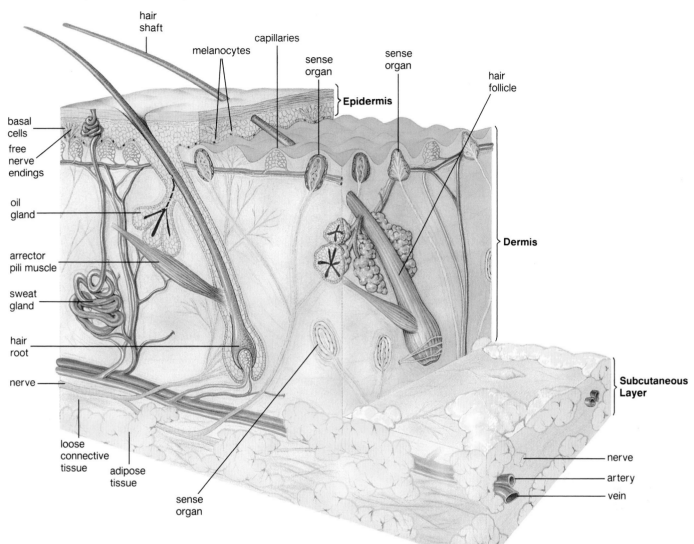

Each animal organ has a structure that suits its function. In human skin, the hardened cells of epidermis provide a protective covering; the dermis contains receptors, blood vessels, and nerves; and the adipose tissue of the subcutaneous layer insulates the body.

Organ Systems

In most animals, individual organs function as part of an *organ system*, the next higher level of animal organization. Figure 33.9 pictorially shows the organ systems within the human body. These same systems are found in all vertebrate animals. The organ systems carry out the life processes that are common to all animals and indeed to all organisms:

Life Processes	Human Systems
Acquire materials and energy (food)	Skeletal system Muscular system Digestive system
Maintain body shape	Muscular system Skeletal system
Exchange gases	Respiratory system
Transport materials	Circulatory system
Excrete wastes	Excretory system
Protect the body from disease	Lymphatic system
Coordinate body activities	Nervous system Endocrine system
Produce offspring	Reproductive system

Animal Organization and Homeostasis

Figure 33.9
Human organ systems and
their functions.

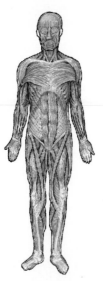 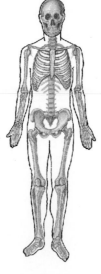 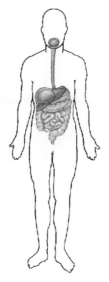 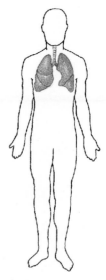 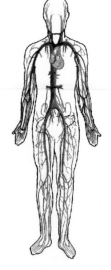

Name of System	Muscular system	Skeletal system	Digestive system	Respiratory system	Circulatory system
Life Process	Acquires materials and energy	Acquires materials and energy	Acquires materials and energy	Exchanges gases	Transports material
Specific Functions	Produces body movements; produces body heat; maintains posture and supports the body	Provides rigid framework for body movement; supports and protects internal parts; produces blood cells; stores minerals	Digests food into small molecules and absorbs these molecules	Exchanges gas between external environment and blood; maintains blood pH by excreting carbon dioxide	Transports nutrients and oxygen to and metabolic wastes from cells; distributes hormones; protects against injury and microbes

Body Cavities

Each organ system has a particular location within the human body. There are 2 main body cavities: the smaller dorsal body cavity and the larger ventral body cavity (fig. 33.10). The brain and the spinal cord are in the dorsal body cavity.

During development, the ventral body cavity develops from the *coelom.* In humans and other mammals, the coelom is divided by a muscular diaphragm that assists breathing. The heart, a pump for the closed circulatory system, and the lungs are located in the upper (thoracic or chest) cavity. The major portion of the digestive system, the entire excretory system, and much of the reproductive system are located in the lower (abdominal) cavity. The major organs of the excretory system are the paired kidneys. The accessory organs of the digestive system are the liver and pancreas. Each sex has characteristic sex organs.

Homeostasis

Claude Bernard pointed out in 1859 that while an animal lives in an external environment, the cells of the body lie within an internal environment. The *internal environment* is a fluid, called tissue fluid, that bathes the cells of the body. Bernard said that the relative stability of the internal environment allows animals to live in an external environment that can vary considerably. Later, Walter Cannon, an American physiologist, introduced the term **homeostasis.** He said there is a dynamic interplay between events that tend to change the internal environment and those that counter this possibility. To achieve homeostasis, the composition of blood and tissue fluid, the body temperature, and the blood pressure, for example, must stay within a normal range.

Animal Structure and Function

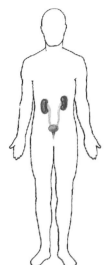

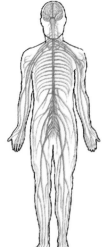

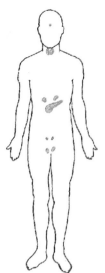

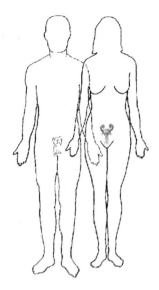

Excretory system

Excretes wastes

Maintains volume and chemical composition of blood and tissue fluid

Lymphatic system

Protects the body from disease

Transports excess tissue fluid to bloodstream; transports fat to blood; helps provide immunity against disease

Nervous system

Coordinates body activities

Along with endocrine system regulates body systems; learning and memory

Endocrine system

Coordinates body activities

Secretes hormones that regulate body metabolism, growth, and reproductive system

Reproductive system

Produces offspring

Male: produces hormones and sperm; transfers sperm to female. Female: produces hormones and egg; provides site for fertilization of egg, implantation, and development of embryo and fetus

---

The internal environment of an animal's body consists of tissue fluid, which bathes the cells.

In the chapters that follow, we will see that most organ systems of the human body contribute to homeostasis (fig. 33.9). The digestive system takes in and digests food, providing nutrient molecules that enter the blood and replace the nutrients that are constantly being used by the body cells. The respiratory system adds oxygen to the blood and removes carbon dioxide. The amount of oxygen taken in and carbon dioxide given off can be increased to meet body needs. The liver and the kidneys contribute greatly to homeostasis. For example, immediately after glucose enters the blood, it can be removed by the liver and stored as glycogen. Later, glycogen is broken down to replace the glucose used by the body

cells; in this way, the glucose composition of the blood remains constant. The hormone insulin, secreted by the pancreas, regulates glycogen storage. The kidneys are also under hormonal control as they excrete wastes and salts, substances that can affect the pH level of the blood.

Although homeostasis is, to a degree, controlled by hormones, it is ultimately controlled by the nervous system. Homeostatic systems have 3 elements: *receptors* (sensors) that react to a stimulus; an *integrating center* that evaluates information from the sensors; and *effectors* (muscles and glands) that carry out a response to the stimulus (fig. 33.11). Maintaining proper temperature and blood pressure levels requires receptors that detect a change and signal an integrating center in the brain. The center then directs the muscles by way of nerves to bring

Figure 33.10

Mammalian body cavities. There is a dorsal cavity, which contains the cranial cavity and the vertebral canal. The brain is in the cranial cavity, and the spinal cord is in the vertebral canal. There is a well-developed ventral cavity, which is divided by the diaphragm into the thoracic cavity and the abdominal cavity. The heart and lungs are in the thoracic cavity, and most other internal organs are in the abdominal cavity.

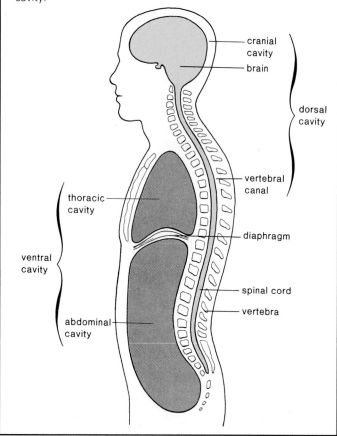

Figure 33.11

Negative feedback control. A receptor (sense organ) is stimulated by a change, such as low temperature, and signals an integrating center (in the brain) that directs effectors (muscles) to react. The response, constriction of blood vessels in the skin, and perhaps shivering, raises the temperature, negating the original stimulus.

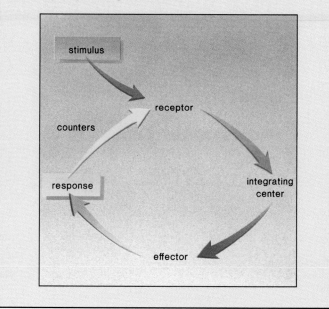

about an adaptive response. Very often homeostatic systems utilize a **negative feedback loop**. For example, if the stimulus (input) was a decrease in body temperature, the response (output) is a rise in body temperature. It is a feedback loop because the output is fed back into the system; it is a negative feedback loop because the output is counter to and cancels the input. **Positive feedback loops** also sometimes occur. In these instances, a choice of events increases the likelihood of a particular response. Blood clotting, for example, requires a number of steps, each leading to the next step until a blood clot forms.

Body Temperature Control

Body temperature control in humans is dependent on a homeostatic system that is self-regulated by a negative feedback loop. This type of homeostatic regulation results in fluctuation between 2 levels (fig. 33.12).

The receptor and the integrating center for body temperature are located in the hypothalamus, a part of the brain. The receptor is sensitive to the temperature of the blood, and when the temperature falls below normal, the receptor signals the integrating center, which directs (via nerve impulses) the muscles (the effectors) of blood vessels in the skin to constrict. This conserves heat. Also, the arrector pili muscles pull hairs erect, and a layer of insulating air is trapped next to the skin. If body temperature falls even lower, the integrating center sends nerve impulses to the skeletal muscles, and shivering occurs. Shivering generates heat, and gradually body temperature rises to 37°C and perhaps higher.

During the period of time the body temperature is normal, the receptor and the integrating center are not active, but once body temperature is higher than normal, they are activated. Now the integrating center directs the blood vessels of the skin to dilate. This allows more blood to flow near the surface of the body, where heat can be lost to the environment. The integrating center also activates the sweat glands, and the evaporation of sweat also helps lower body temperature. Gradually, body temperature decreases to 37°C and perhaps lower. Once body temperature is below normal the cycle begins again.

Homeostasis of internal conditions is a self-regulating mechanism that results in slight fluctuations above and below a mean. For example, body temperature rises above and drops below a normal temperature of 37°C.

Animal Structure and Function

Figure 33.12

Temperature control. When the body temperature rises, the regulator center directs the blood vessels to dilate and the sweat glands to be active. Now, the body temperature lowers. Then, the regulator center directs the blood vessels to constrict, hairs to stand on end, and even shivering to occur if needed. Now, the body temperature rises again. Because the receptors are sensitive only to a change in the internal environment, the body temperature fluctuates above and below normal.

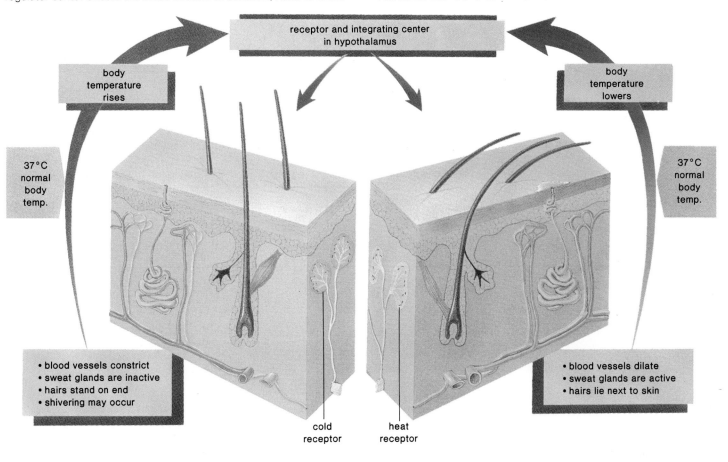

receptor and integrating center in hypothalamus

body temperature rises

body temperature lowers

37°C normal body temp.

37°C normal body temp.

- blood vessels constrict
- sweat glands are inactive
- hairs stand on end
- shivering may occur

- blood vessels dilate
- sweat glands are active
- hairs lie next to skin

cold receptor

heat receptor

Summary

1. Tissues arise from the embryonic germ layers: ectoderm, mesoderm, and endoderm. Ectoderm produces nervous tissue; mesoderm produces muscle and certain systems (circulatory, excretory, reproductive); endoderm produces the linings of the digestive and respiratory tracts and associated organs. Epithelial tissue comes from all 3 layers.

2. Epithelial tissue covers the body and lines cavities. There is squamous, cuboidal, and columnar epithelium. Each type can be simple or stratified, glandular, or have modifications like cilia. Epithelial tissue protects, absorbs, secretes, and excretes.

3. Connective tissue has a matrix between cells. Loose connective tissue and fibrous connective tissue contain fibroblasts and fibers (collagen and elastic). Adipose tissue is a type of loose connective tissue; tendons and ligaments are fibrous connective tissue.

4. Cartilage (protein matrix) and bone (calcium matrix) are rigid connective tissues. Blood is connective tissue in which plasma is the matrix.

5. Muscular (contractile) tissue can be smooth or striated (both skeletal and cardiac). In humans, smooth muscle is in the wall of internal organs; skeletal muscle is attached to bone, and cardiac muscle is in the heart.

6. Nervous tissue contains neurons having 3 parts: dendrites, cell body, and axon. Neuroglial cells support and protect neurons.

7. Organs contain various tissues. Skin is an organ that has 3 tissue layers: epidermis (stratified squamous epithelium); dermis (fibrous connective epithelium); and subcutaneous (loose connective epithelium). Receptors, hair follicles, blood vessels, and nerves are in the dermis.

8. Organ systems contain several organs. Organ systems of humans carry out the life processes that are common to all organisms and also have specific functions (see fig. 33.9).

9. The human body contains 2 main cavities. The dorsal cavity contains the brain and spinal cord. The ventral cavity contains the thoracic (heart and lungs) and abdominal cavity (most other internal organs).

10. Homeostasis refers to the relative constancy of the internal environment, which is very often maintained by negative feedback loops in which the response negates the stimulus. As an example, consider body temperature control in humans. When the body is cold, receptors signal an integrating center that directs the muscles of the blood vessels (effectors) to constrict. When the body is warm, receptors signal an integrating center that directs the blood vessels to dilate. Also, sweat glands are activated. It can be seen that this type of regulation causes fluctuations above and below a mean.

Writing Across the Curriculum

In order to practice writing skills, students should write out the answers to any or all of the study questions and the critical thinking questions. The study questions are sequenced in the same order as the text. Suggested answers to the critical thinking questions are in appendix D.

Study Questions

1. Name the 4 major types of tissues.
2. Describe the structure and functions of 3 types of epithelial tissue.
3. Describe the structure and functions of 6 types of connective tissue.
4. Describe the structure and functions of 3 types of muscular tissue.
5. Nervous tissue contains what types of cells?
6. Describe the structure of skin, and state at least 2 functions of this organ.
7. In general terms, describe the location of the human organ systems.
8. Tell how the various systems of the body contribute to homeostasis.
9. What is the function of receptors, the integrating center, and effectors in a negative feedback loop? Why is it called negative feedback?

Objective Questions

1. Which of these is mismatched?
   a. epithelial tissue—protection and absorption
   b. muscular tissue—contraction and conduction
   c. connective tissue—binding and support
   d. nervous tissue—conduction and message sending
2. Which of these is not epithelial tissue?
   a. simple cuboidal and stratified columnar
   b. bone and cartilage
   c. stratified squamous and simple squamous
   d. All are epithelial tissue.
3. Which tissue is more apt to line a lumen?
   a. epithelial tissue
   b. connective tissue
   c. nervous tissue
   d. muscular tissue
4. Tendons and ligaments are
   a. connective tissue.
   b. associated with the bones.
   c. found in vertebrates.
   d. All of these.
5. Which tissue has cells in lacunae?
   a. epithelial tissue
   b. cartilage
   c. bone
   d. Both b and c.
6. Cardiac muscle is
   a. striated.
   b. involuntary.
   c. smooth.
   d. Both a and b.
7. Which of these components of blood fights infection?
   a. red blood cells
   b. white blood cells
   c. platelets
   d. All of these.
8. Which of these body systems contribute to homeostasis?
   a. digestive system and excretory system
   b. respiratory system and nervous system
   c. nervous system and endocrine system
   d. All of these.
9. With negative feedback,
   a. the output cancels the input.
   b. there is a fluctuation above and below the average.
   c. there is self-regulation.
   d. All of these.
10. When a person is cold, the blood vessels
    a. dilate and the sweat glands are inactive.
    b. dilate and the sweat glands are active.
    c. constrict and the sweat glands are inactive.
    d. constrict and the sweat glands are active.
11. Identify each of these tissues, and tell whether the tissue is a type of epithelial tissue, muscular tissue, nervous tissue, or connective tissue.

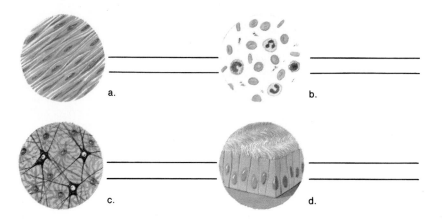

a.

b.

c.

d.

Concepts and Critical Thinking

1. *An animal's body has levels of organization.*

   How is each level of organization more than the sum of its parts?

2. *All organisms carry out certain life processes.*

   Review figure 7.10 and associate parts of the figure with appropriate life processes listed on page 545.

3. *The internal environment of organisms stays relatively constant.*

   How does each of the life processes (except producing offspring) help keep the internal environment relatively constant?

Selected Key Terms

epithelial tissue (ep″ĭ-the′le-al tish′u) 539
connective tissue (kŏ-nek′tiv tish′u) 540
fibroblast (fi′bro-blast) 540
adipose tissue (ad′ē-poz tish′u) 541
tendon (ten′don) 541

ligament (lig′ah-ment) 541
cartilage (kar′tĭ-lij) 541
bone (bōn) 541
blood (blud) 541
muscular (contractile) tissue (mus′ky-lar tish′u) 542

neuron (nu′ron) 543
epidermis (ep″ĭ-der′mis) 544
dermis (der′mis) 544
homeostasis (ho″me-o-sta′sis) 546
negative feedback (neg′ah-tiv fēd′bak) 548

34

Circulatory System

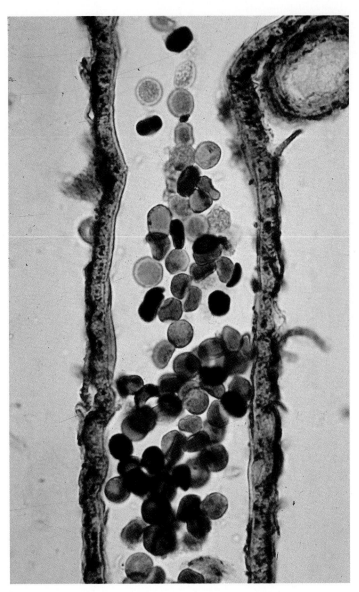

Vertebrates have a closed circulatory system—the blood is always contained within vessels and never runs free. Exchange of gases and nutrients for wastes has to take place across one-cell-thick capillary walls.

Your study of this chapter will be complete when you can

1. compare the circulatory systems of invertebrate and vertebrate animals;
2. compare the structure and function of the 3 types of blood vessels;
3. trace the path of blood through the mammalian heart and about the body;
4. describe how the heart beats, and tell how the heartbeat is controlled;
5. list the factors that control the pressure and velocity of blood in the vessels;
6. list the components of human blood plasma, and describe the function and the source of each;
7. describe the structure and function of human blood cells;
8. list the 3 steps necessary for the clotting of blood;
9. draw and explain a diagram depicting capillary exchange within tissues;
10. list the ABO blood types and describe a laboratory procedure for typing blood.

Every cell requires a supply of oxygen and of nutrient molecules and must rid itself of waste molecules. Single-celled organisms make these exchanges directly with the external environment that surrounds their body (fig. 34.1a). Even some small multicellular animals do not have an internal transport system because their cells can be serviced without one (fig. 34.1b and c). Larger invertebrates and vertebrates usually have a circulatory system that transports oxygen and nutrient molecules to cells and transports waste molecules away from cells.

Invertebrates

A phylogenetic tree of the animal kingdom is depicted in figure 26.2. Only the coelomate animals have a circulatory system. The noncoelomates do not need one, for reasons we will discuss, and the pseudocoelomates (i.e., roundworms) use coelomic fluid to service the cells.

Gastrovascular Cavity

The sac body plan of cnidarians and flatworms makes a circulatory system unnecessary. In cnidarians (see fig. 26.5), cells are either part of an external layer or they line the gastrovascular cavity. In either case, each cell is exposed to water and can independently exchange gases and rid itself of wastes. The cells that line the gastrovascular cavity are specialized to carry out digestion. They pass nutrient molecules to other cells by diffusion. In flatworms, a trilobed gastrovascular cavity ramifies throughout the small and flattened body (fig. 34.1). No cell is very far from one of the 3 digestive branches, so nutrient molecules can diffuse from cell to cell. Similarly, diffusion meets the respiratory and excretory needs of the cells.

Open and Closed Circulatory Systems

Arthropods and most mollusks have an *open circulatory system.* In an open circulatory system, blood is not always contained with blood vessels (fig. 34.2a). A heart pumps blood into

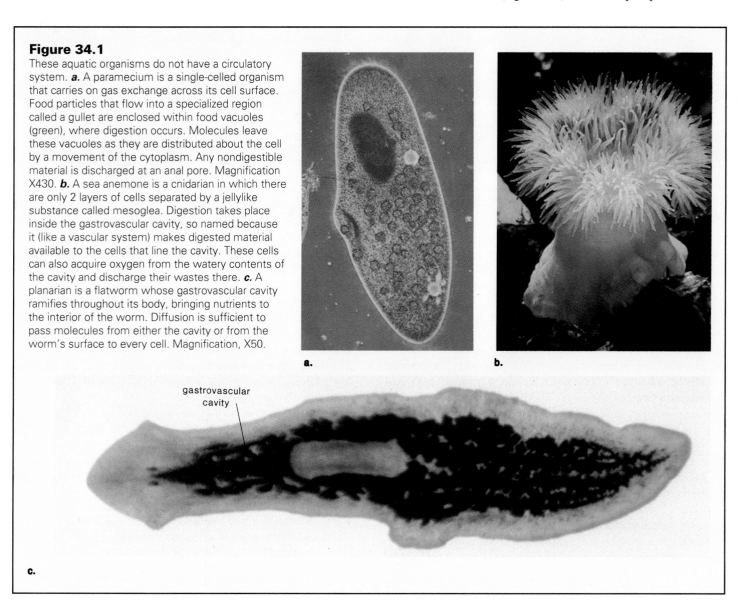

Figure 34.1
These aquatic organisms do not have a circulatory system. *a.* A paramecium is a single-celled organism that carries on gas exchange across its cell surface. Food particles that flow into a specialized region called a gullet are enclosed within food vacuoles (green), where digestion occurs. Molecules leave these vacuoles as they are distributed about the cell by a movement of the cytoplasm. Any nondigestible material is discharged at an anal pore. Magnification X430. *b.* A sea anemone is a cnidarian in which there are only 2 layers of cells separated by a jellylike substance called mesoglea. Digestion takes place inside the gastrovascular cavity, so named because it (like a vascular system) makes digested material available to the cells that line the cavity. These cells can also acquire oxygen from the watery contents of the cavity and discharge their wastes there. *c.* A planarian is a flatworm whose gastrovascular cavity ramifies throughout its body, bringing nutrients to the interior of the worm. Diffusion is sufficient to pass molecules from either the cavity or from the worm's surface to every cell. Magnification, X50.

a.

b.

gastrovascular cavity

c.

Figure 34.2

a. The grasshopper has an open circulatory system. A hemocoel is a body cavity filled with blood, which freely bathes the internal organs. The heart keeps the blood moving, but this open system probably could not supply oxygen to wing muscles rapidly enough. These muscles receive oxygen directly from air tubes. *b.* The earthworm has a closed circulatory system. The dorsal and ventral blood vessels are joined by 5 pairs of anterior hearts and by branch vessels in the rest of the worm.

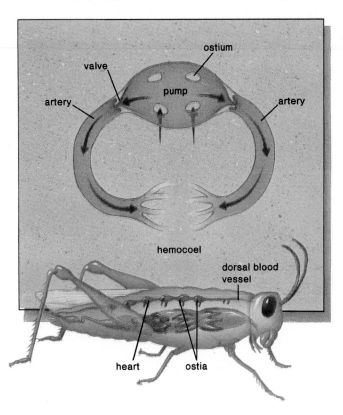

a. Open circulatory system

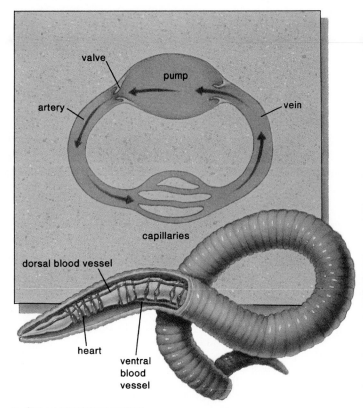

b. Closed circulatory system

vessels, and then these vessels empty either into body cavities, where blood bathes the internal organs, or into sinuses located within the organs themselves. In an open circulatory system, blood ebbs and flows in a sluggish manner.

In the grasshopper, for example, the dorsal heart pumps blood into a dorsal aorta, which empties into a body cavity termed a *hemocoel*. The hemocoel, appropriately named, is filled with blood. When the heart contracts, openings called ostia are closed; when the heart relaxes, the blood is sucked back into the heart by way of the ostia. The blood of a grasshopper is colorless because it does not contain hemoglobin or any other respiratory pigment. It does not carry oxygen but only nutrients. Oxygen is taken to cells and carbon dioxide is removed from them by way of air tubes, which are found throughout the body. Flight muscles require a very efficient means of receiving oxygen, and the open circulatory system probably would not suffice.

Vertebrates and several invertebrates, including annelids, squids, and octopuses, have a *closed circulatory system.* Blood, which usually consists of cells and plasma, is pumped by the heart into a system of blood vessels and the blood never runs free (fig. 34.2*b*). There are valves that prevent the backward flow of blood.

In the segmented earthworm, for example, 5 pairs of anterior lateral vessels pump blood into the ventral blood vessel, which has a branch in every segment of the worm's body. Blood moves through these branches into capillaries, where exchanges with tissue fluid take place. Blood then moves into veins that return it to the dorsal blood vessel. This dorsal blood vessel returns blood to the heart for repumping.

The earthworm has red blood because it contains the respiratory pigment **hemoglobin.** Hemoglobin is dissolved in the blood and is not contained within cells. The earthworm has no specialized boundary for gas exchange with the external environment. Gas exchange takes place across the body wall, which must always remain moist for this purpose.

A gastrovascular cavity in acoelomate aquatic invertebrates performs the functions of an internal circulatory system. Coelomate invertebrates do have a circulatory system, which may be closed or open.

Animal Structure and Function

Figure 34.3

Vertebrate circulatory system. The heart is a pump that keeps the blood moving. **a.** Blood leaving the heart moves from an artery to arterioles to capillaries to venules and then returns to the heart by way of a vein. **b.** Arteries have well-developed walls with a thick middle layer of elastic tissue and smooth muscle. **c.** Capillary walls are one cell thick. Exchanges take place across their thin walls. **d.** Veins have flabby walls, particularly because the middle layer is not thick. Veins have valves that point toward the heart.

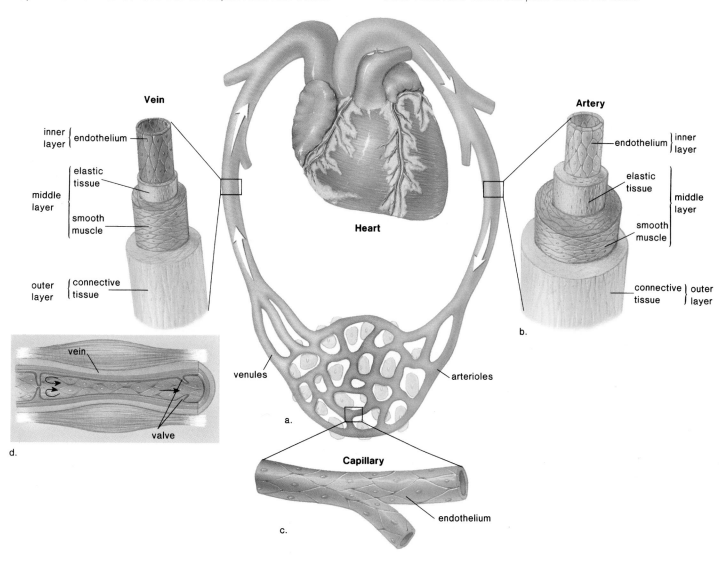

Vertebrates

All vertebrate animals have a closed circulatory system called a **cardiovascular system** (fig. 34.3). It consists of a strong, muscular heart, in which the *atria* (atrium, sing.) primarily receive blood and the muscular *ventricles* pump blood out through the blood vessels. There are 3 kinds of blood vessels: **arteries,** which carry blood away from the heart; **capillaries,** which exchange materials with tissue fluid; and **veins,** which return blood to the heart.

Arteries, the strongest of the blood vessels, are resilient; they are able to expand and constrict because their walls have substantial layers of both elastic tissue and smooth muscle. Arterioles are small arteries whose constriction can be regulated by the nervous system. Arteriole constriction and dilation affect blood pressure. The greater the number of vessels dilated, the lower the blood pressure.

Arterioles branch into capillaries, which are extremely narrow, microscopic tubes with a wall composed of only one layer of cells. Capillary beds (many capillaries interconnected) are present in all regions of the body; consequently, a cut to any body tissue draws blood. Capillaries are the most important part of a closed circulatory system because exchange of nutrient and waste molecules takes place across their thin walls.

The distribution of blood in the various capillary beds is regulated by **sphincters,** circulatory muscles that open and close tubular structures. In this instance, when the sphincter

Figure 34.4

Comparison of circulatory systems in vertebrates. ***a.*** In a fish, the blood moves in a single loop. The heart has a single atrium and ventricle and pumps the blood into the gill region, where gas exchange takes place. Blood pressure created by the pumping of the heart is dissipated after the blood passes through the gill capillaries. This is a disadvantage of this single-loop system. ***b.*** Amphibians have a double-loop system in which the heart pumps blood to both the lungs and the body itself. The system is not very efficient because oxygenated blood and deoxygenated blood mix in the single ventricle. ***c.*** The pulmonary and systemic systems are completely separate in birds and mammals since the heart is divided by a septum into a right and left half. The right side pumps blood to the lungs, and the left side pumps blood to the body proper.

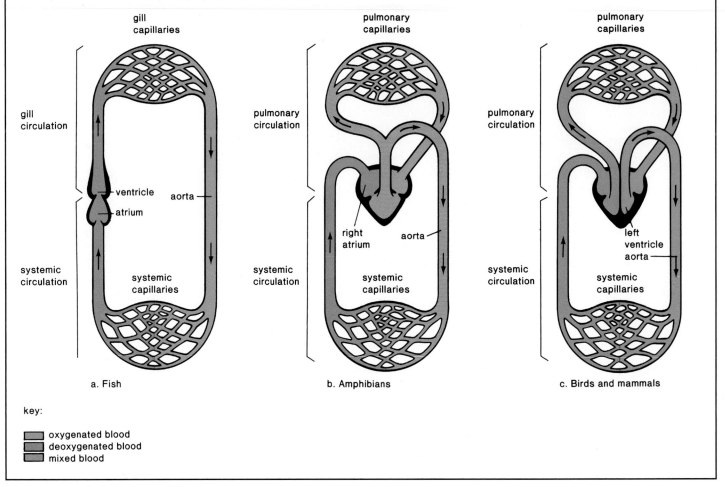

key:

- oxygenated blood
- deoxygenated blood
- mixed blood

relaxes, blood can enter a capillary bed. When the sphincter contracts, a capillary is closed and no blood can enter. Contraction and relaxation of sphincters is controlled by the nervous system. Not all capillary beds are open at the same time. After an animal has eaten, the capillary beds around the digestive tract are usually open, and during muscular exercise, the capillary beds of the skeletal muscles are open.

Venules and veins collect blood from the capillary beds and take it to the heart. First the venules drain the blood from the capillaries, and then they join together to form a vein. The wall of a vein is much thinner than that of an artery because the muscle and elastic fiber layers are poorly developed. Valves within the veins point, or open, toward the heart, preventing a backflow of blood.

Comparison of Circulatory Pathways

Among vertebrate animals, there are 3 different types of circulatory pathways. In fishes, blood follows a one-circuit (single-loop circulatory) pathway through the body. The heart has a single atrium and a single ventricle (fig. 34.4*a*). The pumping action of the ventricle sends blood under pressure only to the gills, where it is oxygenated. After passing through gill capillaries, there is little blood pressure left to distribute the oxygenated blood to the tissues.

In contrast, the other vertebrates have a 2-circuit (double-loop circulatory) pathway. The heart pumps blood to the lungs, called *pulmonary circulation,* and pumps blood to the tissues, called *systemic circulation.* This double pumping action assures adequate blood pressure and flow to both circulatory loops.

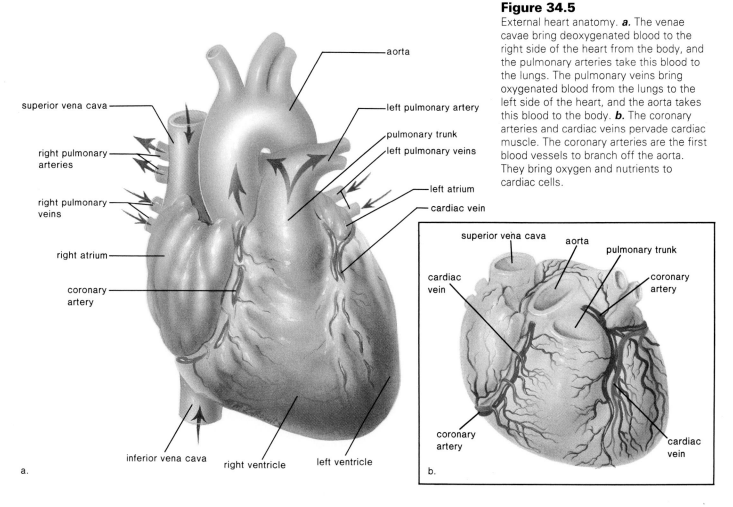

Figure 34.5
External heart anatomy. **a.** The venae cavae bring deoxygenated blood to the right side of the heart from the body, and the pulmonary arteries take this blood to the lungs. The pulmonary veins bring oxygenated blood from the lungs to the left side of the heart, and the aorta takes this blood to the body. **b.** The coronary arteries and cardiac veins pervade cardiac muscle. The coronary arteries are the first blood vessels to branch off the aorta. They bring oxygen and nutrients to cardiac cells.

In amphibians, the heart has 2 atria, but there is only a single ventricle (fig. 34.4*b*). The hearts of other vertebrates are partially (most reptiles) or completely (some reptiles, all birds, and mammals) divided into right and left halves (fig. 34.4*c*). The right ventricle pumps blood to the lungs, and the left ventricle pumps blood to the rest of the body. This arrangement increases the likelihood of adequate blood pressure for both the pulmonary and systemic circulations.

> In mammals, birds, and reptiles, the heart is divided into a right side (pumping deoxygenated blood) and a left side (pumping oxygenated blood); this ensures adequate blood pressure for both the pulmonary and systemic circulations.

Human Circulation

William Harvey discovered in the seventeenth century that the blood circulates in humans and other animals (see following reading). He realized that human circulation resembles that of other mammals (fig. 34.4*c*). The functions of the circulatory system include (1) transporting gases, nutrients, and wastes about the body, (2) clotting to prevent loss of blood from injured blood vessels, and (3) fighting infection or invasion of the body by microorganisms.

The Human Heart

The heart is a cone-shaped muscular organ about the size of a fist (fig. 34.5). It is located between the lungs directly behind the sternum and is tilted so that the apex is directed to the left. The major portion of the heart is called the *myocardium* and consists largely of cardiac muscle tissue (fig. 33.6*c*). Epithelial tissue lines the heart chambers, and the organ is surrounded by a membranous *pericardial sac*. Normally, this sac contains a small quantity of liquid to lubricate the heart.

Internally, the right and left sides of the heart are separated by a *septum* (fig. 34.6). The heart has valves that direct the flow of blood and prevent its backward flow. The valves that lie between the atria and the ventricles are called the *atrioventricular valves*. These valves are supported by strong fibrous strings called *chordae tendineae*. There are also *semilunar valves*, which resemble half moons, between the ventricles and their attached vessels.

Figure 34.6

Internal heart anatomy. ***a.*** The venae cavae empty into the right atrium, and the pulmonary trunk leaves from the right ventricle. The pulmonary veins enter the left atrium, and the aorta leaves from the left ventricle. Note that the muscles attaching the chordae tendineae chordae tendineae to the right ventricular wall have been cut. ***b.*** A diagrammatic representation that allows you to trace the path of blood through the heart.

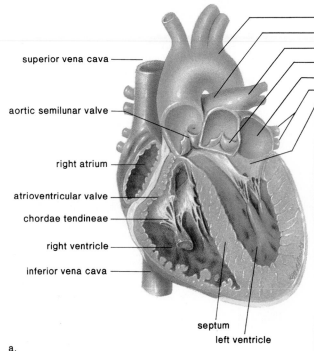

- aorta
- pulmonary trunk
- left pulmonary artery
- pulmonary semilunar valve
- left atrium
- left pulmonary veins
- atrioventricular valve

superior vena cava

aortic semilunar valve

right atrium

atrioventricular valve

chordae tendineae

right ventricle

inferior vena cava

septum

left ventricle

a.

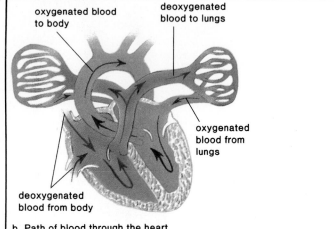

b. Path of blood through the heart

- The superior (anterior) vena cava and the inferior (posterior) vena cava carrying deoxygenated blood (low in oxygen and high in carbon dioxide) enter the right atrium.
- The right atrium sends blood through an atrioventricular valve (the tricuspid valve) to the right ventricle.
- The right ventricle sends blood through the pulmonary semilunar valve into the pulmonary trunk and the pulmonary arteries to the lungs.
- The pulmonary veins carrying oxygenated blood (high in oxygen and low in carbon dioxide) from the lungs enter the left atrium.
- The left atrium sends blood through an atrioventricular valve (the bicuspid, or mitral, valve) to the left ventricle.
- The left ventricle sends blood through the aortic semilunar valve into the aorta to the body proper.

The Heartbeat

The heart contracts, or beats, about 70 times a minute, and each heartbeat lasts about 0.85 sec. Each heartbeat, or *cardiac cycle,* consists of the following elements (the word **systole** refers to contraction of heart muscle, and the word **diastole** refers to relaxation of heart muscle):

Cardiac Cycle		
Time	Atria	Ventricles
0.15 sec	Systole	Diastole
0.30 sec	Diastole	Systole
0.40 sec	Diastole	Diastole

First the atria contract (while the ventricles relax), and then the ventricles contract (while the atria relax) and then all chambers rest. The short systole of the atria is appropriate since the atria send blood only into the ventricles. It is the muscular ventricles that actually pump blood out into the circulatory system proper. When the word *systole* is used alone, it usually refers to the left ventricular systole.

When the heart beats, the familiar lub-DUPP sound is heard as the valves of the heart close. The *lub* is caused by vibrations of the heart when the atrioventricular valves close, and the *DUPP* is heard when vibrations occur due to the closing of the semilunar valves. These valves aid circulation of the blood through the heart because they permit only one-way flow and do not allow a backward flow of blood. The *pulse* is a wave of fibrillation, which passes down the walls of the arterial blood vessels when the aorta expands following ventricle systole. Because there is one arterial pulse per ventricular systole, the arterial pulse rate can be used to determine the heart rate.

The contraction of the heart is intrinsic to the heart; it does not require outside nervous stimulation. The reason for this is a unique type of tissue called nodal tissue, with both muscular

Animal Structure and Function

Figure 34.7

Control of the cardiac cycle. ***a.*** The SA node sends out a stimulus that causes the atria to contract. When this stimulus reaches the AV node, it signals the ventricles to contract by way of the Purkinje fibers. ***b.*** A normal ECG indicates that the heart is functioning properly. The P wave occurs as the atria contract; the QRS wave occurs as the ventricles contract; and the T wave occurs when the ventricles are recovering from contraction. ***c.*** Abnormal ECGs: sinus tachycardia is an abnormally fast heartbeat due to a fast pacemaker; ventricular fibrillation is irregular heartbeat due to irregular simulation of the ventricles; and mitral stenosis occurs because the bicuspid (mitral) valve is obstructed.

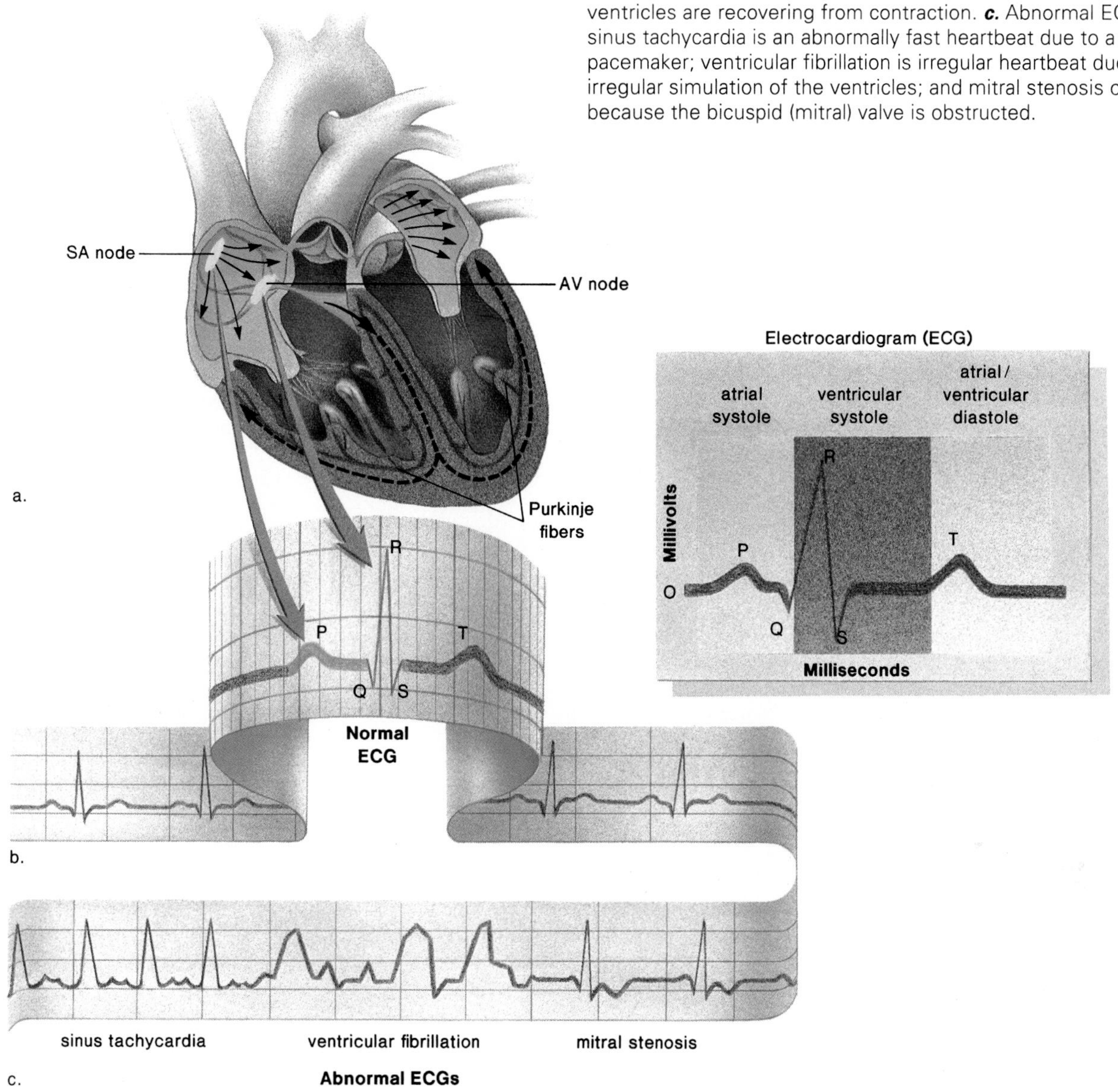

and nervous characteristics, located in 2 regions of the heart. The first of these, the *SA (sinoatrial) node,* is found in the upper dorsal wall of the right atrium; the other, the *AV (atrioventricular) node,* is found in the base of the right atrium very near the septum (fig. 34.7). The SA node initiates the heartbeat and automatically sends out an excitation impulse every 0.85 sec to cause the atria to contract. When the impulse reaches the AV node, it signals the ventricles to contract by way of specialized fibers called *Purkinje fibers.* The **SA node** is called the *pacemaker* because it usually keeps the heartbeat regular.

The rate of the heartbeat is also under nervous control. A cardiac center in the medulla oblongata of the brain can alter the beat of the heart by way of the autonomic nervous system. This system is made up of 2 divisions: the parasympathetic system, which promotes the functions we associate with normal activities, and the sympathetic system, which brings about the responses we associate with times of stress. The parasympathetic system slows the heartbeat, and the sympathetic system increases the heartbeat. Various factors, such as the relative need for oxygen or the level of blood pressure, determine which of these systems is activated.

illiam Harvey (1578-1657) was the first to offer proof that the blood circulates in the body of humans and other animals. Harvey was an English scientist of the seventeenth century, a time of renewed interest in the collection of facts, the use of the hypothesis, experimentation, and respect for mathematics. The seventeenth century was the time of the scientific revolution.

After many years of research and study, Harvey hypothesized that the heart is a pump for the entire circulatory system and that blood flows in a circuit. In contrast to former anatomists, Harvey dissected not only dead but also live organisms and observed that when the heart beats, it contracts, forcing blood into the aorta. Had this blood come from the right side of the heart? To do away with the complication of the lungs (pulmonary circulation), Harvey turned to fishes and noticed that the heart first receives and then pumps the blood forward. He observed that blood in the fetus passes directly from the right side of the heart through the septum to the left side. He felt confident that in mature higher organisms, all blood moves from the right to the left side of the heart by way of the lungs.

Harvey then wanted to show an intimate connection between the arteries and

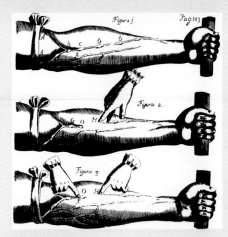

Figure 34.A

To show that the veins return blood to the heart, Harvey tied a ligature (*Figure 1*) above the elbow to observe the accumulation of blood in the veins. Blood can be forced past a valve from H to O (*Figure 2*), but not in the opposite direction (*Figure 3*). Therefore, it can be deduced that blood ordinarily moves toward the heart in the veins.

veins in the tissues of the body. Again using live lower organisms, he demonstrated that if an artery is slit, the whole blood system empties, including arteries and veins. He measured the capacity of the

left ventricle in humans and found it to be 2 oz. Since the heart beats 72 times a minute, in one hr the left ventricle forces into the aorta no less than 72 X 60 X 2 = 8,640 oz = 540 lb, or 3 times the weight of a heavy man! Could so much blood be created and consumed every hour? The same blood must return again and again to the heart.

Harvey also studied the valves in the veins and suggested their true purpose. By the use of ligatures, he demonstrated that a tight ligature on the arm causes the artery to swell on the side of the heart, and a slack ligature causes the vein to swell on the opposite side (fig. 34.A). He said, "This is an obvious indication that the blood passes from the arteries into the veins . . . . and there is an anastomosis of the 2 orders of vessels."

Harvey's methods showed how fruitful research might be done. He established that physical and mechanical evidence could provide data for a theory of circulation. He erred, however, when he speculated on the function of the heart and lung. He thought the heart heated the blood, and the lung served to cool it or control the degree of heat. His basic method, however, contributed to the scientific revolution and set an example for others to follow.

The conduction system of the heart includes the SA node, the pacemaker, which results in atrial contraction, and also the AV node and the Purkinje fibers, which result in ventricular contraction.

Vascular Pathways

Human circulation includes 2 circular pathways, the **pulmonary circuit** and the **systemic circuit** (fig. 34.8).

Pulmonary Circuit

Deoxygenated blood from all regions of the body collects in the right atrium and then passes into the right ventricle, which pumps it into the pulmonary trunk. The pulmonary trunk divides into the pulmonary arteries, which carry blood to arterioles and capillaries in the lungs. After passing through lung capillaries located around the alveoli (see fig. 37.8), oxygenated blood returns to the left atrium of the heart through pulmonary venules and veins.

Systemic Circuit

The **aorta** and the **venae cavae** (vena cava, sing.) serve as the major pathways for blood in the systemic circuit. To trace the

path of blood to any organ in the body, you need only mention the aorta, the proper branch of the aorta, the organ, and the vein returning blood to the vena cava. In the systemic circuit, arteries contain oxygenated blood and have a bright red color, but veins contain deoxygenated blood and appear in purplish color.

The *coronary arteries* are extremely important because they serve the heart muscle itself (fig. 34.5). (The heart is not nourished by the blood in its chambers.) The coronary arteries arise from the aorta just above the aortic semilunar valve. They lie on the exterior surface of the heart, where they branch into arterioles and then capillaries. The capillary beds enter venules, which join to form the cardiac veins, and these empty into the right atrium. The coronary arteries have a very small diameter and may become blocked, causing a heart attack, as discussed in the reading, "Cardiovascular Disease."

A **portal system** is one that begins and ends in capillaries. One place in the human body where a portal system is found is between the small intestine and the liver. Blood passes from the capillaries of intestinal villi into venules that join to form the hepatic portal vein, a vessel that connects the intestine with the liver. The hepatic vein leaves the liver and enters the inferior vena cava.

Figure 34.8

Blood vessels in the pulmonary and systemic circuits. The blue-colored vessels carry deoxygenated blood, and the red-colored vessels carry oxygenated blood; the arrows indicate the flow of blood. In order to trace blood from the right to the left side of the heart, you must begin at the lung capillaries. In order to trace blood from the gut capillaries to the right atrium, you must consider the hepatic portal vein and hepatic vein.

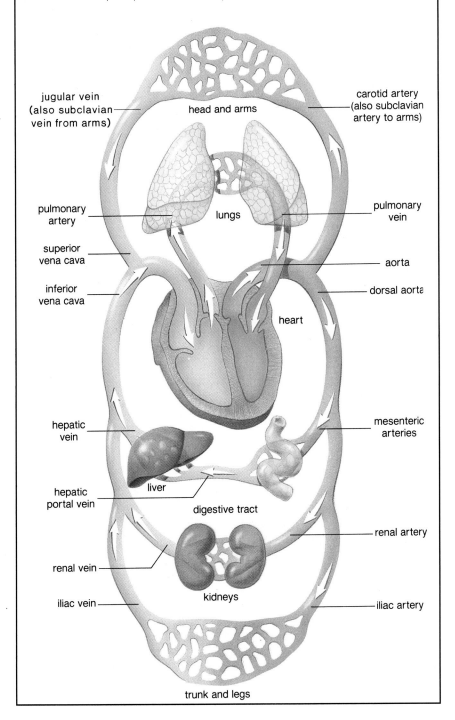

jugular vein (also subclavian vein from arms)
head and arms
carotid artery (also subclavian artery to arms)
pulmonary artery
lungs
pulmonary vein
superior vena cava
aorta
inferior vena cava
dorsal aorta
heart
hepatic vein
mesenteric arteries
hepatic portal vein
liver
digestive tract
renal artery
renal vein
kidneys
iliac vein
iliac artery
trunk and legs

The pulmonary circuit takes deoxygenated blood to the lungs and oxygenated blood to the heart. The systemic circuit takes blood through the aorta to the vena cavae by numerous routes.

Systemic Blood Flow

When the left ventricle contracts, blood is forced into the arteries under pressure. *Systolic pressure* results from blood being forced into the arteries during ventricular systole, and *diastolic pressure* is the pressure in the arteries during ventricular diastole. Human **blood pressure** is measured with a sphygmomanometer, which has a pressure cuff that permits measurement of the amount of pressure required to stop the flow of blood through an artery. Blood pressure is normally measured on the brachial artery, which is in the upper arm, and is stated in millimeters of mercury (mm Hg). A blood pressure reading consists of 2 numbers, for example, 120/80—which represents systolic and diastolic pressures, respectively. For a man under age 45, a reading above 130/90 indicates *hypertension* (high blood pressure) and beyond age 45, a reading above 140/95 is considered hypertensive. High blood pressure is associated with cardiovascular disease, as discussed on page 566.

As blood flows from the aorta into the various arteries and arterioles, blood pressure falls (fig. 34.9). Also, the difference in pressure between systolic and diastolic gradually decreases until it disappears in the capillaries, where there is an even, steady flow of blood. Blood pressure in the venules is very low (5–10 mm Hg) and even lower in the vena cava (2 mm Hg). This low a pressure cannot move blood back to the heart, especially from the limbs of the body. When skeletal muscles near veins contract, they put pressure on the collapsible walls of the veins and on blood contained in these vessels. Veins, however, have *valves* that prevent the backward flow of blood, and therefore pressure from muscle contraction is sufficient to move blood through veins toward the heart (fig. 34.10). *Varicose veins,* abnormal dilations in superficial veins, develop when the valves of the veins become weak and ineffective due to a backward pressure of the blood.

In the human circulatory system, the beat of the heart supplies the pressure that keeps blood moving in the arteries; pressure drops off in the capillaries; and skeletal muscle contraction pushes blood in the veins back to the heart.

Human Blood

Blood has 2 main portions: the liquid portion, called plasma, and cells. **Plasma** contains many types of molecules, including nutrients, wastes, salts, and proteins (table 34.1). The salts and proteins buffer the blood, effectively keeping the pH near 7.4. They also maintain

Figure 34.9
Blood pressure in different parts of the human circulatory system. The pulse pressure (includes systolic and diastolic pressures) decreases as blood flows into smaller arteries and disappears as blood enters capillaries. The slow and even movement of the blood through the capillaries facilitates exchange of molecules there. The blood pressure in the veins is so low that it cannot account for the movement of blood in these vessels. Skeletal muscle contraction pushes the blood along from valve to valve in the veins.

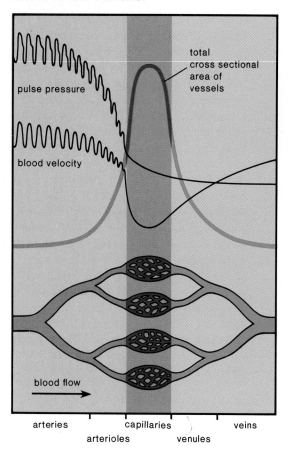

Figure 34.10
Cross section of a valve. *a.* When a valve is open, a result of pressure on the veins exerted by skeletal muscle contraction, the blood flows toward the heart in veins. *b.* Valves close when external pressure is no longer applied to them. Closure of the valves prevents the blood from flowing in the opposite direction.

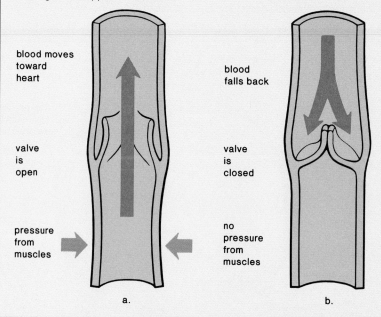

Table 34.1
Components of Plasma

Component	Function	Source
Water	Maintains blood volume and transports molecules	Absorbed from intestine
Proteins	Maintain blood osmotic pressure and pH	
Albumin	Transports organic molecules	Liver
Fibrinogen	Clotting	Liver
Globulins	Fight infection	Lymphocytes
Gases		
Oxygen	Cellular respiration	Lungs
Carbon dioxide	End product of metabolism	Tissues
Nutrients		
Fats, glucose, amino acids, etc.	Food for cells	Absorbed from intestinal villi
Salts	Maintain blood osmotic pressure and pH; aid metabolism	Absorbed from intestinal villi
Wastes		
Urea and uric acid	End products of metabolism	Liver
Hormones, vitamins, etc.	Aid metabolism	Varied

Plasma is 90%–92% water, 7%–8% plasma proteins, and not quite 1% salts. All other components are present in even smaller amounts.

Animal Structure and Function

blood's osmotic pressure so that water has an automatic tendency to enter blood capillaries. The plasma proteins assist in transporting large organic molecules in blood. For example, the lipoproteins that transport cholesterol are globulins, and bilirubin, a breakdown product of hemoglobin, is transported by albumin.

Types of Blood Cells

Blood cells are of 3 types: red blood cells, or **erythrocytes**; white blood cells, or **leukocytes**; and platelets, or **thrombocytes.**

Red Blood Cells

Red blood cells are small biconcave disks that lack a nucleus and contain the respiratory pigment hemoglobin. There are 4–6 million red blood cells per mm^3 of whole blood, and each one of these cells contains about 200 million hemoglobin molecules. **Hemoglobin** contains 4 globin protein chains, each associated with heme, an iron-containing group. Iron combines loosely with oxygen, and in this way oxygen is carried in the blood. If there is an insufficient number of red blood cells or the cells do not have enough hemoglobin, the individual suffers from anemia and has a tired, run-down feeling.

Red blood cells are manufactured continuously in the red bone marrow of the skull, ribs, vertebrae, and ends of the long bones. The growth factor erythropoietin, which stimulates the production of red blood cells, is now a biotechnology product (p. 284). It is helpful to persons with anemia and is also sometimes abused by athletes who want to increase performance.

Before they are released from the bone marrow into blood, red blood cells lose their nucleus and acquire hemoglobin (fig. 34.11a). Possibly because they lack a nucleus, red blood cells live about 120 days. They are destroyed chiefly in the liver and the spleen, where they are engulfed by large phagocytic cells. When red blood cells are destroyed, hemoglobin is released. The iron is recovered and is returned to the red bone marrow for reuse. The heme portions of the molecules undergo chemical degradation and are excreted by the liver as bile pigments in the bile. The bile pigments are primarily responsible for the color of feces.

> Oxygen is transported to the tissues in combination with hemoglobin, a pigment found in red blood cells.

White Blood Cells

White blood cells differ from red blood cells in that they are usually larger and have a nucleus, they lack hemoglobin, and without staining, they appear white in color. With staining, white blood cells appear light blue unless they have granules that bind with certain stains. The *granular leukocytes* have a lobed nucleus and may be neutrophils, which have granules that do not take up a dye; eosinophils, which have granules that take up the red dye eosin; and basophils, which have granules that take up a basic dye staining them a deep blue. *Agranular leukocytes* (with no granules) have a circular or indented nucleus. There are 2 types—the larger monocytes and the smaller lymphocytes (fig. 34.11a and b). The newly discovered stem cell growth factor (SGF) can be used to increase the production of all white blood cells, and there are also various

specific stimulating factors such as GM-CSF (granulocyte-macrophage colony-stimulating growth factor) that can be used to stimulate the production of specific stem cells. These growth factors should be helpful to patients with low immunity, such as AIDS patients.

When microorganisms enter the body due to an injury, the response is called an *inflammatory reaction* because there is swelling and reddening at the injured site. The damaged tissue has released kinins, which cause vasodilation, and histamines, which cause increased capillary permeability. **Neutrophils,** which are amoeboid, squeeze through the capillary wall and enter the tissue fluid, where they phagocytize foreign material. Monocytes appear and are transformed into **macrophages,** which are large phagocytizing cells that release white blood cell growth factors (see fig. 35.4). Soon there is an explosive increase in the number of leukocytes. The thick, yellowish fluid called pus contains a larger proportion of dead white blood cells that have fought the infection.

Lymphocytes also play an important role in fighting infection. Certain lymphocytes called T cells attack cells that contain viruses. Other lymphocytes called B cells produce antibodies. Each B cell produces just one type of antibody, which is specific for one type of antigen. An **antigen,** which is most often a protein but sometimes a polysaccharide, is found in the outer covering of a parasite or is present in its toxin. When **antibodies** combine with antigens, the complex is often phagocytized by a macrophage. An individual is actively immune when a large number of B cells are all producing the specific antibody needed for a particular infection. Chapter 35 deals with immunity and explores this topic in detail.

> The inflammatory response is a "call to arms"—it marshals white blood cells to the site of invasion by microorganisms. Neutrophils and monocytes (macrophages) are phagocytic; B lymphocytes produce antibodies to combine with disease-causing antigens.

Platelets

Strictly speaking, platelets (thrombocytes) are not cells. Platelets result from fragmentation of certain large cells, called megakaryocytes, in the bone marrow (fig. 34.11). **Platelets** are involved in the process of blood clotting, or coagulation. Also necessary to the process are fibrinogen and prothrombin, proteins manufactured and deposited in blood by the liver. If clotting occurs in a test tube, a fluid called *serum* collects above the clot. Serum has the same composition as plasma except that it lacks prothrombin and fibrinogen.

Blood Clotting When a blood vessel in the body is damaged, platelets clump at the site of the puncture and partially seal the leak. They and the injured tissues release a clotting factor called *prothrombin activator* that converts prothrombin to thrombin. This reaction requires calcium (Ca^{++}) ions. Thrombin, in turn, acts as an enzyme that severs 2 short amino acid chains from each fibrinogen molecule. These activated fragments then join end to end, forming

Figure 34.11

a. Blood cell formation in red bone marrow. Multipotent stem cells give rise to specialized stem cells that produce the various types of blood cells. Each stem cell is under the control of specific growth factors. **b.** Human blood cells, their abundance, size, and function. (See next page.)

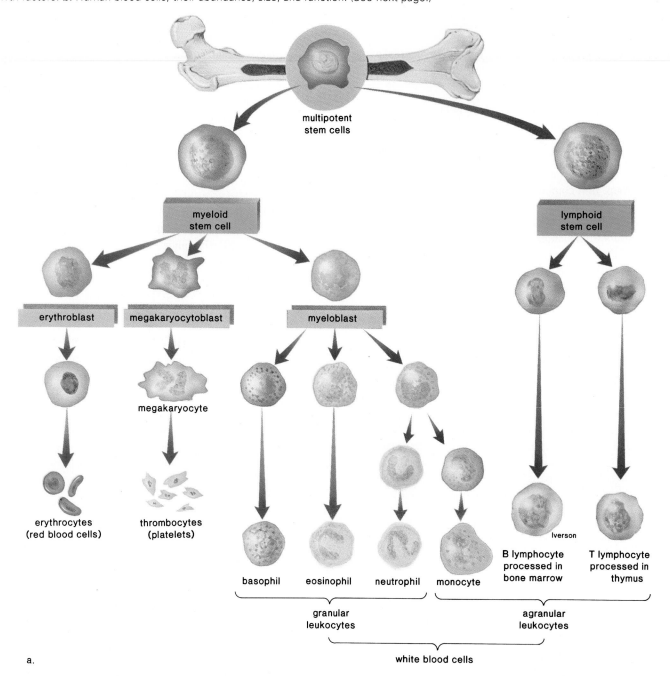

a.

long threads of *fibrin*. Fibrin threads wind around the platelet plug in the damaged area of the blood vessel and provide the framework for the clot. Red blood cells also are trapped within the fibrin threads; these cells make a clot appear red.

A fibrin clot is only temporarily present. As soon as blood vessel repair is initiated, an enzyme called plasmin destroys the fibrin network and restores the fluidity of plasma. This is a protective measure because a blood clot can act as a thrombus or an embolus. In either case, it interferes with circulation and even can cause death of tissues in the area.

The steps necessary for blood clotting upon injury are quite complex, but they can be summarized as in figure 34.12.

A blood clot consists of platelets and red blood cells entangled within fibrin threads.

Human Blood Cells			
Component	Number (per mm³ of blood)	Size	Function
Erythrocyte (red blood cell)	5,000,000	7.5 μm	Transport oxygen; assists CO_2 transport
Leukocyte (white blood cell)	7,000	8–20 μm	Fight infection
Thrombocytes (platelets)	200,000–400,000	2–4 μm	Aid blood clotting
White Blood Cell Types	% of Total White Blood Cells	Size	Function
Granular leukocytes			
Neutrophils	55–65	9–12 μm	Phagocytize primarily bacteria
Eosinophils	2–3	9–12 μm	Phagocytize and destroy antigen-antibody complexes
Basophils	0.5	9–12 μm	Congregate in tissues; release histamine when stimulated
Agranular leukocytes			
Lymphocytes	25–33	8–10 μm	B cell produces antibodies in blood and lymph; T cell destroys virus infected cells
Monocytes	4–7	12–20 μm	Become macrophages— phagocytize bacteria and viruses

b.

Figure 34.12

Blood clotting. **a.** Platelets and damaged tissue cells release prothrombin activator, which acts on prothrombin in the presence of calcium to produce thrombin, which acts on fibrinogen to form fibrin. **b.** A scanning electron micrograph showing a red blood cell caught in the fibrin threads of a clot. Magnification, X3,000.

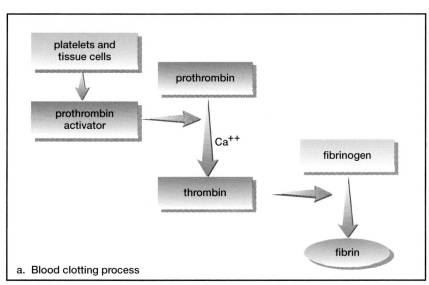

a. Blood clotting process

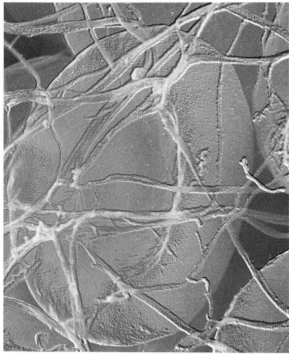

b. Blood clot

Circulatory System

565

Cardiovascular Disease

Because sudden cardiac death happens once every 72 sec in the United States, it is well to identify the factors that predispose an individual to cardiovascular disease. The risk factors for cardiovascular disease include the following:

Male sex
Family history of heart attack under age 55
Smoking more than 10 cigarettes a day
Severe obesity (30% or more overweight)
Hypertension
Unfavorable HDL and LDL cholesterol blood levels
Impaired circulation to the brain or the legs
Diabetes mellitus

Hypertension (high blood pressure) is well recognized as a major factor in cardiovascular disease, and 2 controllable behaviors contribute to hypertension. Smoking cigarettes, including filtered cigarettes, causes hypertension, as does obesity. It is best to never take up the habit of smoking cigarettes, but most of the detrimental effects can be reversed if you stop smoking. Since it is very difficult for obese individuals to lose weight, it is recommended that weight control be a lifelong endeavor.

Investigators have identified several behaviors that may help to reduce the possibility of heart attack and stroke. Exercise seems to be critical. Sedentary individuals have a risk of cardiovascular disease that is about double that of those who are very active. One physician, for example, recommends that his patients walk for one hour, 3 times a week. Stress reduction also is desirable. The same investigator recommends everyday mediation and yogalike stretching and breathing exercises to reduce stress.

Another behavior that is now well known is the adoption of a diet that is low in saturated fats and cholesterol because such a diet is believed by many to protect against the development of cardiovascular disease. Cholesterol is ferried in the blood by 2 types of plasma proteins called LDL (low-density lipoprotein) and HDL (high-density lipoprotein). LDL (called "bad" lipoprotein) takes cholesterol to the tissues from the liver, and HDL (called "good" lipoprotein) transports cholesterol out of the tissues to the liver. When the LDL level in blood is abnormally high or the HDL level is abnormally low, cholesterol accumulates in the cells. When cholesterol-laden cells line the arteries, plaque develops, which interferes with circulation (fig. 34.B).

Cholesterol guidelines have been established by the National Heart, Lung, and Blood Institute. According to the institute, everyone should know his or her cholesterol blood level. Individuals with a borderline high cholesterol blood level (200–239 mg/100 ml) should be tested further if they already have heart disease or if they have 2 known risk factors of cardiovascular disease (see list). Individuals with a high cholesterol blood level (240 mg/100 ml) always should be tested further. Persons with an LDL cholesterol level of over 130 mg/100 ml should be treated if they have other risk factors, and those with an LDL cholesterol level of 160 mg/100 ml should be treated even if this is the only risk factor.

Persons with a total-to-HDL cholesterol ratio higher than 4.5 also are considered to be at risk. Heart attack has occurred in individuals who have a normal total cholesterol level but who also have an unfavorable total-to-HDL cholesterol ratio. For example, if a person's total cholesterol blood level is 200, but the HDL level is only 25 mg/100 ml, then the total-to-HDL cholesterol ratio is 8.0, and circulatory difficulties most likely will develop.

First and foremost, treatment for unfavorable cholesterol levels consists of adopting a diet that is low in saturated fat and cholesterol (see p. 599). Although the prescribed diet does not lower cholesterol blood level in all persons, it is expected to do so for most individuals. If diet alone does not bring down the cholesterol blood level, drugs can be prescribed. Some of the drugs act in the intestine to remove cholesterol, and others act in the body to prevent its production. These drugs reduce the blood level of cholesterol, but their long-term side effects are not completely known and may be serious. Considering this, some investigators do not recommend these drugs to lower cholesterol blood levels.

A diet low in saturated fat and cholesterol can lower the total cholesterol blood level and the LDL level of some individuals, but this diet most likely will not raise the

Capillary Exchange

No cell is far away from one or more capillaries because capillaries are found throughout every tissue in the body. The cross-sectional area of capillaries is so large that blood flows very slowly through the capillaries (fig. 34.9). This slow rate allows adequate time for exchange of materials between blood and tissue fluid. **Tissue fluid** is the internal environment of cells, and the circulatory system performs an important homeostatic function by providing nutrients to and removing wastes from tissue fluid.

Much of the exchange between blood and tissue fluid occurs by diffusion through the one-cell-thick capillary walls. While lipid-soluble substances can pass freely through the plasma membrane of cells, water-soluble substances usually diffuse through pores present in junctions between the cells of the capillary walls.

Two forces control movement of fluid through the capillary wall: osmotic pressure, which tends to cause water to move from tissue fluid to blood, and blood pressure, which tends to cause water to move in the opposite direction.

As figure 34.13 illustrates, blood pressure is higher than osmotic pressure at the arterial end of a capillary. This tends to force fluid out through the pores in the capillary wall. Midway along the capillary, where blood pressure is lower, the 2 forces essentially cancel one another, and there is no net movement of water. Solutes now diffuse according to their concentration gradient, however, nutrients (glucose and oxygen) diffuse out of the capillary, and wastes (carbon dioxide) diffuse into the capillary. Since proteins are too large to pass out of the capillary, tissue fluid tends to contain all components of plasma except proteins. At the

Animal Structure and Function

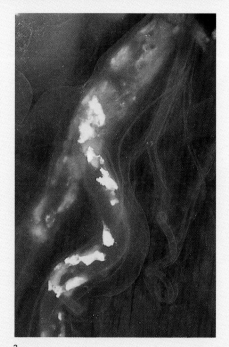

a.

smooth
muscle cells

lumen
of vessel

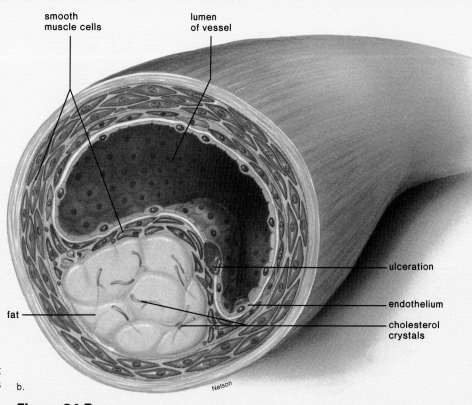

ulceration

endothelium

cholesterol
crystals

fat

b.

Nelson

Figure 34.B

a. Plaques (*yellow*) in the coronary artery of a heart patient. **b.** Cross section of an artery that is partially blocked by plaque. The plaque bulges out into the lumen of the artery, obstructing blood flow. **b.** From Kent M. Van De Graaff and Stuart I. Fox, *Concepts of Human Anatomy and Physiology,* 2d ed. Copyright © 1989 Wm. C. Brown Publishers, Dubuque, Iowa. All Rights Reserved. Reprinted by permission.

HDL level. Aside from certain drugs that apparently can raise HDL level, exercise is sometimes effective.

There is nothing that can be done about some of the cardiovascular risk factors, such as male gender and family history. Other risk factors, however, likely can be controlled if the individual believes it is worth the effort. It is clear that the 4 great admonitions for a healthy life—heart-healthful diet, regular exercise, proper weight maintenance, and refraining from smoking—all contribute to acceptable blood pressure and cholesterol blood levels.

venule end of a capillary, where blood pressure has fallen, osmotic pressure is greater than blood pressure, and water tends to move into the capillary. Almost the same amount of fluid that left the capillary returns to it, although there is always some excess tissue fluid collected by the lymphatic capillaries (fig. 34.14).

> Oxygen and nutrient molecules exit a capillary near the arterial end; carbon dioxide and waste molecules enter a capillary near the venous end.

Blood Typing

Although there are at least 12 well-known blood type identification systems, the ABO system and the RH system are most often used to determine blood type.

The ABO System

Before the twentieth century, blood transfusions were attempted, although sometimes the results were dire and even caused the death of the recipient. Concerned about these occurrences, a newly established physician in Vienna, Karl Lansteiner, began to examine the effect of mixing different samples of blood. In the end, he and his associates determined that there are 4 major blood groups among humans. They designated the types of blood as A, B, AB, and O. Table 34.2 shows that blood type is dependent upon the presence of antigens, specific glycoproteins, on red blood cells. Type O blood has neither the A antigen nor the B antigen on red blood cells; the other types of blood do have one or both of the antigen(s) present. These are antigens only to a recipient and not to the person with that type blood.

Figure 34.13

Diagram of a capillary exchange. At the arterial end of a capillary, the blood pressure is higher than the osmotic pressure. Therefore, water exits here. At the venous end of a capillary, the osmotic pressure is higher than the blood pressure. Therefore, water enters here. In between the 2 ends of the capillary, nutrients, gases, and wastes diffuse in or out according to their concentration gradients.

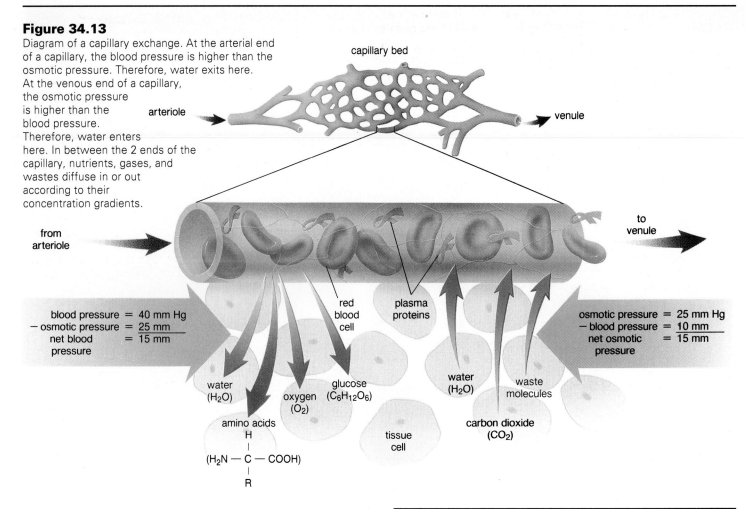

blood pressure = 40 mm Hg
− osmotic pressure = 25 mm
net blood = 15 mm
pressure

osmotic pressure = 25 mm Hg
− blood pressure = 10 mm
net osmotic = 15 mm
pressure

water (H_2O)

oxygen (O_2)

glucose ($C_6H_{12}O_6$)

water (H_2O)

waste molecules

amino acids
H
|
(H_2N — C — COOH)
|
R

carbon dioxide (CO_2)

tissue cell

red blood cell

plasma proteins

Within the plasma, there are antibodies to the antigens that are not present on the red blood cells. Therefore, for example, type A has an antibody called anti-B in the plasma. This is reasonable because if the A antigen and anti-A antibody are present, for example, **agglutination,** or clumping of red blood cells, occurs. For a recipient to receive blood from a donor, the recipient's plasma must not have an antibody that causes the donor's cells to agglutinate. For this reason it is important to determine each person's blood type. Figure 34.15 demonstrates a way to use the antibodies derived from plasma to determine the blood type. If clumping occurs after a sample of blood is exposed to a particular antibody, the person has that type of blood.

The Rh System

Another important antigen in matching blood types is the Rh factor. Persons with this particular antigen on the red blood cells are Rh positive (Rh+); those without it are Rh negative (Rh−). Rh-negative individuals do not normally make antibodies to the Rh factor, but they do make them when exposed to the Rh factor. It is possible to extract these antibodies and use them for blood type testing (fig. 34.15).

The Rh factor is particularly important during pregnancy. If the mother is Rh negative and the father is Rh positive, the child may be Rh positive. The child's Rh-positive red blood cells begin to

Figure 34.14

Lymphatic vessels. Arrows indicate that lymph is formed when lymphatic capillaries take up excess tissue fluid. Lymphatic capillaries lie near blood capillaries.

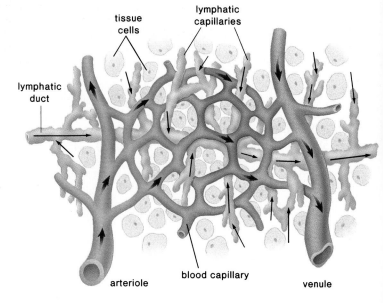

Table 34.2
Blood Groups

Blood Type	Antigen or Red Blood Cells	Antibody in Plasma	% U.S.* Black	% U.S.* Caucasian
A	A	Anti-B	25	41
B	B	Anti-A	20	7
AB	A,B	None	4	2
O	None	Anti-A and anti-B	51	50

*Blood type frequency for other races is not available.

leak across the mother's circulatory system because placental tissues normally break down before and at birth. This causes the mother to produce Rh antibodies. If the mother becomes pregnant with another Rh-positive baby, Rh antibodies may cross the placenta and destroy this child's red blood cells. This is called *hemolytic disease of the newborn,* and the problem is solved by giving Rh-negative women an Rh-immune globulin injection just after the birth of an Rh-positive child. This injection contains Rh-antibodies, which attack the baby's red blood cells before these cells stimulate the mother's immune system to produce its own antibodies.

Human blood consists of liquid plasma, which transports nutrient and waste molecules, and cells. Red blood cells carry oxygen, and white blood cells fight infection. Blood clotting involves the platelets, and blood typing is based on types of antigens on the red blood cells.

Figure 34.15
The standard test to determine ABO and Rh blood type consists of putting a drop of anti-A antibodies, anti-B antibodies, and anti-Rh antibodies on a slide. To each of these a drop of the person's blood is added. *a.* A possible reaction to any one of these is shown. Top, no reaction; therefore, no antigen to that antibody is present. Bottom, agglutination occurs; therefore, antigen is present. *b.* Several possible results. For example, in the top row there is no agglutination with either anti-A or anti-B, but there is agglutination with anti-Rh. Therefore, the person's blood type is O⁺.

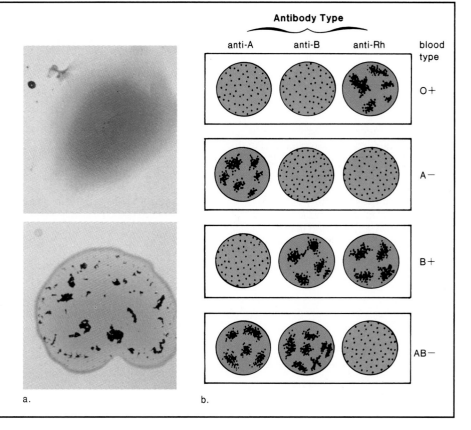

a.   b.

Summary

1. Some invertebrates do not have a transport system. The presence of a gastrovascular cavity allows diffusion alone to supply the needs of cells in cnidarians and flatworms.
2. Other invertebrates do have a transport system. Insects have an open system and earthworms have a closed one.
3. Vertebrates have a closed system in which arteries carry blood away from the heart to capillaries, where exchange takes place, and veins carry blood to the heart.
4. Fishes have a single circulatory loop because the heart with the single atrium and ventricle pumps blood only to the gills. The other vertebrates have both a pulmonary and systemic circulation. Amphibians have 2 atria but a single ventricle. Birds and mammals, including humans, have a heart with 2 atria and 2 ventricles, in which oxygenated blood is always separate from deoxygenated blood.

5. The heartbeat in humans begins when the SA node (pacemaker) causes the 2 atria to contract, sending blood through the atrioventricular valves to the 2 ventricles. The AV node causes the 2 ventricles to contract, sending blood through the semilunar valves to the pulmonary trunk and the aorta. Now all chambers rest. The heart sounds, lub-DUPP, are caused by the closing of the valves.

6. Blood pressure created by the beat of the heart accounts for the flow of blood in the arteries, but skeletal muscular contraction is largely responsible for the flow of blood in the veins, which have valves preventing a backward flow.

7. Blood has 2 main parts: plasma and cells. Plasma contains mostly water (92%) and proteins (8%) but also nutrients and wastes.

8. The red blood cells contain hemoglobin and function in oxygen transport.

9. Defense against disease depends on the various types of leukocytes. Neutrophils and monocytes are phagocytic and are especially responsible for the inflammatory reaction. Lymphocytes are involved in the development of immunity to disease.

10. The platelets and 2 plasma proteins, prothrombin and fibrinogen, function in blood clotting, an enzymatic process that results in fibrin threads.

11. Blood clotting is a complex process that includes 3 major events: platelets and injured tissue release prothrombin activator, which enzymatically changes prothrombin to thrombin; thrombin is an enzyme that causes fibrinogen to be converted to fibrin threads.

12. When blood reaches a capillary, water moves out at the arterial end, due to blood pressure. At the venule end, water moves in, due to osmotic pressure. In between, nutrients diffuse out and wastes diffuse in.

13. In the ABO blood system there are 4 types of blood: A, B, AB, and O, depending on the type of antigen present on the red blood cells. If the red blood cells have a particular antigen, they clump when exposed to the corresponding antibody. Table 34.2 tells which types of antibody are present in the plasma for the various ABO blood groups.

Study Questions

1. Describe transport in those invertebrates that have no circulatory system; those that have an open circulatory system; and those that have a closed circulatory system.
2. Compare the circulatory systems of a fish, an amphibian, and a mammal.
3. Trace the path of blood in humans from the right ventricle to the left atrium; from the left ventricle to the kidneys and return to the right atrium; from the left ventricle to the small intestine and return to the right atrium.
4. Define these terms: pulmonary circulation, systemic circulation, portal system.

5. Describe the beat of the heart, mentioning all the factors that account for this repetitive process. Describe how the heartbeat affects blood flow; what other factors are involved in blood flow?
6. Discuss the life cycle and function of red blood cells.
7. How are white cells classified? What is the function of neutrophils, monocytes, and lymphocytes?
8. Name the steps that take place when blood clots. Which substances are present in blood at all times, and which appear during the clotting process?

9. What forces operate to facilitate exchange of molecules across the capillary wall?
10. What are the 4 ABO blood types in humans? For each, state the antigen(s) on the red blood cells and the antibody(ies) in the plasma.
11. Problems can arise during childbearing if the mother is which Rh type and the father is which Rh type? Explain why this is so.

Objective Questions

1. Which one of these would you expect to be part of a closed, but not an open, circulatory system?
   a. ostia
   b. capillary beds
   c. hemocoel
   d. heart
2. In a one-circuit circulatory system, blood pressure
   a. is constant throughout the system.
   b. drops significantly after gas exchange has taken place.

   c. is higher at the intestinal capillaries than at the gill capillaries.
   d. cannot be determined.
3. In which of the following animals is the blood entering the aorta incompletely oxygenated?
   a. frog
   b. chicken
   c. monkey
   d. fish
4. Which of these factors has little effect on blood flow in arteries?
   a. the heartbeat
   b. blood pressure

   c. total cross-sectional area of vessels
   d. skeletal muscle contraction
5. In humans, blood returning to the heart from the lungs returns to the
   a. right ventricle.
   b. right atrium.
   c. left ventricle.
   d. left atrium.
6. Systole refers to the contraction of the
   a. major arteries.
   b. SA node.
   c. atria and ventricles.
   d. All of these.

Animal Structure and Function

7. A baby born to which of these couples is most likely to suffer from hemolytic disease of newborn.

   a. Rh-positive mother and Rh-negative father

   b. Rh-negative mother and Rh-negative father

   c. Rh-positive mother and Rh-positive father

   d. Rh-negative mother and Rh-positive father

8. During blood typing, agglutination indicates that the

   a. plasma contains certain antibodies.

   b. red blood cells carry certain antigens.

   c. plasma contains certain antigens.

   d. red blood cells carry certain antibodies.

9. Water enters capillaries on the venule side as a result of

   a. active transport from tissue fluid.

   b. an osmotic pressure gradient.

   c. increased blood pressure on this side.

   d. higher red blood cell concentration on this side.

10. Which of these associations is incorrect?

   a. white blood cells—infection fighting

   b. red blood cells—blood clotting

   c. plasma—water, nutrients, and wastes

   d. platelets—blood clotting

11. The last step in blood clotting

   a. requires calcium ions.

   b. occurs outside the bloodstream.

   c. converts prothrombin to thrombin.

   d. converts fibrinogen to fibrin.

12. Label this diagram of the heart:

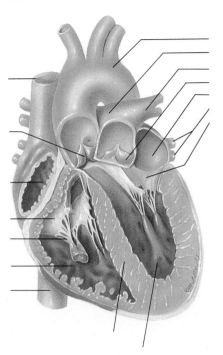

Concepts and Critical Thinking

1. *The internal environment of most animals remains relatively constant.*

How does the composition of tissue fluid stay relatively constant in vertebrates?

2. *A circulatory system is necessary to the life of a complex animal.*

Why is a circulatory system imperative for complex animals?

3. *Animals are physical and chemical machines; for example, energy must be exerted in order to keep the blood circulating.*

What type of energy keeps the blood circulating? How do animals acquire the necessary energy?

Selected Key Terms

hemoglobin (he″mo-glo′bin) 554
artery (ar′ter-e) 555
capillary (kap′ĭ-lar″e) 555
vein (vān) 555
systole (sis′to-le) 558
diastole (di-as′to-le) 558

pulmonary circuit (pul′mo-ner″e ser′kut) 560
systemic circuit (sis-tem′ik ser′kut) 560
aorta (a-or′tah) 560
vena cava (ve′nah ka′vah) 560
plasma (plaz′mah) 562
erythrocyte (ĕ-rith′ro-sīt) 563

leukocyte (lu′ko-sīt) 563
thrombocyte (throm′bō-sīt) 563
neutrophil (nu′tro-fil) 563
macrophage (mak′ro-fāj) 563
lymphocyte (lim′fo-sīt) 563
tissue fluid (tish′u floo′id) 566

35

Lymphatic System and Immunity

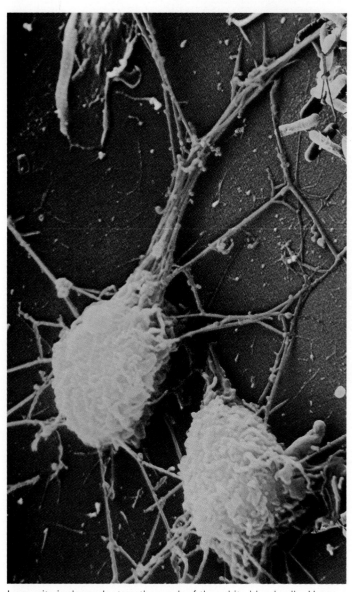

Immunity is dependent on the work of the white blood cells. Here the largest of the white blood cells, called a macrophage, is about to engulf some *E. coli* bacteria. Macrophages can consume a bacterium in less than 1/100 of a second.

Your study of this chapter will be complete when you can

1. name 3 functions of the lymphatic system;
2. describe the structure and the function of the lymphatic vessels and the lymphoid organs;
3. name 3 general ways the body defends itself against infections, and give examples of each;
4. contrast the maturation, the structure, and the function of B and T lymphocytes;
5. explain the clonal selection theory as it pertains to B cells and T cells;
6. describe the structure and the function of an antibody;
7. describe the different types of T cells, and describe the function of each type;
8. contrast active immunity with passive immunity, and tell why the former lasts longer;
9. tell how monoclonal antibodies are produced, and list ways in which they are used today;
10. name and discuss 3 types of immunological side effects.

The Lymphatic System

he mammalian **lymphatic system** consists of lymphatic vessels and the lymphoid organs. This system, which is closely associated with the cardiovascular system, has 3 main functions: (1) lymphatic vessels take up excess tissue fluid and return it to the bloodstream; (2) lymphatic capillaries absorb fats at the intestinal villi and transport them to the bloodstream (p. 596); and (3) the lymphatic system helps to defend the body against disease.

Lymphatic Vessels

Lymphatic vessels are quite extensive; every region of the body is supplied richly with lymphatic capillaries (fig. 35.1). The construction of the larger lymphatic vessels is similar to that of cardiovascular veins, including the presence of valves. Also, the movement of lymph within these vessels is dependent upon skeletal muscle contraction. When the muscles contract, the lymph is squeezed past a valve that closes, preventing the lymph from flowing backwards.

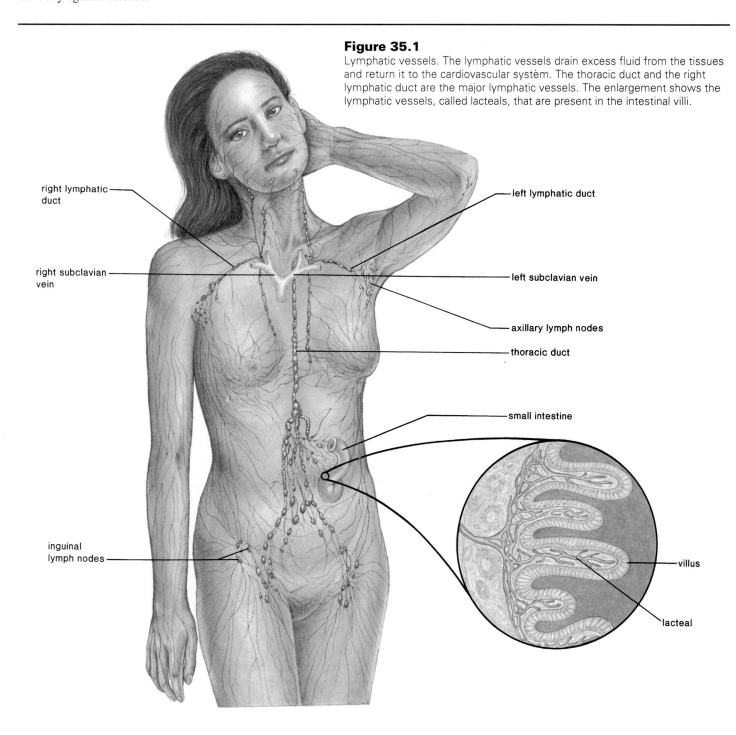

Figure 35.1
Lymphatic vessels. The lymphatic vessels drain excess fluid from the tissues and return it to the cardiovascular system. The thoracic duct and the right lymphatic duct are the major lymphatic vessels. The enlargement shows the lymphatic vessels, called lacteals, that are present in the intestinal villi.

right lymphatic duct

right subclavian vein

inguinal lymph nodes

left lymphatic duct

left subclavian vein

axillary lymph nodes

thoracic duct

small intestine

villus

lacteal

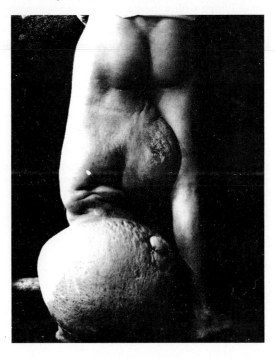

Figure 35.2
Elephantiasis. When the lymphatic vessels are blocked due to an infection by filarial worms, extreme edema results. Because tissue fluid is not drained away by the lymphatic vessels, the leg is swollen and the skin is thickened.

The lymphatic system begins with lymphatic capillaries that lie near blood capillaries. These capillaries take up fluid that has diffused from and has not been reabsorbed by the blood capillaries (fig. 34.14). Once tissue fluid enters the lymphatic vessels, it is called **lymph.** The lymphatic capillaries join to form lymphatic vessels that merge before entering one of 2 ducts: the thoracic duct or the right lymphatic duct.

The *thoracic duct* is much larger than the right lymphatic duct. It serves the lower extremities, the abdomen, the left arm, and the left side of the head and the neck. In the thorax, the left thoracic duct enters the left subclavian vein. The *right lymphatic duct* serves only the right arm and the right side of the head and the neck. It enters the right subclavian vein.

A malfunction in the lymphatic system can result in edema, which is localized swelling caused by the accumulation of tissue fluid. In the tropics, infection of lymphatic vessels by a parasitic worm causes elephantiasis, a condition in which a limb swells and supposedly resembles the limb of an elephant (fig. 35.2).

> The lymphatic system is a one-way system. Lymph flows from a capillary to ever-larger lymphatic vessels and finally to a lymphatic duct, which enters a subclavian vein.

Lymphoid Organs

The lymphoid organs include the bone marrow, the lymph nodes, the spleen, and the thymus (fig. 35.3).

Bone Marrow

In the adult, bone marrow is present only in the bones of the skull, the sternum, the ribs, the clavicle, the spinal column, and the ends of the femur and the humerus (p. 666). Bone marrow contains connective tissue cells called reticular cells, which produce a network of reticular fibers. Reticular cells and developing blood cells are packed about thin-walled venous sinuses. Mature blood cells enter the bloodstream at these sinuses.

Radioactive tracer studies have shown that the bone marrow is the site of origination for all the types of blood cells (see fig. 34.11), including both the granular and agranular leukocytes. In other words, the bone marrow contains lymphoid tissue that produces lymphocytes. The B (for bone marrow) lymphocytes mature in the bone marrow, and the T (for thymus) lymphocytes mature in the thymus gland. The structure and the function of B and T lymphocytes are discussed in the following section.

The bone marrow also contains monocytes that have developed into resident macrophages. These large phagocytic cells help to cleanse the marrow and the adjacent blood sinuses.

Lymph Nodes

At certain points along lymphatic vessels, small (about 2.5 cm), ovoid or round structures called lymph nodes occur. A lymph node has a fibrous connective tissue capsule. Connective tissue also divides a node into nodules. Each nodule contains a sinus (open space) filled with many lymphocytes and macrophages. As lymph passes through the sinuses, it is purified of infectious organisms and any other debris.

While nodules usually occur within lymph nodes, they also can occur singly or in groups. The *tonsils* are composed of partly encapsulated lymph nodules. There are also nodules called *Peyer's patches* within the intestinal wall.

The lymph nodes occur in groups in certain regions of the body. For example, the inguinal nodes are in the groin and the axillary nodes are in the arm pits.

> Lymph nodes are divided into sinus-containing lobules, where the lymph is cleansed by phagocytes.

The Spleen

The spleen is located in the upper left abdominal cavity just beneath the diaphragm. The spleen is constructed the same as a lymph node. Outer connective tissue divides the organ into lobules that contain sinuses. In the spleen, however, the sinuses are filled with blood instead of lymph. Especially since the blood vessels of the spleen can expand, this organ serves as a blood reservoir and makes blood available in times of low pressure or when the body needs extra oxygen in the blood.

A spleen nodule contains red pulp and white pulp. Red pulp contains red blood cells, lymphocytes, and macrophages. The white pulp contains only lymphocytes and macrophages. Both types of pulp help to purify the blood that passes through the spleen. If the spleen ruptures due to injury, it can be removed. Although the functions of the spleen are duplicated by other organs, the individual is expected to be slightly more susceptible to infections and may have to take antibiotic therapy indefinitely.

Animal Structure and Function

Figure 35.3

Lymphoid organs. The bone marrow is the site of lymphocyte and monocyte (macrophage) production and B cell maturation. The thymus is the site of T cell maturation. (A child is shown because the thymus is larger in children than in adults.) The lymph node is the site of lymphocyte and macrophage accumulation (the tonsils are modified lymph nodes). The spleen is the site of lymphocyte, macrophage, and red blood cell accumulation.

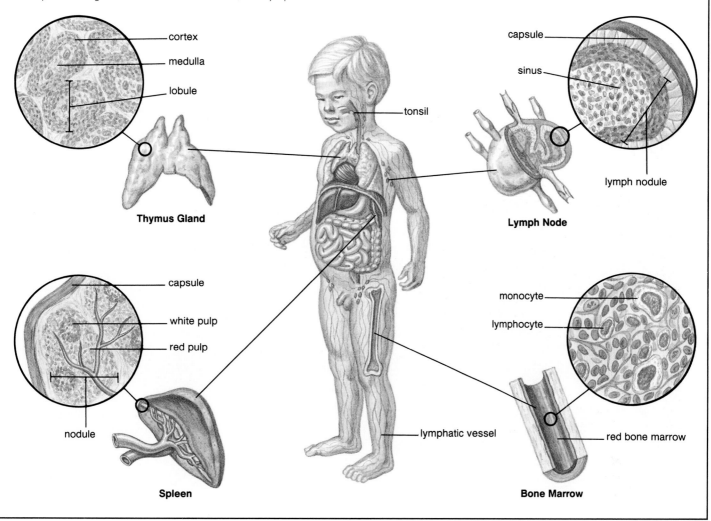

The spleen is divided into sinus-containing lobules, where the blood is cleansed by phagocytes.

The Thymus

The thymus is located along the trachea behind the sternum in the upper thoracic cavity. This gland varies in size, but it is larger in children than in adults. The thymus also is divided into lobules by connective tissue. The T lymphocytes mature in these lobules.

The thymus secretes thymosin, a molecule that is believed to be an inducing factor; that is, it causes pre-T cells to become T cells. Thymosin also may have other functions in immunity.

The thymus is divided into lobules, where T lymphocytes mature.

Immunity

Immunity is the ability of the body to protect itself from foreign substances and cells, including infectious microbes. The *first line of defense* is available immediately because it involves mechanisms that are nonspecific. The *second line of defense* takes a little longer to act because it is highly specific and contains mechanisms that are tailored to a particular threat.

General Defense

The environment contains many types of organisms that are able to invade and to infect the body. There are 3 general defense mechanisms that are useful against all types of organisms: barriers to entry, phagocytic white blood cells, and protective proteins.

Figure 35.4

Macrophage (red) engulfing bacteria (green). Monocyte-derived macrophages are the body's scavengers. They engulf microbes and debris in the body's fluids and tissues.

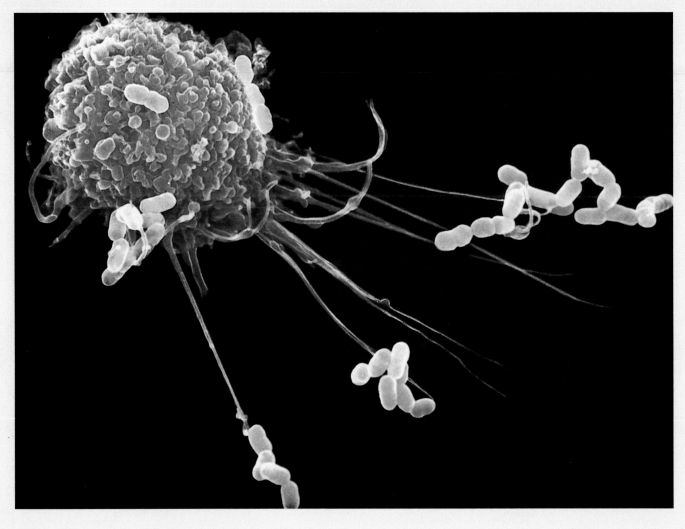

Barriers to Entry

The skin, along with the mucous membrane lining the respiratory and digestive tracts, is a mechanical barrier to entry by bacteria and viruses. The secretions of the oil glands in the skin contain chemicals that weaken or kill bacteria. The respiratory tract is lined by cells that sweep mucus and trapped particles up into the throat, where they can be swallowed. The stomach has an acidic pH that inhibits the growth of many types of bacteria. A mix of bacteria that normally reside in the intestine and other organs, such as the vagina, prevent pathogens from taking up residence.

Phagocytic White Blood Cells

If microbes do gain entry to the body, the inflammatory reaction (p. 563) and other nonspecific forces come into play. For example, neutrophils and monocyte-derived macrophages are phagocytic white blood cells that engulf some bacteria upon contact (fig. 35.4). Infections may be accompanied by fever, which is a protective response because phagocytes function better at a higher-than-normal body temperature.

Protective Proteins

The **complement system** is a series of proteins produced by the liver that are present in the plasma. When the first protein is activated, a cascade of reactions occurs. Every protein molecule in the series activates another in a predetermined sequence. Certain complement proteins form pores in bacterial cell walls and membranes. Then fluids and salts enter the bacterium until it bursts (fig. 35.5). Some simply coat bacteria while others attract phagocytes to the scene. Although complement is a general defense mechanism, it also plays a role in specific defense, as we will see later.

Animal Structure and Function

Figure 35.5
Action of the complement system. The complement system is a number of proteins always present in the plasma. When activated, some of these form pores in bacterial cell wall membranes, allowing fluids and salts to enter, until the cell eventually bursts.

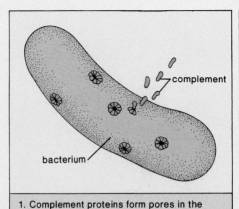

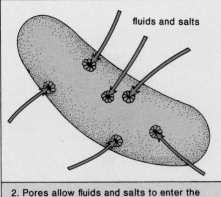

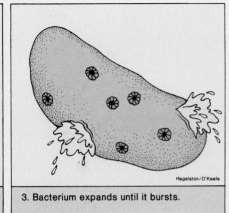

| 1. Complement proteins form pores in the bacterial cell wall and plasma membrane. | 2. Pores allow fluids and salts to enter the bacterium. | 3. Bacterium expands until it bursts. |

When most viruses infect a tissue cell, the infected cell produces and secretes interferon. **Interferon** binds to receptors on noninfected cells, and this action causes these cells to prepare for possible attack by producing substances that interfere with viral replication.

A cell that has bound interferon is protected against any type of virus; therefore, interferon should be very useful in preventing viral infection. Interferon is specific to the species, however; only human interferon can be used in humans. It used to be quite a problem to collect enough interferon for clinical and research purposes, but now interferon is a biotechnology product.

> The first line of defense against disease is nonspecific. It consists of barriers to entry, phagocytic white blood cells, and protective proteins.

Specific Defense

Sometimes, we are threatened with an invasion by microorganisms that cannot be counteracted successfully by the general defense mechanisms. In such cases, the immune system is activated to provide a specific defense. The **immune system** consists of lymphocytes and monocytes and also the lymphoid organs and lymphatic vessels where these white blood cells are found in high concentration.

The immune system allows us to develop an immunity against a specific antigen. **Antigens** are usually protein (or polysaccharide) molecules that specific lymphocytes recognize as foreign to the body. Antigens occur on bacteria and viruses, but they also can be part of a foreign cell or a cancerous cell. Ordinarily, we do not become immune to our body's own normal cells; therefore, it is said that the immune system is able to distinguish self from nonself. Immunity is primarily the result of the action of B

lymphocytes and T lymphocytes, which have different functions. **B lymphocytes,** also called B cells, become plasma cells that produce **antibodies,** proteins that are capable of combining with and inactivating antigens. These antibodies are secreted into the blood and the lymph. In contrast, **T lymphocytes,** also called T cells, do not produce antibodies. Instead, certain T cells directly attack cells bearing antigens they recognize. Other T cells regulate the immune response.

Lymphocytes are capable of recognizing an antigen because they have receptor molecules on their surface. The shape of the receptors on any particular lymphocyte are complementary to the shape of one specific antigen. It is often said that the receptor and the antigen fit together like *a lock and a key.* It is estimated that during our lifetime, we encounter a million different antigens, so we need the same number of different lymphocytes for protection against those antigens.

> There are 2 types of lymphocytes. B cells produce and secrete antibodies that combine with antigens. Certain T cells directly attack antigen-bearing cells, and others regulate the immune response.

The Action of B Cells

The receptor on a B cell is called a membrane-bound antibody because it is structured like an antibody. When a B cell encounters a bacterial cell or a toxin bearing an appropriate antigen, it is activated; that is, it has the potential to produce many **plasma cells** that will secrete antibodies against this antigen (fig. 35.6). (For the B cell to realize this potential, it must be stimulated by a helper T cell [p. 583].) All of the plasma cells derived from one parent lymphocyte are called a clone, and a clone produces the same type of antibody. Notice that a B cell does not clone until its antigen is present. The *clonal selection theory* states that the antigen selects which B cell will produce a clone of plasma cells.

Figure 35.6

Clonal selection theory as it applies to B cells. Antigens have different shapes. In this diagram the shape of the antigen causes it to attach to the receptor of the middle B cell and not those of the B cells on the right or the left. The antigen and receptor fit together in a lock-and-key manner because their shapes are complementary. The B cell will divide if stimulated by a helper T cell (fig. 35.10) and produce many plasma cells that secrete specific antibodies against this antigen by the fifth day. Memory cells that retain the ability to secrete these specific antibodies at a future time are also produced.

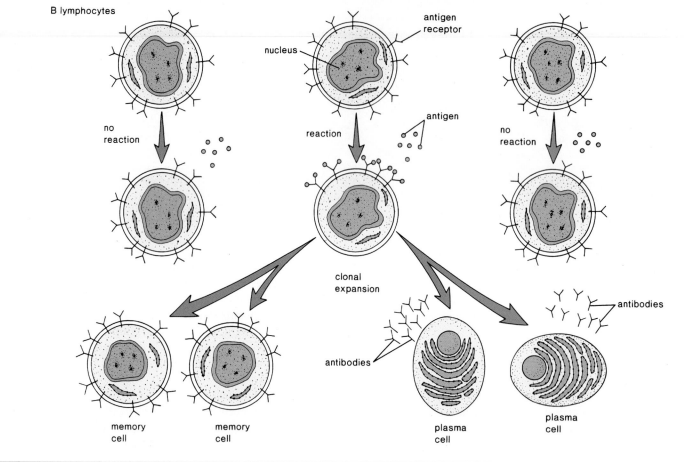

Once antibody production is sufficient, the antigen disappears from the system, and the development of plasma cells ceases. Some members of a clone do not participate in antibody production, however; instead, they remain in the bloodstream as **memory B cells.** Memory B cells are capable of producing the antibody specific to a particular antigen for some time. As long as these cells are present, the individual is said to be actively immune: future antibody production is possible because the memory cells can produce more plasma cells if the same antigen invades the system again.

Defense by B cells is called *antibody-mediated immunity* because B cells produce antibodies. It also is called humoral immunity because these antibodies are present in the bloodstream.

> B cells are responsible for antibody-mediated immunity. After they recognize an antigen, they (if stimulated by a helper T cell) divide to produce both antibody-secreting plasma cells and memory B cells.

Antibodies The most common type of antibody (IgG) is a Y-shaped protein molecule having 2 arms. Each arm has a long "heavy" chain and a short "light" chain of amino acids. These chains have *constant regions,* where the sequence of amino acids is set, and *variable regions,* where the sequence of amino acids varies to produce a particular shape (fig. 35.7). The antigen binds to the antibody at the variable regions of one arm in a lock-and-key manner. In other words, the variable regions form an antibody-binding site that is specific for a particular antigen.

The constant regions are not identical among all the antibodies. Instead, they are the same for different classes of antibodies. Most antibodies found in the blood belong to the class IgG (immunoglobulin G) (table 35.1).

The antigen-antibody reaction can take several forms, but quite often the antigen-antibody reaction produces complexes of antigens combined with antibodies. Such an antigen-antibody complex, sometimes called the immune complex, marks the anti-

Animal Structure and Function

Figure 35.7

Structure of the most common antibody (IgG). **a.** Computer model of an antibody. **b.** Schematic drawing. Each arm of an antibody contains a light (short) amino acid chain and a heavy (long) amino acid chain. In the variable regions, the amino acid sequence varies between antibodies so that the binding sites have a shape that is complementary to the shape of a particular antigen. An antigen binds to a variable region in a lock-and-key manner.

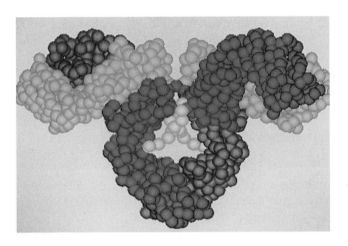

a.

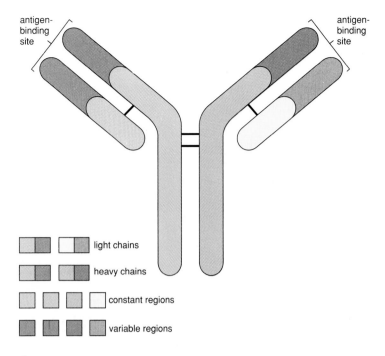

b.

gen for destruction by other forces. For example, the complex may be engulfed by neutrophils or macrophages, or it may activate a portion of blood serum called the complement system, page 576.

> An antibody combines with its antigen in a lock-and-key manner. When several antigens bind with antibodies, an antigen-antibody complex forms.

The Actions of T Cells

There are 4 different types of T cells: cytotoxic T cells, helper T cells, memory T cells, and suppressor T cells. All 4 types look alike but can be distinguished by their functions.

Cytotoxic T (T_C) cells sometimes are called **killer T cells**. In immune individuals, they attack and destroy cells bearing a foreign antigen, such as virus-infected cells or cancerous cells. These T cells have storage vacuoles that contain a chemical called perforin because it perforates cell membranes. The perforin molecules form a pore in the membrane that allows water and salts to enter. The cell under attack then swells and eventually bursts (fig. 35.8).

It often is said that T cells are responsible for *cell-mediated immunity,* characterized by destruction of antigen-bearing cells. Of all the T cells, only T_C cells are involved in this type of immunity.

Helper T (T_H) cells regulate immunity by enhancing the response of other immune cells. In response to an antigen, they enlarge and secrete lymphokines, including interferon and the interleukins.[1] **Lymphokines** are stimulatory molecules that cause

[1]For example, the lymphokine that stimulates B cell to secrete antibodies is called interleukin-4.

Table 35.1
Types of Antibodies

Classes	Description
IgG	Main antibody type in circulation; attacks microorganisms and their toxins
IgA	Main antibody type in secretions, such as saliva and milk; attacks microorganisms and their toxins
IgE	Antibody responsible for allergic reactions
IgM	Antibody type found in circulation; largest antibody, with 5 subunits
IgD	Antibody type found primarily as a membrane-bound immunoglobulin

T_H cells to clone and other immune cells to perform their functions. For example, T_H cells stimulate macrophages to phagocytize antibodies and B cells to manufacture antibodies. Because the AIDS virus attacks T_H cells, it inactivates the immune response. This is discussed in chapter 43.

When an activated T_H cell divides, the clone contains **suppressor T (T_S) cells** and **memory T (T_M) cells.** Once there is a sufficient number of T_S cells, the immune response ceases. Following suppression, a population of T_M cells persists, perhaps for life. These cells are able to secrete lymphokines and to stimulate macrophages and B cells whenever the same antigen enters the body once again.

Figure 35.8

Cell-mediated immunity. Scanning electron micrograph showing cytotoxic T cells destroying a cancer cell. During the killing process, vacuoles in a T cell fuse with the plasma membrane and release pore-forming proteins. Fluids and salts enter through the pores, and the target cell eventually bursts.

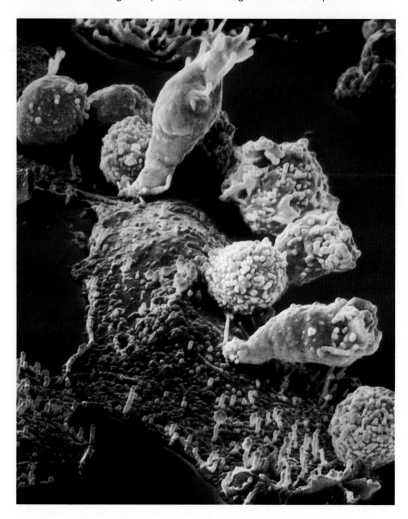

T_C cells are responsible for cell-mediated immunity; T_H cells promote the immune response; T_S cells suppress the immune response; and T_M cells maintain immunity.

The Activation of Cytotoxic and Helper T Cells T cells have receptors just as B cells do. Unlike B cells, however, T_C cell and T_H cells are unable to recognize an antigen that simply is present in lymph or blood. Instead, the antigen must be presented to them by an antigen-presenting cell (APC). When an APC, such as a macrophage, engulfs a bacterium or a virus, the APC enzymatically breaks it down to peptide fragments that are antigenic (have the properties of an antigen). The antigenic peptide fragment is linked to an **MHC** (major histocompatibility complex) **protein,** and together they are displayed to a T cell at the plasma membrane. The importance of MHC proteins first was recognized when it was discovered that they contribute to the specificity of tissues and make it difficult to transplant tissue from one person to another. In other words, the donor and the recipient must be histo-(tissue) compatible (the same or nearly so) for a transplant to be successful without the administration of immunosuppressive drugs.

There are 2 types of MHC proteins, known as MHC I and MHC II. Most cells of the body display MHC I–type proteins and only cells of the immune system, that is, macrophages, B cells, and some T cells display MHC II-type proteins. This allows cells of the immune system to recognize one another.

Figure 35.9*a* shows a macrophage and a B cell presenting an antigen to a T_H cell. Once a T_H cell recognizes an antigen, it undergoes clonal expansion, producing T_S cells and T_M cells that also can recognize this same antigen. While well-known APCs are macrophages and B cells, actually any cell in the body can be an APC. For example, figure 35.9*b* shows a virus-infected cell presenting an antigen to a T_C cell. Once a T_C cell recognizes an antigen, it attacks and destroys any cell that is infected with the same virus.

In order for T_C and T_H cells to recognize an antigen, the antigen, along with an MHC protein, must be presented to them by an APC.

Table 35.2 and figure 35.10 summarize our discussion of B cells and T cells.

Immunotherapy

The immune system can be manipulated to help people avoid or recover from diseases. Some of the techniques to do this have been utilized for a long time, and some are relatively new.

Induced Immunity

Active immunity, which provides long-lasting protection against a disease-causing organism, develops after an individual is infected with a virus or a bacterium. In many instances today, however, it is not necessary to suffer an illness to become immune because it is possible to be artificially immunized against a disease using vaccines. **Vaccines** traditionally are bacteria and viruses (antigens) that have been treated so that they are no longer virulent (able to cause disease), but new methods of producing vaccines are being developed. For example, it is possible to use the recombinant DNA technique to mass-produce a protein that can be used as a vaccine. This method is being used to prepare a vaccine against hepatitis B.

After a vaccine is given, it is possible to determine the amount of antibody present in a sample of serum—this is called the *antibody titer.* After the first exposure to an antigen, a primary response occurs. For a period of several days, no antibodies are present; then, there is a slow rise in the titer, followed by a gradual

How Do You Test for an HIV Infection?

I t is now well recognized that acquired immune deficiency syndrome (AIDS) is caused primarily by a human immunodeficiency virus (HIV-1), which infects T_H lymphocytes. Even though the virus attacks the immune system, B cells do begin to produce antibodies within weeks or months after the infection begins. All of the commonly used tests detect the presence of these antibodies by having them react to their antigens. They use plastic beads that are coated with either inactivated virus or just specific viral envelope proteins.

At the start of the test, an antigen coated bead is placed in a small tube along with the specimen (human serum or plasma) to be tested (fig. 35.A). If the specimen contains HIV-1 antibodies, they will bind to the bead and remain bound when the bead is washed and rinsed. These antibodies are designated as Ab_1 because they are the first antibodies attached to the bead.

The next step is to get a color reaction that can be read photometrically by a laboratory instrument. First, a conjugate designated as Ab_2 – Enz is added to the tube. The conjugate is made up of antihuman globulin attached to an enzyme. The Ab_2 – Enz will bind to Ab_1 on the bead. After washing and rinsing, a substrate is added that will react with the enzyme and cause a color change. For example, one of the commonly used enzymes is peroxidase derived from horseradish. When hydrogen peroxide and o-phenylenediamine are added, the hydrogen peroxide releases oxygen that causes the o-phenylenediamine to go from a clear to a yellow-orange color. The intensity of the color reaction is compared to a known negative test result (no HIV-1 antibody is present in the specimen) and a known positive test result (HIV-1 antibody is present in the specimen).

The test we have described is called an HIV-1 antibody *enzyme-linked immunoassay* (ELISA). Note that since this is a test for HIV-1 antibody, it cannot detect persons who are infectious but have not yet produced HIV-1 antibodies. Nor can it reliably detect persons infected with HIV-2, another HIV that can cause AIDS. On the other hand, false positive results are possible in persons suffering from diseases other than AIDS. To guard against false positive results, it is common practice to do a more specialized and expensive test on specimens that give a positive ELISA test. (This test resembles the procedure described for DNA fingerprinting on page 592.) Patients who give a positive result with both the HIV-1 antibody ELISA and with the confirmatory test are told they have an HIV infection and that they should seek immediate treatment.

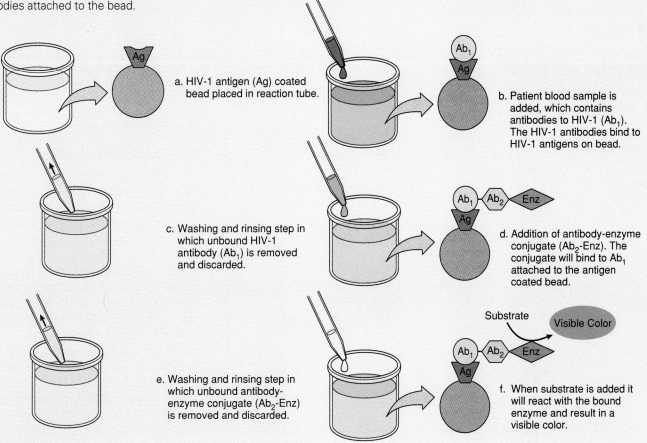

a. HIV-1 antigen (Ag) coated bead placed in reaction tube.

b. Patient blood sample is added, which contains antibodies to HIV-1 (Ab_1). The HIV-1 antibodies bind to HIV-1 antigens on bead.

c. Washing and rinsing step in which unbound HIV-1 antibody (Ab_1) is removed and discarded.

d. Addition of antibody-enzyme conjugate (Ab_2-Enz). The conjugate will bind to Ab_1 attached to the antigen coated bead.

e. Washing and rinsing step in which unbound antibody-enzyme conjugate (Ab_2-Enz) is removed and discarded.

f. When substrate is added it will react with the bound enzyme and result in a visible color.

Figure 35.A
A positive test for an HIV infection has these steps. In the first step, HIV-1 antibodies (Ab_1) in the specimen bind to HIV-1 antigens on a bead. In the second step, antibody-enzyme conjugate (Ab_2) binds to Ab_1. The enzyme brings about a color change that signifies the test is positive.

Figure 35.9

T cell activation. **a.** Either a macrophage or a B cell presents an antigen to a helper T cell. To accomplish this, the antigen has to be digested to peptides that are combined with an MHC II protein. The complex is presented to the T cell. In return, the helper T cell produces and secretes lymphokines that stimulate T cells and other immune cells. **b.** Cells infected with a virus present one of the viral proteins along with an MHC I protein to a cytotoxic T cell. This causes the cytotoxic T cell to attack and to destroy any cell infected with the same virus (see fig. 35.8).

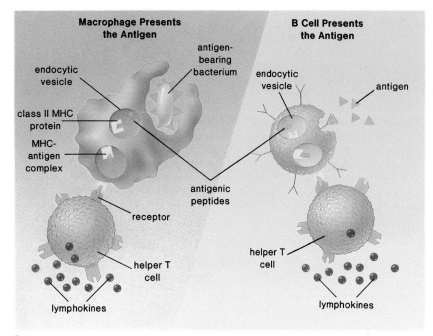

a.

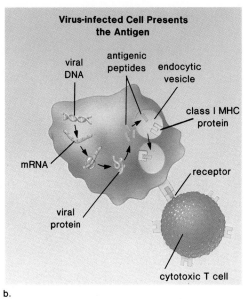

b.

decline (fig. 35.11). After a second exposure, a secondary response may occur. If so, the titer rises rapidly to a level much greater than before. The second exposure in that case often is called the *"booster"* because it boosts the antibody titer to a high level. A good secondary response can be related to the number of plasma cells and memory cells in the serum. Upon the second exposure, these cells already are present, and antibodies can be produced rapidly.

Passive immunity occurs when an individual is given antibodies (immunoglobulins) to combat a disease. Since these antibodies are not produced by the individual's B cells, passive immunity is short-lived. For example, newborn infants possess passive immunity because antibodies have crossed the placenta from the mother's blood. These antibodies soon disappear, however, so that within a few months, infants become more susceptible to infections. Breast-feeding prolongs the passive immunity an infant receives from the mother because there are antibodies in the mother's milk.

Vaccines can be used to make people actively immune. Passive immunity is short-lived because the antibodies are administered to and not made by the individual.

Monoclonal Antibodies

Every plasma cell derived from the same B cell secretes antibodies against a specific antigen, as previously discussed. These are *monoclonal antibodies* because all of them are the same type (mono) and because they are produced by plasma cells derived from the same B cell (clone). Monoclonal antibodies can be produced in vitro (in laboratory glassware). B lymphocytes are removed from the body (today, usually a mouse) and are exposed to a particular antigen. Then they are fused with a myeloma cell (a malignant plasma cell) because these cells, unlike normal plasma cells, live and divide indefinitely. The fused cells are called hybridomas; *hybrid* because they result from the fusion of 2 different cells and *oma* because one of the cells is a cancer cell.

Figure 35.10

Summary of B cell and T cell functions. Both cells are produced in the bone marrow, but only B cells mature there. T cells mature in the thymus. An antigen activates the appropriate B cell, which divides and differentiates to give a clone of antibody-producing plasma cells and memory cells. An antigen activates a T cell (either cytotoxic T or helper T) if it is presented with an MHC II protein by an APC such as a macrophage. Cytotoxic T cells then attack and destroy antigen-bearing cells; helper T cells divide to give a clone of helper T cells that produce lymphokines and also suppressor T and memory T cells.

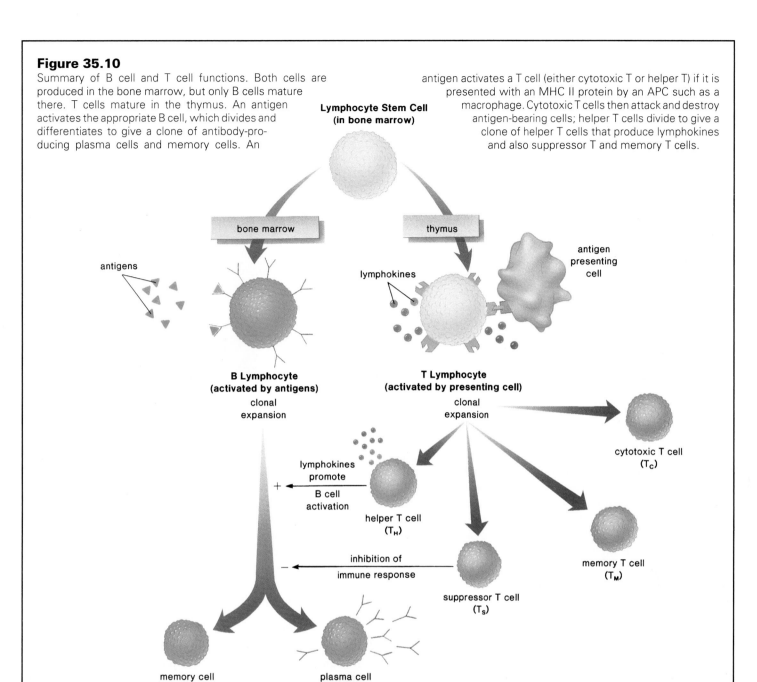

Table 35.2

Some Properties of B Cells and T Cells

Property	B Cells	T Cells	
		Cytotoxic	Helper
Type of immunity	Antibody mediated	Cell mediated	_____
Antigen recognition	Direct recognition	Must be presented by APC	Must be presented by APC
Response	Become antibody-producing plasma cells	Search and destroy antigen-bearing cells	Secrete lymphokines; stimulate other immune cells
Final response	Memory cells	_____	Suppressor and memory cells

Figure 35.11

Development of active immunity due to immunization. The primary response after the first exposure of a vaccine is minimal, but the secondary response that may occur after the second exposure shows a dramatic rise in the amount of antibody present in serum.

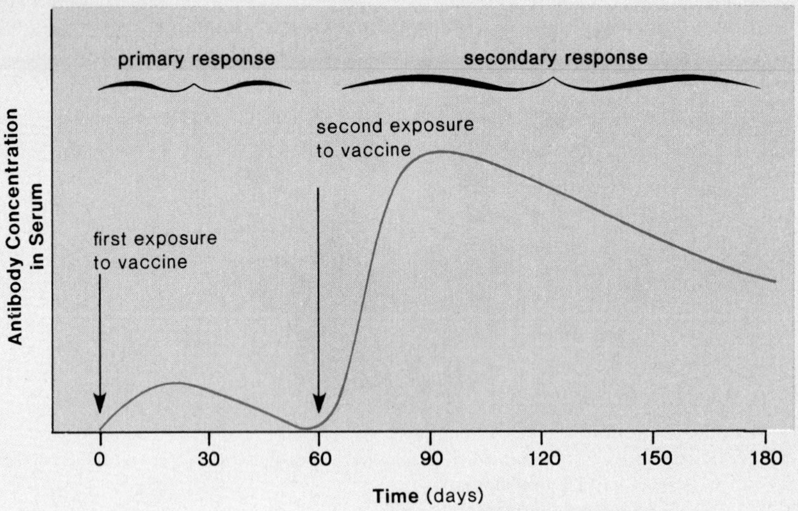

Figure 35.12

Monoclonal antibody use. These cells are infected with herpes simplex type 1 virus (HSV-1), which can be detected by a monoclonal antibody specific for the virus and tagged with a fluorescent dye. When the cells fluoresce, it shows that the virus is inside the cells.

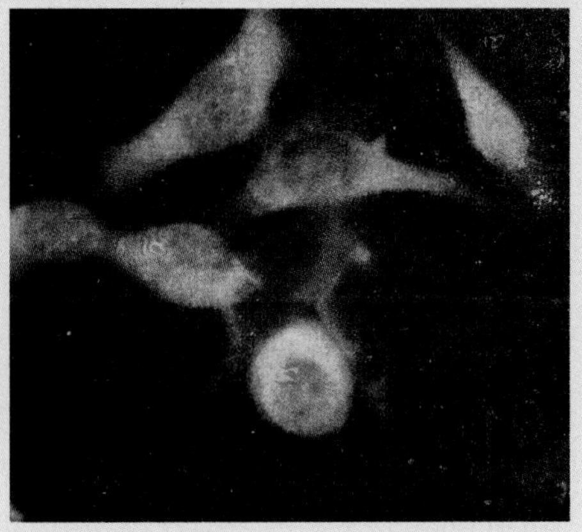

At present, monoclonal antibodies are used for quick and certain diagnosis of various conditions. For example, a particular hormone is present in the urine of a pregnant woman. A monoclonal antibody can be used to detect the hormone and so indicate that the woman is pregnant. Monoclonal antibodies also are used for identifying infections (fig. 35.12). They are so accurate they even can sort out the different types of T cells in a blood sample. Because monoclonal antibodies can distinguish between cancerous and normal tissue cells, they are used to carry radioactive isotopes or toxic drugs only to tumors so these can be destroyed selectively.

Monoclonal antibodies are produced in pure batches—they react specifically with just one type of molecule (antigen); therefore, they can distinguish one cell, or even one molecule, from another.

Immunological Side Effects and Illnesses

The immune system protects us from disease because it can tell self from nonself. Sometimes, however, the immune system is underprotective, as when an individual develops cancer, or is overprotective, as when an individual has allergies.

Allergies

Of the 5 varieties of antibodies (table 35.1)—IgA, IgD, IgE, IgG, and IgM—IgE antibodies cause allergies. IgE antibodies are found in the bloodstream, but they, unlike the other types of antibodies, also reside in the membrane of *mast cells* found in the tissues. Some investigators contend that mast cells are basophils that have left the bloodstream and have taken up residence in the tissues. In any case, when the *allergen,* an antigen that provokes an allergic reaction, attaches to the IgE antibodies on mast cells, these cells release histamine and other substances that cause mucus secretion and airway constriction, resulting in the characteristic symptoms of allergy. On occasion, basophils and other white blood cells release these chemicals into the bloodstream. The increased capillary permeability that results from this can lead to fluid loss and shock.

Tissue Rejection

Both T_C cells and/or antibodies can bring about disintegration of foreign tissue introduced into the body. Organ rejection can be controlled in 2 ways: careful selection of the organ to be transplanted and the administration of immunosuppressive drugs. It is best if the transplanted organ has the same type of MHC proteins as those of the recipient, because T_C cells learn to recognize foreign MHC proteins. The immunosuppressive drug cyclosporine has been in use for some years. An experimental drug, FK-506, eventually may replace cyclosporine as the drug of choice for transplant patients. In more than 100 patients taking FK-506, the rate of organ rejection was one-sixth that of cyclosporine.

Autoimmune Diseases

Certain human illnesses are due to the production of antibodies that act against an individual's own tissues. In myasthenia gravis, autoantibodies attack the neuromuscular junctions so that the muscles do not obey nervous stimuli. Muscular weakness results. In multiple

Animal Structure and Function

sclerosis (MS), antibodies attack the myelin of nerve fibers, causing various neuromuscular disorders. A person with systemic lupus erythematosus (SLE) forms various antibodies to different constituents of the body, including the DNA of the cell nucleus. The disease sometimes results in death, usually due to kidney damage. In rheumatoid arthritis, the joints are affected. When an autoimmune disease occurs, a viral infection of tissues often has set off an immune reaction to the body's own tissues. There is evidence to suggest that type I diabetes is the result of this sequence of events, as well as heart damage following rheumatic fever.

Summary

1. The lymphatic system has various functions. Lymphatic capillaries collect excess tissue fluid, which moves in lymphatic veins to blood circulatory veins. Lacteals absorb the products of fat digestion, and lymphoid organs (bone marrow, thymus, lymph nodes, and spleen) help defend the body from disease.
2. The general defense of the body consists of barriers to entry, phagocytic white blood cells, and protective proteins.
3. The immune response is specific to a particular antigen and requires 2 types of lymphocytes, both of which are produced in the bone marrow. B cells mature in the bone marrow, and T cells mature in the thymus.
4. B cells are responsible for antibody-mediated immunity. When the shape of an antigen fits the shape of a B-cell receptor, that B cell undergoes clonal expansion, producing antibody-secreting plasma cells and memory cells. In keeping with the clonal selection theory, the antigen selects the B cell that proliferates.
5. An antibody is a Y-shaped molecule that has 2 binding sites. Each antibody is specific for a particular antigen.
6. For a T cell to recognize an antigen, the antigen must be presented by an APC along with an MHC protein. (MHC I in macrophages and MHC II for virus-infected cells.)
7. There are 4 types of T cells: T_C cells kill cells on contact; T_H cells stimulate other immune cells; T_S cells suppress the immune response; T_M cells are memory cells.
8. AIDS is an HIV infection of T_H cells, and therefore it destroys the cells that are needed to mount an immune response.
9. Immunity can be fostered by immunotherapy. Vaccines are available to promote active immunity, and antibodies sometimes are available to provide an individual with short-term passive immunity. Monoclonal antibodies are used for various purposes.
10. Immunity has certain undesirable side effects. Allergies are due to an overactive immune system that forms antibodies to substances not normally recognized as foreign. T_C cells attack transplanted organs. Autoimmune illnesses occur when antibodies against the body's own cells form.

Writing Across the Curriculum

In order to practice writing skills, students should write out the answers to any or all of the study questions and the critical thinking questions. The study questions are sequenced in the same order as the text. Suggested answers to the critical thinking questions are in appendix D.

Study Questions

1. What is the lymphatic system, and what are its 3 functions?
2. Describe the structure and the function of the bone marrow, lymph nodes, the spleen, and the thymus.
3. Distinguish between the general defense and the specific defense of the body against disease.
4. Contrast B and T cells in as many ways as possible.
5. What is the clonal selection theory?
6. Explain the process that allows a T cell to recognize an antigen.
7. List the 4 types of T cells, and state their functions.
8. Discuss allergies, tissue rejection, and autoimmune disease as they relate to the immune system.
9. Relate active immunity to the presence of plasma cells and memory cells.
10. How is active immunity achieved? How is passive immunity achieved?

Objective Questions

1. Complement
   a. is a general defense mechanism.
   b. is a series of proteins present in the plasma.
   c. plays a role in destroying bacteria.
   d. All of these.
2. Which one of these does not pertain to T cells?
   a. have specific receptors
   b. have cell-mediated immunity
   c. stimulate antibody production by B cells
   d. have no effect on macrophages
3. Which one of these does not pertain to B cells?
   a. have passed through the thymus
   b. specific receptors
   c. antibody-mediated immunity
   d. synthesize and liberate antibodies

4. The clonal selection theory says that
   a. an antigen selects out certain B cells and suppresses them.
   b. an antigen stimulates the multiplication of B cells that produce antibodies against it.
   c. T cells select out those B cells that should produce antibodies regardless of antigens.
   d. T cells suppress all those B cells except the ones that should multiply and divide.

5. Plasma cells are
   a. the same as memory cells.
   b. formed from blood plasma.
   c. B cells that are actively secreting antibody.
   d. inactive T cells carried in the plasma.

6. For a T cell to recognize an antigen, it must interact with
   a. complement.
   b. a macrophage.
   c. a B cell.
   d. All of these.

7. Antibodies combine with antigens
   a. at variable regions.
   b. at constant regions.
   c. only if macrophages are present.
   d. Both a and c.

8. Which one of these is mismatched?
   a. Helper T cells—help complement react
   b. Killer T cells—active in tissue rejection
   c. Suppressor T cells—shut down the immune response

   d. Memory T cells—long-living line of T cells

9. Vaccines are
   a. the same as monoclonal antibodies.
   b. treated bacteria or viruses or one of their proteins.
   c. MHC proteins.
   d. All of these.

10. The theory behind the use of lymphokines in cancer therapy is
    a. if cancer develops, the immune system has been ineffective.
    b. lymphokines stimulate the immune system.
    c. cancer cells bear antigens that should be recognizable by T_K cells.
    d. All of these.

Concepts and Critical Thinking

1. *The body defends its integrity.*

   How does your study of immunity support this concept?

2. *Multitiered mechanisms maintain homeostasis.*

   How does your study of immunity support this concept?

3. *Organs belong to organ systems.*

   Argue that red bone marrow is a part of the skeletal, circulatory, and lymphatic systems.

Selected Key Terms

lymphatic system (lim-fat'ik sis'tem) 573
immunity (i-mūn-i-tee) 575
interferon (in"ter-fer'on) 577
antigen (an'tī-jen) 577
B lymphocyte (be lim'fo-sīt) 577

antibody (an'tī-bod"e) 577
T lymphocyte (te lim'fo-sīt) 577
plasma cell (plaz'mah sel) 577
memory B cell (mem'o-re be sel) 578
cytotoxic T (T_c) cell (si"to-tok'sik te sel) 579
helper T (T_H) cell (hel'per te sel) 579

suppressor T (T_s) cell (sŭ-pres'or te sel) 579
memory T (T_M) cell (mem'o-re te sel) 579
MHC protein (em ăch se pro'tein) 580
vaccine (vak'sēn) 580

36

Digestive System and Nutrition

A long-nosed tree snake is feeding on an anole. Snakes are discontinuous eaters—few eat more than once a week; many eat about once a month. Snakes can eat large prey because their jaws and their entire digestive tract are expandable. A snake's curved teeth hold onto the prey animal and pull it back into the esophagus.

Your study of this chapter will be complete when you can

1. contrast the dentition and digestive tracts of mammalian herbivores and mammalian carnivores;
2. contrast the characteristics of a continuous and a discontinuous feeder;
3. contrast the characteristics of an incomplete and complete digestive tract;
4. name the parts of the human digestive system and describe, in general, their structure and their function;
5. outline the digestion of carbohydrates (e.g., starch), proteins, and fats in humans by listing the names and functions of the appropriate enzymes and by telling where in the human digestive tract these enzymes function;
6. list the contributions of the liver and pancreas to the digestive process in humans;
7. name 6 functions of the liver;
8. give the names and the functions of hormones that affect the flow of digestive juices;
9. in general, tell how an animal's body makes use of the products of digestion;
10. describe a balanced diet; describe how to achieve a balanced diet and how to avoid too much fat, sugar, and salt in the diet.

A nimals are heterotrophic organisms that must take in preformed food. The many adaptations animals exhibit to acquire and to digest food are extremely varied and can be related to a wide variety of available food. Necessarily, we will be able to discuss only a few examples. When animals take in food, they acquire the energy needed to carry on daily activities and the building blocks needed to grow and/or repair tissues.

Animal Feeding Modes

Some animals are omnivores; they eat both plants and animals. Others are herbivores; they feed only on plants. Still others are carnivores; they eat only other animals (fig. 36.1).

Among mammals, the dentition differs according to mode of nutrition (fig. 36.2). Humans, as well as raccoons, rats, and brown bears, are omnivores, in which dentition is nonspecialized and adequate to deal with both a vegetable diet and a meat diet. An adult human has 32 teeth. One-half of each jaw has teeth of 4 different types: 2 chisel-shaped *incisors* for biting; one pointed *canine* for tearing; 2 fairly flat *premolars* for grinding; and 3 *molars*, well-flattened for crushing.

Among herbivores, the koala of Australia is famous for its diet of only eucalyptus leaves, and likewise many other mammals are browsers, feeding off bushes and trees. Grazers, like the horse, feed off grasses. The horse has sharp, even incisors for neatly clipping off blades of grass and large, flat premolars and molars for grinding and crushing up the grass. Extensive grinding and crushing disrupts plant cell walls, allowing bacteria located in the cecum

Figure 36.1

Feeding modes of invertebrates. *a.* Some invertebrates are omnivores, as are earthworms. As an earthworm burrows, it takes up soil and organic material by the joint action of its liplike prostomium and muscular pharynx. *b.* Some invertebrates are herbivores, as are grasshoppers. There are mouthparts for chewing the food before it enters the pharynx. *c.* Some invertebrates are carnivores, as are spiders. Wolf spiders, jumping spiders, and tarantulas hunt for prey, which they seize and immobilize with their fangs and pedipalps, anterior appendages. Other spiders trap prey in webs made of silk. Digestive juices are secreted into the body of an insect and then it is sucked dry.

a.

b.

c.

Animal Structure and Function

Figure 36.2

Composition of the teeth of several mammals. **a.** Horses are herbivores that graze on grasses. Note the sharp incisors, reduced canines, and large, flat premolars and molars. **b.** Lions are carnivores that prey on other animals. Note the pointed incisors, enlarged canines, and jagged premolars and molars. **c.** Humans are omnivores that have nonspecialized teeth.

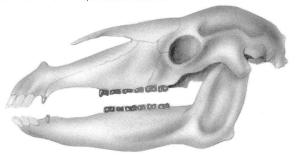

a. Horse

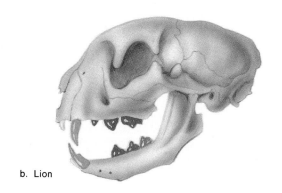

b. Lion

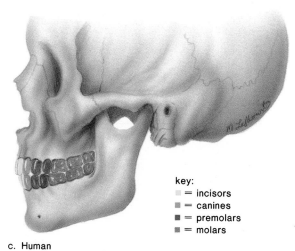

c. Human

key:
= incisors
= canines
= premolars
= molars

to get at and digest cellulose. Other mammalian grazers, like the cow and deer, are ruminants. In contrast to horses, they graze quickly and swallow partially chewed grasses into a special part of the stomach called a rumen. Here, microorganisms start the digestive process and the result, called cud, is regurgitated at a later and more convenient time. The cud is chewed before being swallowed for complete digestion.

Many mammals, including dogs, toothed whales, and polar bears, are carnivores. A carnivore, like a lion, uses pointed incisors and enlarged canine teeth to tear off pieces small enough to be quickly swallowed. Meat is rich in protein and fat and easier to digest than plant material. Therefore, the digestive system of carnivores is shorter and doesn't have the specialization seen in herbivores.

> The dentition of a mammal reflects its type of diet. Omnivores have nonspecialized teeth. Herbivores have large, flat molars that grind food; carnivores have incisors and canines to tear off chunks of food.

Continuous Feeders versus Discontinuous Feeders

The clam is a continuous feeder, which is often called a *filter feeder* (fig. 36.3*a*). Water is always moving into the mantle cavity by way of the incurrent siphon (slitlike opening) and depositing particles on the gills. The size of the incurrent siphon permits the entrance of only small particles, which adhere to the gills. Ciliary action moves suitably sized particles to the labial palps, which force them through the mouth into the stomach. Digestive enzymes are secreted by a large digestive gland, but amoeboid cells present throughout the tract are believed to complete the digestive process by intracellular digestion.

Marine fanworms (p. 435) are sessile filter feeders that live in a tube and extend feathery tentacles to capture food. The baleen whale is an active filter feeder. The baleen, a curtainlike fringe, hangs from the roof of the mouth and filters small shrimp called krill from the water. The baleen whale filters up to a ton of krill in a single feeding.

The squid is an example of a discontinuous feeder (fig. 36.3*b*). The body of a squid is streamlined, and the animal moves rapidly through the water using jet propulsion (forceful expulsion of water from a tubular funnel). The head of a squid is surrounded by 10 arms, 2 of which have developed into long, slender tentacles whose suckers have toothed, horny rings. These tentacles seize prey (fishes, shrimps, and worms) and bring it to the squid's bearlike jaws, which bite off pieces pulled into the mouth by the action of a radula, a toothy tongue. An esophagus leads to a stomach and cecum, where digestion occurs. The stomach, supplemented by the cecum, holds food until digestion is complete. Discontinuous feeders, whether they are carnivores or herbivores, require such a storage area.

> Continuous feeders, such as sessile filter feeders, do not need a storage area for food; discontinuous feeders, such as herbivores and carnivores, do need a storage area for food.

Digestive Tracts

Generally, animals have some sort of digestive tract or gut. The gut can be simple or complex. A simple gut has few, if any, specialized parts, and a complex gut does have specialized parts.

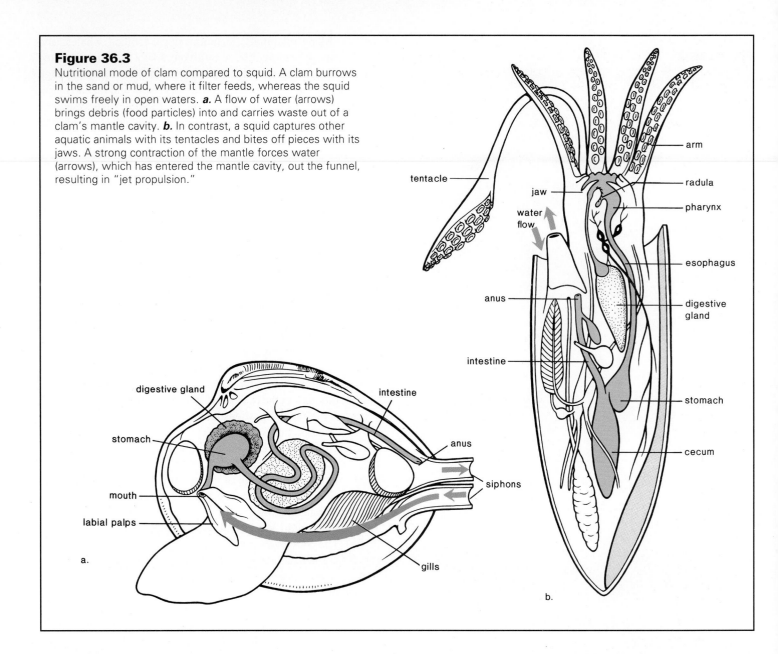

Figure 36.3
Nutritional mode of clam compared to squid. A clam burrows in the sand or mud, where it filter feeds, whereas the squid swims freely in open waters. ***a.*** A flow of water (arrows) brings debris (food particles) into and carries waste out of a clam's mantle cavity. ***b.*** In contrast, a squid captures other aquatic animals with its tentacles and bites off pieces with its jaws. A strong contraction of the mantle forces water (arrows), which has entered the mantle cavity, out the funnel, resulting in "jet propulsion."

Incomplete Gut versus Complete Gut

An *incomplete gut* has a single opening, usually called a mouth. The planarian is an example of an animal with an incomplete gut (fig. 36.4). It is carnivorous and feeds largely on smaller aquatic animals. Its digestive system contains only a mouth, a pharynx, and an intestine. When the worm is feeding, the pharynx actually extends beyond the mouth. It wraps its body about the prey and uses its muscular pharynx to suck up minute quantities at a time. Digestive enzymes present in the *gastrovascular cavity* allow some extracellular (outside the cells) digestion to occur. Digestion is finished intracellularly by the cells that line the cavity, which branches throughout the body. No cell in the body is far from the intestine, and therefore diffusion alone is sufficient to distribute nutrient molecules.

The digestive system of a planarian is notable for its lack of specialized parts. It is saclike because the pharynx serves not only as an entrance for food but also as an exit for nondigestible material. Specialization of parts does not occur under these circumstances.

Recall that the planarian has some parasitic relatives. The tapeworm (see fig. 26.8), for example, has no digestive system at all—it simply absorbs nutrient molecules from the intestinal juices that surround its body. The body wall is highly modified for this purpose: it has millions of microscopic fingerlike projections that increase the surface area for absorption.

In contrast to the planarian, an earthworm has a complete gut. A *complete gut* has a mouth and anus. Earthworms feed mainly on decayed organic matter in dirt (fig. 36.5). The muscular *pharynx* draws in food with a sucking action. The *crop* is a storage area that has expansive thin walls. The *gizzard* has thick muscular walls for churning and grinding the food. Digestion is extracellular—in the intestine. The surface area of digestive tracts is often increased for absorption of nutrient molecules, and in the earthworm there is an intestinal fold called the typhlosole. Undigested remains pass out of the body at the anus.

Specialization of parts is obvious in the earthworm because the pharynx, crop, gizzard, and intestine each has a particular function in the digestive process. A complete gut leads to a digestive system with specialization of parts.

> In contrast to the incomplete, saclike gut, the complete gut with both a mouth and an anus has many specialized parts.

Human Digestive Tract

The human gut is complete, and there is a complex digestive system (fig. 36.6). Each part of the system has a specific function (table 36.1).

Mouth

Food is chewed in the mouth, and it is also mixed with saliva. We have already mentioned that human dentition is nonspecialized because humans are omnivores (fig. 36.2*c*). There are 3 pairs of *salivary glands* that send their juices by way of ducts to the mouth. Saliva contains the enzyme *salivary amylase,* which begins the process of starch digestion. The disaccharide maltose is a typical end product of salivary amylase digestion:

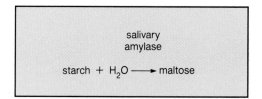

$$\text{starch} + H_2O \xrightarrow{\text{salivary amylase}} \text{maltose}$$

While in the mouth, food is manipulated by a muscular tongue, which has touch and pressure receptors similar to those in the skin. *Taste buds,* chemical receptors that are stimulated by the chemical composition of food, are also found primarily on the tongue and also on the surface of the mouth and the

Figure 36.4

Planarians are rather simple aquatic organisms. *a.* They have a very extensive nonspecialized digestive tract, the gastrovascular cavity. Incomplete digestive tracts have little specialization of parts because the single opening serves as both entrance and exit. *b.* The planarian relies on intracellular digestion to complete the digestive process. Phagocytosis produces a vacuole, which joins with an enzyme-containing lysosome. Notice that the food particles are always surrounded by membrane; therefore, they are never part of the cytoplasm. The digested products pass from the vacuole before any nondigestible material is eliminated at the plasma membrane.

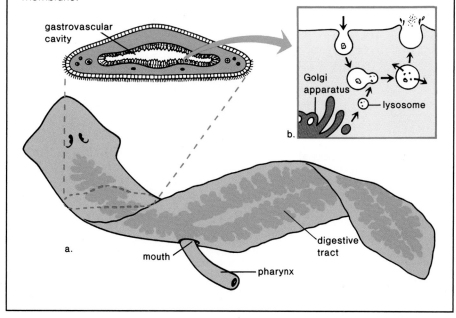

Figure 36.5

The earthworm has a complete digestive tract with both a mouth and an anus and many other specialized parts. The absorptive surface of the intestine is increased by an internal fold called the typhlosole.

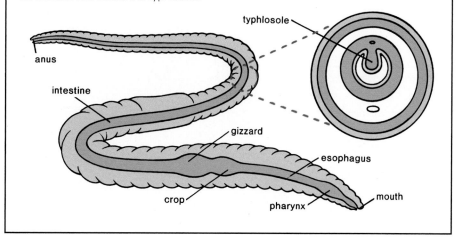

Figure 36.6

Human digestive system. The enlargement shows the layers that make up the wall of the small intestine. The *mucous membrane layer* is made up of columnar epithelium overlying loose connective tissue. The epithelium has deep folds called villi (fig. 36.10), which increase the absorptive surface of the small intestine. The *submucosal layer* is a dense connective tissue layer containing nerve fibers, blood vessels, and lymphatic vessels. The products of digestion enter these vessels. The *smooth muscle layer* has both circular and longitudinal muscular tissue. Rhythmic contraction of these muscles brings about peristalsis. In the *serous membrane layer* a thin sheet of connective tissue underlies a thin outermost sheet of squamous epithelium. This membrane is part of the peritoneum, which lines the entire abdominal cavity.

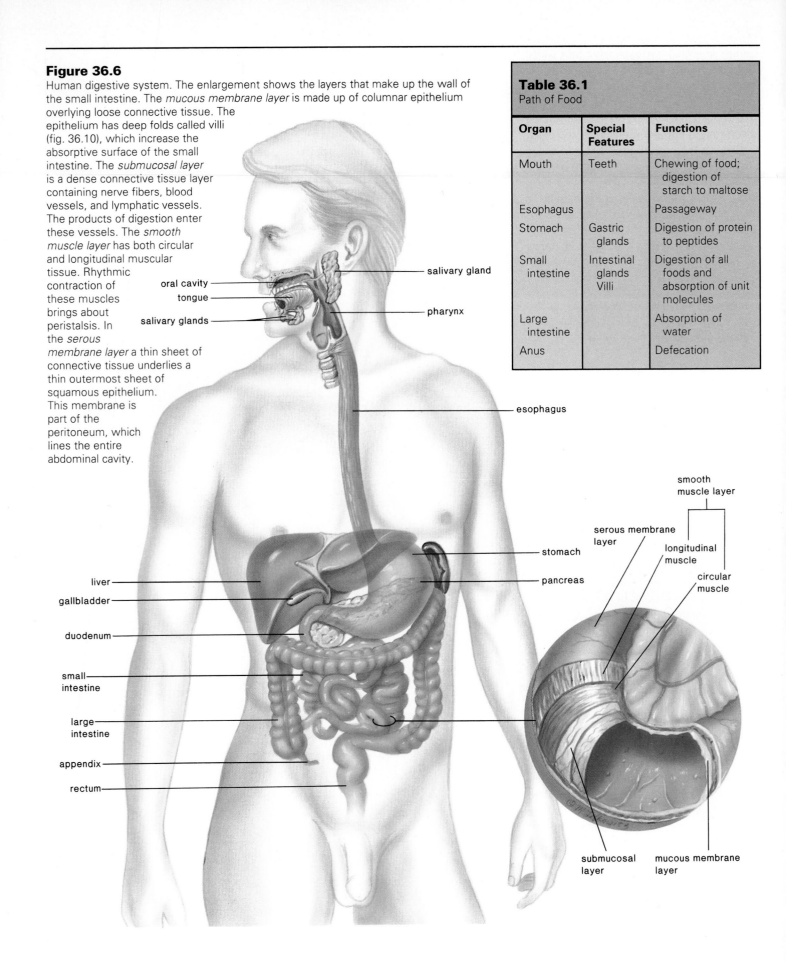

Table 36.1
Path of Food

Organ	Special Features	Functions
Mouth	Teeth	Chewing of food; digestion of starch to maltose
Esophagus		Passageway
Stomach	Gastric glands	Digestion of protein to peptides
Small intestine	Intestinal glands Villi	Digestion of all foods and absorption of unit molecules
Large intestine		Absorption of water
Anus		Defecation

Animal Structure and Function

Figure 36.7

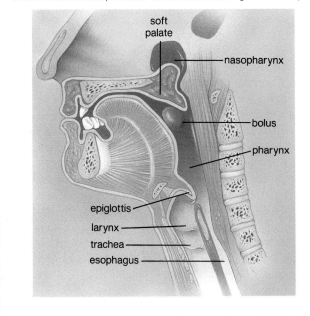

Swallowing. Respiratory and digestive passages diverge in the pharynx. During swallowing, the epiglottis covers the opening to the trachea and prevents food from entering this airway.

soft palate
nasopharynx
bolus
pharynx
epiglottis
larynx
trachea
esophagus

Figure 36.8

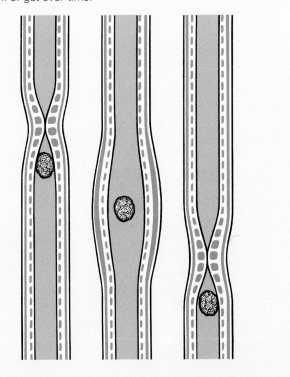

Peristalsis in the digestive tract. Rhythmic waves of muscle contraction move material along the digestive tract. The 3 drawings show how a peristaltic wave moves through a single section of gut over time.

wall of the pharynx. After the food has been thoroughly chewed and mixed with saliva, the tongue starts the process of swallowing by pushing the food (bolus) back to the pharynx.

Pharynx and Esophagus

The digestive and respiratory passages cross in the pharynx (fig. 36.7). Therefore, during swallowing, the path of air to the lungs could be blocked if food enters the trachea (windpipe). Normally, however, a flap of tissue called the **epiglottis** covers the opening into the **trachea** as muscles move the bolus through the pharynx into the esophagus, a tubular structure that takes food to the stomach. When food enters the **esophagus,** the rhythmic contraction of the gut wall, **peristalsis,** begins (fig. 36.8). Peristalsis pushes the bolus down the esophagus to the stomach.

Stomach

Usually, the stomach stores up to 2 liters of partially digested food. Therefore, humans can periodically eat relatively large meals and spend the rest of their time at other activities. But the stomach is much more than a mere storage organ, as was discovered by William Beaumont in the mid-nineteenth century. Beaumont, an American doctor, had a French Canadian patient, Alexis St. Martin. St. Martin had been shot in the stomach, and when the wound healed, he was left with a fistula, or opening, that allowed Beaumont to collect gastric (stomach) juices produced by *gastric glands* and to look inside the stomach to see what was going on there. Beaumont was able to determine that the muscular walls of the stomach contract vigorously and mix food with juices that are secreted whenever food enters the stomach (fig. 36.9). He found that *gastric juice* contains *hydrochloric acid* (HCl) and a substance

active in digestion. (This substance was later identified as pepsin.) He also found that the gastric juices are produced independently of the protective mucous secretions of the stomach. Beaumont's work, which was very carefully and painstakingly done, pioneered the study of the physiology of digestion.

So much hydrochloric acid is secreted by the stomach that it routinely has a pH of about 2.0. Such a high acidity usually is sufficient to kill bacteria and other microorganisms that might be in food. This low pH also stops the activity of salivary amylase, which functions optimally at the near-neutral pH of saliva, but it promotes the activity of pepsin. *Pepsin* is a hydrolytic enzyme that acts on protein to produce peptides:

$$\text{protein} + H_2O \xrightarrow{\text{pepsin}} \text{peptides}$$

By now the stomach contents have a thick, soupy consistency and are called *chyme.* At the base of the stomach is a narrow opening controlled by a *sphincter.* Whenever the sphincter relaxes, a small quantity of chyme squirts through the opening into the **duodenum,** the first part of the *small intestine* (fig. 36.6). When chyme enters the duodenum, it sets off a neural reflex that causes

Figure 36.9

This scanning electron micrograph provides a greatly enlarged view of what Beaumont saw when he looked through the opening in St. Martin's stomach. The wall of the stomach has folds called rugae (Ru), which disappear when the stomach is full. The arrows indicate the openings to the gastric glands, which produce gastric secretions, including hydrochloric acid and pepsinogen, a precursor that becomes pepsin in the stomach. The wall of the stomach has the same 4 layers as the small intestine, but with modifications. Notice how large the mucous membrane layer (Mu = mucous membrane layer) is because of the presence of mucosal muscle (MM). Su = submucosal layer; ML = smooth muscle layer; Se = serous membrane layer. Magnification, X55.

Kessel, R. G., and Kardon, R. H.: *Tissues and Organs: A Text-Atlas of Scanning Electron Microscopy.* © 1979 by W. H. Freeman and Co.

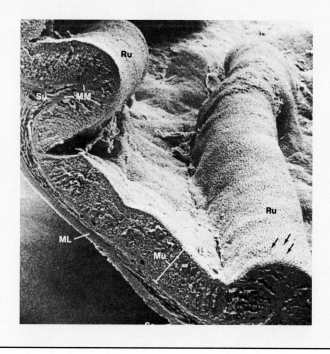

Small Intestine

The human *small intestine* is a coiled tube about 3 m long in a living person. The mucous membrane layer has ridges and furrows that give it an almost corrugated appearance. On the surface of these ridges and furrows are small, fingerlike projections called **villi.** Cells on the surfaces of the villi have minute projections called microvilli. The villi and microvilli greatly increase the effective surface area of the small intestine. If the small intestine were simply a smooth tube, it would have to be 500 m–600 m long to have a comparable surface area.

When the chyme enters the duodenum, proteins and carbohydrates are only partly digested, and fat digestion still needs to be carried out. Considerably more digestive activity is required before these nutrients can be absorbed through the intestinal wall. Two important accessory glands, the liver and the pancreas, send secretions to the duodenum (fig. 36.6). The liver produces **bile,** which is stored in the **gallbladder** and sent to the duodenum by way of a duct. Bile looks green because it contains pigments that are products of hemoglobin breakdown. This green color is familiar to anyone who has observed the color changes of a bruise. Hemoglobin within the bruise is breaking down into the same types of pigments found in bile. Bile also contains bile salts, which are *emulsifying* agents that break up fat into fat droplets so that they mix with the water:

$$\text{fat} \xrightarrow{\text{bile salts}} \text{fat droplets}$$

Emulsified fat is more easily acted on by enzymes.

The pancreas sends pancreatic juice into the duodenum, also by way of a duct. While the pancreas is an *endocrine* gland when it produces and secretes insulin into the bloodstream, it is an *exocrine* gland when it produces and secretes pancreatic juice into the duodenum. Pancreatic juice contains sodium bicarbonate ($NaHCO_3$), which neutralizes the chyme and makes the pH of the small intestine slightly basic. Pancreatic juice also contains digestive enzymes that act on every major component of food. *Pancreatic amylase* digests starch to maltose; *trypsin* and other enzymes digest protein to peptides; and *lipase* digests fat droplets to glycerol and fatty acids:

$$\text{starch} + H_2O \xrightarrow{\text{pancreatic amylase}} \text{maltose}$$

$$\text{protein} + H_2O \xrightarrow{\text{trypsin}} \text{peptides}$$

$$\text{fat droplets} + H_2O \xrightarrow{\text{lipase}} \text{glycerol} + \text{fatty acids}$$

the muscles of the sphincter to contract vigorously and to close the opening temporarily. Then the sphincter relaxes again and allows more chyme to squirt through. The slow manner in which chyme enters the small intestine allows digestion to be more thorough than it otherwise would be.

Normally, a thick layer of mucus protects the wall of the stomach and first part of the duodenum. But if by chance gastric juice does penetrate the mucus, pepsin starts to digest the stomach or duodenal lining and an *ulcer* results. An ulcer is an open sore in the wall caused by the gradual disintegration of tissues. It is believed that the most frequent cause of an ulcer is oversecretion of gastric juice due to too much nervous stimulation. Evidence is growing, however, that suggests a bacterial infection may be involved.

Before food enters the small intestine, it passes through the mouth, esophagus, and stomach. In the mouth, salivary amylase digests starch to maltose; in the stomach, pepsin digests protein to peptides.

Table 36.2
Digestive Enzymes

Reaction	Enzyme	Produced by	Site of Occurrence
Starch + H_2O → maltose	a. Salivary amylase b. Pancreatic amylase	a. Salivary glands b. Pancreas	a. Mouth b. Small intestine
Maltose + H_2O → glucose*	Maltase	Intestinal cells	Small intestine
Protein + H_2O → peptides	a. Pepsin b. Trypsin	a. Gastric glands b. Pancreas	a. Stomach b. Small intestine
Peptides + H_2O → amino acids*	Peptidases	Intestinal cells	Small intestine
Fat + H_2O → glycerol + fatty acids*	Lipase	Pancreas	Small intestine

*Absorbed by villi.

Figure 36.10
Anatomy of intestinal lining. *a.* The products of digestion are absorbed by villi, fingerlike projections of the intestinal wall. *b.* Each villus contains blood vessels and a lacteal.

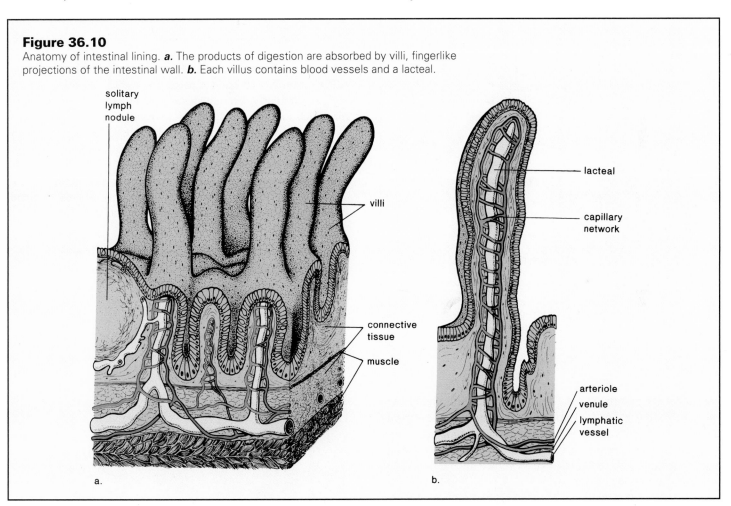

The epithelial cells of the villi produce *intestinal enzymes,* which remain attached to the plasma membrane of microvilli. These enzymes complete the digestion of peptides and sugars. Peptides, which result from the first step in protein digestion, are digested by peptidases to amino acids. Maltose, which results from the first step in starch digestion, is digested by maltase to glucose:

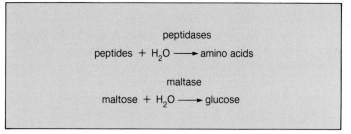

Digestive System and Nutrition

Figure 36.11

Hormonal control of digestive gland secretions. Especially after eating a protein-rich meal, gastrin produced by the lower part of the stomach enters the bloodstream and thereafter stimulates the upper part of the stomach to produce more digestive juices. Acid chyme from the stomach causes the duodenum to secrete secretin and CCK-PZ. Secretin and CCK-PZ stimulate the pancreas to release pancreatic juice, and CCK-PZ alone stimulates the gallbladder to release bile.

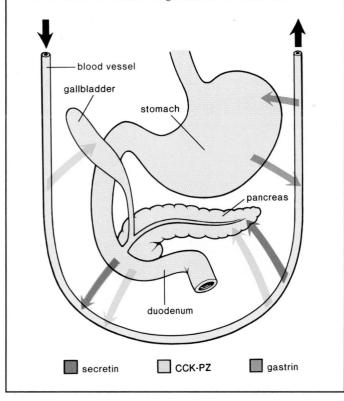

The wall of the small intestine is lined by villi. The epithelial cells of the villi produce intestinal enzymes that finish the digestive process. Sugars and amino acids enter blood vessels, and reformed fats enter the lacteals of the villi.

Control of Secretions Under normal circumstances, the digestive secretions are released *only* when food is present in the digestive tract. Saliva in the mouth and gastric juices in the stomach flow primarily in response to the taste, smell, or sometimes even the thought of food. Also, food in the stomach stimulates sense receptors that signal the brain to stimulate, by way of nerves, the gastric glands in the stomach.

Perhaps the major factor in gastric secretory regulation, however, is the hormone *gastrin*, which is produced by the stomach itself. When proteins contact the stomach mucosa, gastrin-producing cells are stimulated to release gastrin into the bloodstream (fig. 36.11). As soon as gastrin circulates through blood vessels and reaches the acid- and enzyme-secreting cells of the stomach lining, those cells respond by secreting large quantities of HCl and pepsin. As the stomach empties, both the neural reflexes and gastrin release subside, and less HCl and pepsin are secreted.

Similarly, some duodenal cells produce the hormone *secretin*, which stimulates the pancreas to release pancreatic juice, especially the sodium bicarbonate component. Other duodenal cells release a hormone called *CCK-PZ* (cholecystokinin-pancreozymin), which stimulates the release of bile. CCK-PZ stimulates the gallbladder to empty its contents through a duct into the duodenum. This same hormone also stimulates the pancreas, especially pancreatic enzyme secretion, as its full name suggests (fig. 36.11)

Liver

Blood vessels from the large intestine and the small intestine merge to form the hepatic portal vein, which leads to the liver:

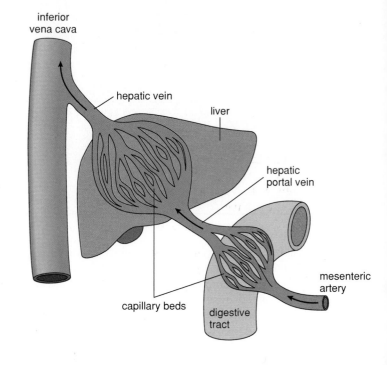

Other disaccharides, each of which is acted upon by a specific enzyme, are digested in the small intestine. Table 36.2 reviews reactions required for the digestion of food and the various digestive enzymes discussed in this chapter.

Absorption The small intestine is specialized for absorption. Molecules are absorbed by the huge number of villi that line the intestinal wall (fig. 36.10). Each villus contains blood vessels and a lymphatic vessel called a **lacteal.** Sugars and amino acids enter villi cells and then are absorbed into the bloodstream. Glycerol and fatty acids enter villi cells and are reassembled into fat molecules, which move into the lacteals.

Absorption continues until almost all products of digestion have been absorbed. Absorption involves active transport and requires an expenditure of cellular energy.

Food is largely made up of carbohydrate (starch), protein, and fat. These very large macromolecules are broken down by digestive enzymes to small molecules that can be absorbed by intestinal villi. This table indicates the steps needed for carbohydrate digestion (starch and maltose), protein digestion (protein and peptides), and fat digestion (fat) and shows that they are all hydrolytic reactions.

The liver has numerous functions, including the following:

1.   Detoxifies the blood by removing and metabolizing poisonous substances.
2.   Makes the blood proteins.
3.   Destroys old red blood cells and converts hemoglobin to the breakdown products in bile (bilirubin and biliverdin).
4.   Produces bile, which is stored in the gallbladder before entering the small intestine, where it emulsifies fats.
5.   Stores glucose as glycogen and breaks down glycogen to glucose between meals to maintain a constant glucose concentration in the blood.
6.   Produces urea from amino groups and ammonia.

Here we are interested in the last 2 functions listed. The liver helps maintain the glucose concentration in blood at about 0.1% by removing excess glucose from the hepatic portal vein and storing it as glycogen. Between meals, glycogen is broken down and glucose enters the hepatic vein. Glycogen is sometimes called *animal starch* because both starch and glycogen are made up of glucose molecules joined together (see fig. 4.7). If by chance the supply of glycogen and glucose runs short, the liver converts amino acids to glucose molecules. Amino acids contain nitrogen in the form of amino groups, whereas glucose contains only carbon, oxygen, and hydrogen. Therefore, before amino acids can be converted to glucose molecules, *deamination,* or the removal of amino groups from the amino acids, must take place. By an involved metabolic pathway, the liver converts these amino groups to urea, the most common nitrogenous waste product of humans, and after its formation in the liver, it is transported by the blood-stream to the kidneys, where it is excreted.

Blood from the small intestine enters the hepatic portal vein, which goes to the liver, a vital organ that has numerous important functions as listed.

Liver Disorders   Jaundice, hepatitis, and cirrhosis are 3 serious diseases that affect the entire liver and hinder its ability to repair itself. When a person is jaundiced, there is a yellowish tint to the skin due to an abnormally large amount of bilirubin in the blood. In *hemolytic jaundice,* red blood cells are broken down in abnormally large amounts; in *obstructive jaundice,* there is an obstruction of the bile duct or damage to the liver cells. Obstructive jaundice often occurs when crystals of cholesterol precipitate out of bile and form gallstones.

Jaundice can also result from *viral hepatitis,* a term that includes hepatitis A, hepatitis B, and hepatitis C. Hepatitis A is most often caused by eating shellfish from polluted waters. Hepatitis B and C are commonly spread by blood transfusions, kidney dialysis, and injection with inadequately sterilized needles. All 3 types of hepatitis can be spread by sexual contact.

Cirrhosis is a chronic disease of the liver in which the organ first becomes fatty. Liver tissue is then replaced by inactive fibrous scar tissue. In alcoholics, who often get cirrhosis of the liver, the condition most likely is caused by the excessive amounts of alcohol the liver is forced to break down.

Large Intestine

About 1.5 liters of water enter the digestive tract daily as a result of eating and drinking. An additional 8.5 liters enter the digestive tract each day carrying the various substances secreted by the digestive glands. About 95% of this water is reabsorbed into cells of the *large intestine,* or *colon.* This water reabsorption is essential. Failure to reabsorb water can result in *diarrhea,* which can lead to serious dehydration and ion loss, especially in children.

The large intestine functions both in water conservation and ion regulation. The colon absorbs some sodium and other ions from the material passing through it. At the same time, colon cells excrete metallic ions into the wastes leaving the body. The role that the large intestine plays in excretion is discussed on page 620. *Vitamin K,* produced by intestinal bacteria, is also absorbed in the colon.

The last 20 cm of the large intestine is the rectum, which terminates in the anus, an external opening. Digestive wastes (feces) eventually leave the body through the rectum and anus. Feces are about 75% water and 25% solid matter. Almost one-third of this solid matter is made up of intestinal bacteria. The remainder is undigested plant material, fats, waste products (such as bile pigments), inorganic material, mucus, and dead cells from the intestinal lining.

Appendicitis and Polyps   Two serious medical conditions are associated with the large intestine. The small intestine joins the large intestine in such a way that there is a blind end on one side (fig. 36.6). This blind sac, or cecum, has a small projection about the size of the little finger, called the appendix. In the case of *appendicitis,* the appendix becomes infected and so filled with fluid that it may burst. The appendix should be removed before it bursts, because a burst appendix can lead to a serious, generalized infection of the abdominal lining, called peritonitis.

The colon is subject to the development of polyps, which are small growths arising from the epithelial lining. Polyps, whether they are benign or cancerous, can be removed individually. The cause of colon cancer is not known, but a low-fat, high-fiber diet, which promotes regularity, is recommended as a protective measure.

Nutrition

To be sure that the essential nutrients are included in the diet, it is necessary to eat a balanced diet. A balanced diet is obtained by eating a variety of foods from the 4 food groups pictured in figure 36.12. Some authorities maintain there are only 3 food groups because milk products and meats provide the same type of nutrients.

Food largely consists of proteins, carbohydrates, and fats. Therefore, we will begin by considering these substances. Vitamins and minerals are also important nutrients and will be discussed later in this chapter.

Proteins

Foods rich in protein include red meat, fish, poultry, dairy products, legumes, nuts, and cereals. Following digestion of protein, amino acids enter the bloodstream and are transported to the tissues. Most of these amino acids are incorporated into structural proteins found in muscles, skin, hair, and nails. Others are used to synthesize such proteins as hemoglobin, plasma proteins, enzymes, and hormones.

Protein formation requires 20 different types of amino acids. Of these, 9 are required from the diet because the body is unable to produce them. These are termed the **essential amino acids.** The body produces the other 11 amino acids by simply transforming one type into another type. Some protein sources, such as meat, are *complete;* they provide all types of amino acids. Vegetables and grains supply us with amino acids, but each vegetable or grain alone is an *incomplete* protein source because at least one of the essential amino acids is absent. It is possible to combine foods, however, to acquire all of the essential amino acids. For example, the combinations of cereal with milk or beans with rice provide all the essential amino acids.

> A complete source of protein is absolutely necessary to ensure a sufficient supply of the essential amino acids.

Even though in this country we emphasize protein intake, it does not take very much protein to meet the daily requirement. In the United States, the *required dietary (daily) allowances (RDAs)* are determined by the National Research Council, a part of the National Academy of Sciences. The RDA for the reference woman (120 lb) is 44 g of protein a day. For the reference man (154 lb), it is 56 g of protein a day. A single serving of roast beef (3 oz) provides 25 g of protein, and a cup of milk provides 8 g.

Unfortunately, in this country, the manner in which we meet our daily requirement for protein is not always the most healthful way. The foods that are richest in protein are apt to be richest in fat.

> Foods richest in protein also tend to be richest in fat.

While it is very important to meet the RDA for protein, consuming more actually can be detrimental. Calcium loss in the urine has been noted when dietary protein intake is over twice the RDA. Everything considered, it is probably a good idea to depend on protein from plant origins (e.g., whole-grain cereals, dark breads, legumes) to a greater extent than is the custom in this country.

Carbohydrates

The quickest, most readily available source of energy for the body is carbohydrates, which can be complex, as in breads and cereals, or simple, as in candy, ice cream, and soft drinks. As

Figure 36.12

The 4 food groups. A variety of foods from each group should be eaten daily.

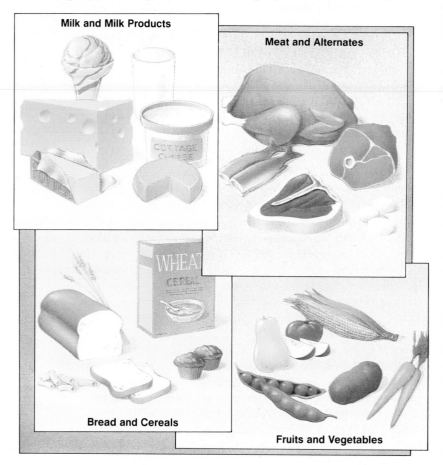

mentioned previously, starches are digested to glucose, which is stored by the liver in the form of glycogen. Between eating, blood glucose is maintained at about 0.1% by the breakdown of glycogen or by the conversion of amino acids to glucose. If necessary, these amino acids are taken from the muscles, even from heart muscle. To avoid this situation, it is suggested that the daily diet contain at least 100 g of carbohydrate. As a point of reference, a slice of bread contains approximately 14 g of carbohydrate.

> Carbohydrates are needed in the diet to maintain the blood glucose level.

Actually, the dietary guidelines produced jointly by the U.S. Department of Agriculture and the Department of Health and Human Services recommend that we increase the proportion of carbohydrates per total energy content of the diet from 46% to 58%. Further, it is assumed that these carbohydrates are complex and not simple. Simple carbohydrates (e.g., sugars) are labeled "empty calories" by some dieticians because they contribute to energy needs and weight gain and are not a part of food

Animal Structure and Function

that supply other nutritional requirements. Table 36.3 gives suggestions on how to cut down on your consumption of dietary sugars (simple carbohydrates).

In contrast to simple sugars, complex carbohydrates are likely to be accompanied by a wide range of other nutrients and by fiber, which is nondigestible plant material. Insoluble fiber, such as that found in wheat bran, has a laxative effect and therefore may reduce the risk of colon cancer. Soluble fiber, such as that found in oat bran, may possibly reduce cholesterol in the blood because it combines with cholesterol in the gut and prevents it from being absorbed.

While the diet should have an adequate amount of fiber, a high-fiber diet can be detrimental. Some evidence suggests that the absorption of iron, zinc, and calcium is impaired by a high-fiber diet.

> Complex carbohydrates along with fiber are considered beneficial to health.

Carbohydrates provide most of the dietary Kcalories.[1] There are only 4 Kcalories per gram of carbohydrate as compared to 9 Kcalories per gram of fat, but since carbohydrates are the bulk of the diet, they provide the most calories.

Lipids

Our discussion of lipids is divided into 2 parts: fats and cholesterol.

Fats

Fats are present not only in butter, margarine, and oils but also in foods high in protein. After being absorbed, the products of fat digestion are transported by the lymph and the blood to the tissues. The liver can alter ingested fats to suit the body's needs, except it is unable to produce the fatty acid linoleic acid. Since this is required for phospholipid production, linoleic acid is considered an *essential fatty acid.*

> Fats have the highest Kcaloric content, but they should not be avoided entirely because they contain the essential fatty acid linoleic acid.

While we need to ingest some fat to satisfy our need for linoleic acid, recent dietary guidelines suggest that we should reduce the amount of fat per total energy content of the diet from 40% to 30%. Dietary fat has been implicated in cancer of the colon, the pancreas, the prostate, and the breast (fig. 36.13). Many animal studies have shown that a high-fat diet stimulates the development of mammary tumors, while a low-fat diet does not. It also has been found that women who have a high-fat diet are more likely to develop breast cancer. Surprisingly, it has been discovered that a reduction in the amount of linoleic acid in the diet helps to prevent breast cancer. Linoleic acid is found in corn, safflower, sunflower, and other common plant oils but is not abundant in olive oil or in fatty fishes and marine animals.

> There is very strong evidence that women who have a diet high in fat are more apt to develop breast cancer.

[1]The amount of heat required to raise 1 kg of water 1°C

Table 36.3
Recommendations for Improved Diet

The Less-Fat Recommendations:

1. Choose lean meat, poultry, fish, dry beans, and peas. Trim fat off.
2. Eat eggs and such organ meats as liver in moderation. (Actually, these are high in cholesterol rather than fat.)
3. Broil, boil, or bake, rather than fry.
4. Limit your intake of butter, cream, hydrogenated oils, shortenings, coconut oil.

The Less-Salt Recommendations:

1. Learn to enjoy unsalted food flavors.
2. Add little or no salt to foods at the table and add only small amounts of salt when you cook.
3. Limit your intake of salty prepared foods such as pickles, pretzels, and potato chips.

The Less-Sugar Recommendations:

1. Eat less sweets such as candy, soft drinks, ice cream, and pastry.
2. Eat fresh fruit or canned fruit without heavy syrup.
3. Use less sugar—white, brown, raw—and less honey and syrups.

Source: American Dietetic Association, based on *Dietary Guidelines for Americans 1980*, U.S. Departments of Agriculture and Health, Education, and Welfare.

Fat is the component of food that has the highest energy content (9.3 Kcal/g compared to 4.1 Kcal/g for carbohydrate). Raw potatoes, which contain roughage, have about 0.9 Kcalories per gram, but when they are cooked in fat, the number of Kcalories jumps to 6 Kcalories per gram. Another problem for those trying to limit their Kcaloric intake is that fat is not always highly visible; for example, butter melts on toast or potatoes. Table 36.3 gives suggestions for cutting down on the amount of fat in the diet.

As a nation, we have increased our consumption of fat from vegetable sources and have decreased our consumption from animal sources, such as red meat and butter. Most likely, this is due to recent studies linking diets high in saturated fats and cholesterol to hypertension and heart attack.

Cholesterol

The risk of cardiovascular disease includes many factors, which are discussed on page 566. One of these factors, according to the National Heart, Lung, and Blood Institute, is a cholesterol blood level of 240 mg/100 ml or higher. If the cholesterol level is this high, additional testing can determine how much of each of 2 important subtypes of cholesterol is in the blood. Cholesterol is carried from the liver to the cells (including the endothelium of the arteries) by plasma proteins called *low-density lipoprotein (LDL)* and is carried away from the cells to the liver by *high-density lipoprotein (HDL)*. LDL cholesterol apparently contributes to the formation of plaque, which can clog the arteries, while HDL cholesterol protects against the development of clogged arteries. It has been found that a diet low in saturated fats prevents the

Figure 36.13

Diet and cancer. Evidence is growing to suggest that the noted dietary factors can contribute to the development of cancer.

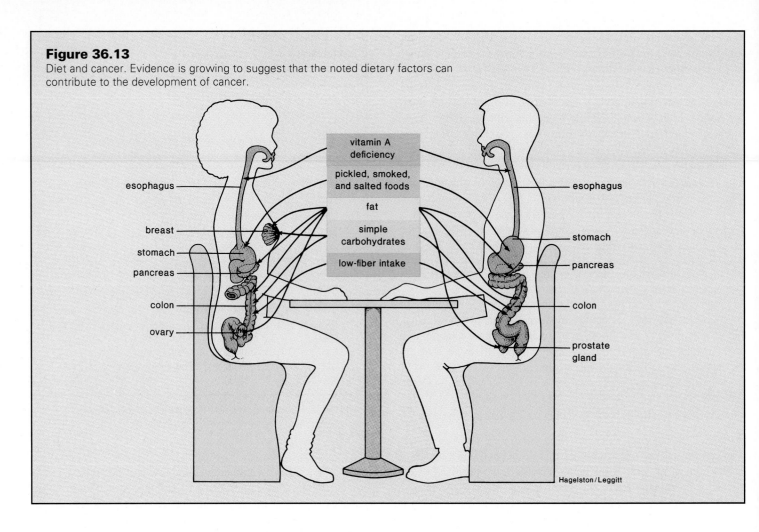

esophagus

breast

stomach

pancreas

colon

ovary

vitamin A
deficiency

pickled, smoked,
and salted foods

fat

simple
carbohydrates

low-fiber intake

esophagus

stomach

pancreas

colon

prostate
gland

Hagelston/Leggitt

---

accumulation of LDL cholesterol in the body. Therefore, the suggestions in regard to fat in table 36.3 are also recommended for cholesterol management. It may also be helpful to restrict the intake of cholesterol itself. To do this, white fish, poultry, and shellfish are recommended; cheese, egg yolks, and liver are not recommended. Egg whites can be substituted for egg yolks for both cooking and eating.

Vitamins and Minerals

Vitamins

Vitamins are organic compounds (other than carbohydrates, fats, and proteins) that the body is unable to produce but needs for metabolic purposes. Therefore, vitamins must be present in the diet; if they are lacking, various symptoms develop (fig. 36.14). Many substances are advertised as vitamins, but in reality there are only 13 vitamins (table 36.4). The table also gives the RDA vitamin values for a "typical" female and a "typical" male. In general, carrots, squash, turnip greens, and collards are good sources for vitamin A. Citrus fruits, other fresh fruits, and vegetables are natural sources of vitamin C. Fortified milk is the primary source of vitamin D, and whole grains are a good source of B vitamins.

It is not difficult to acquire the RDAs for vitamins if your diet is balanced (fig. 36.12), because each vitamin is needed in small amounts only. Many vitamins are portions of coenzymes. For example, niacin is part of the coenzyme NAD+, and riboflavin is a part of FAD. Coenzymes are needed in only small amounts because each one can be used repeatedly.

The National Academy of Sciences suggests that we eat more fruits and vegetables to acquire a good supply of vitamins C and A because these 2 vitamins may help guard against the development of cancer. Nevertheless the intake of excess vitamins by way of pills is discouraged because this practice can possibly lead to illness. For example, excess vitamin C can cause kidney stones and can also be converted to oxalic acid, a molecule that is toxic to the body. Vitamin A taken in excess can cause peeling of the skin, bone and joint pains, and liver damage. Excess vitamin D can cause vomiting, diarrhea, and kidney damage. Megavitamin therapy, therefore, should always be supervised by a physician.

A properly balanced diet includes all the vitamins and minerals needed by most individuals to maintain health.

Do Well on Tests and Exams
Get Your Own Tutor!

Wm. C. Brown Publishers has now developed a take-home tutor that will help you get better grades in the classroom and prepare for careers in the nursing/allied health sciences. The ***Biology Learning Support Package*** is a complete set of study aids designed to facilitate learning and achievement. The package includes:

STUDENT STUDY GUIDE
ISBN 0-697-12384-7

For each text chapter there is a corresponding study guide chapter that includes suggestions for review, objective questions, and essay questions that will review the major chapter concepts.

• • •

HOW TO STUDY SCIENCE
ISBN 0-697-14474-7

This excellent workbook offers you helpful suggestions for meeting the considerable challenges of a college science course. It offers tips on how to take notes, how to get the most out of laboratories, as well as how to overcome science anxiety. The book's unique design helps you develop critical thinking skills, while facilitating careful note taking.

• • •

THE LIFE SCIENCE LEXICON
ISBN 0-697-12133-X

This portable, inexpensive reference will help you quickly master the vocabulary of the life sciences. Not a dictionary, it carefully explains the rules of word construction and derivation, in addition to giving complete definitions of all important terms.

• • •

BIOLOGY STUDY CARDS
ISBN 0-697-03069-5

This boxed set of 300 two-sided study cards provides a quick, yet thorough visual synopsis of all key biological terms and concepts in the general biology curriculum. Each card features a masterful illustration, pronunciation guide, definition, and description in context.

• • •

IS YOUR MATH READY FOR BIOLOGY?
ISBN 0-697-22677-8

This unique booklet provides a diagnostic test that measures your math ability. Part II of the booklet provides helpful hints on the necessary math skills needed to successfully complete a biological science course.

To get your own take-home tutor, just ask your bookstore manager, or if unavailable through your bookstore, call toll-free 1-800-338-5578 and order using your credit card.

Figure 36.14
Illnesses due to vitamin deficiency. *a.* Bowing of bones (rickets) due to vitamin D deficiency. *b.* Dermatitis of areas exposed to light (pellagra) due to niacin deficiency. *c.*
Bleeding of gums (scurvy) due to vitamin C deficiency. *d.* Fissures of lips (cheilosis) due to
riboflavin deficiency.

a.

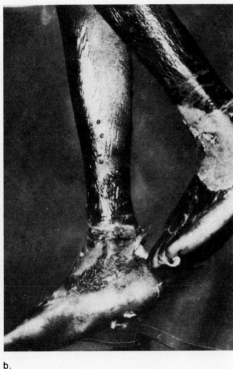

b.

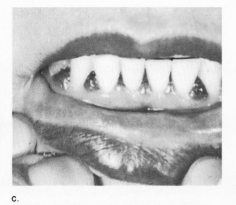

c.

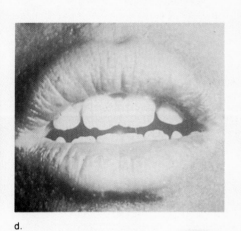

d.

Minerals

In addition to vitamins, various minerals are also required by the
body. Some minerals (calcium, phosphorus, potassium, sulfur,
sodium, chlorine, and magnesium) are recommended in amounts
more than 100 mg per day. These macrominerals serve as constituents of cells and body fluids and as structural components of
tissues. For example, calcium is needed for the construction of
bones and teeth and also for nerve conduction and muscle contraction. Other minerals (iron, manganese, copper, iodine, cobalt, and
zinc) are recommended in amounts of less than 20 mg per day.

These microminerals are more likely to have very specific
functions. For example, iron is needed for the production of
hemoglobin, and iodine is used in the production of thyroxin, a
hormone produced by the thyroid gland. As research continues,
more and more elements have been added to the list of those
considered essential. During the past 3 decades, very small amounts
of molybdenum, selenium, chromium, nickel, vanadium, silicon,
and even arsenic have been found to be essential to good health.

Some individuals do not receive enough iron (especially
women), calcium, magnesium, or zinc in their diets. Adult females
need more iron in their diet than males (RDA of 18 mg compared
to 10 mg) because they lose hemoglobin each month during
menstruation. Stress can bring on a magnesium deficiency, and a
vegetarian diet may lack zinc, which is usually obtained from meat.
A varied and complete diet, however, usually supplies the RDAs
for minerals.

Calcium

There is much interest in calcium supplements to counteract the
possibility of developing *osteoporosis,* a degenerative bone disease that afflicts an estimated one-fourth of older men and one-half
of older women in the United States. These individuals have porous
bones that tend to break easily because they lack sufficient calcium.
In 1984, a National Institute of Health conference on osteoporosis
advised postmenopausal women to increase their intake of calcium
to 1,500 mg and all others to 1,000 mg (compared with the RDA
of 800 mg).

Table 36.4
Vitamins: Their Role in the Body and Food Sources

Vitamins	Role in Body*	Good Food Sources*	Deficiency[†]	Excess[†]
Fat-Soluble Vitamins				
Vitamin A	Assists in the formation and maintenance of healthy skin, hair, and mucous membranes; aids in the ability to see in dim light (night vision); essential for proper bone growth, tooth development, and reproduction	Deep yellow/orange and dark green vegetables and fruits (carrots, broccoli, spinach, cantaloupe, sweet potatoes); cheese, milk, and fortified margarines	Blindness, especially night blindness	Headache, vomiting, peeling of skin, liver damage, anorexia, swelling of long bones
Vitamin D	Aids in the formation and maintenance of bones and teeth; assists in the absorption and use of calcium and phosphorus	Milks fortified with vitamin D; tuna, salmon, or cod liver oil; also made in the skin when exposed to sunlight	Bone deformities, rickets in children	Vomiting, diarrhea, weight loss, kidney damage
Vitamin E	Protects vitamin A and essential fatty acids from oxidation; prevents cell membrane damage	Vegetable oils and margarine, nuts, wheat germ and whole grain breads and cereals, green leafy vegetables	None known in humans	Relatively nontoxic
Vitamin K	Aids in synthesis of substances needed for clotting of blood; helps maintain normal bone metabolism	Green leafy vegetables, cabbage, and cauliflower; also made by bacteria in intestines of humans, except for newborns	Possibly severe bleeding	Synthetic forms at high doses may cause jaundice
Water-Soluble Vitamins				
Vitamin C	Important in forming collagen, a protein that gives structure to bones, cartilage, muscle, and vascular tissue; helps maintain capillaries, bones, and teeth; aids in absorption of iron; helps protect other vitamins from oxidation	Citrus fruits, berries, melons, dark green vegetables, tomatoes, green peppers, cabbage, and potatoes	Scurvy (degeneration of skin, teeth, blood vessels)	Possibility of kidney stones
Thiamin	Helps in release of energy from carbohydrates; promotes normal functioning of nervous system	Whole-grain products, dried beans and peas, sunflower seeds, nuts	Beriberi (nerve changes, edema, heart failure)	None reported
Riboflavin	Helps body transform carbohydrate, protein, and fat into energy	Nuts, yogurt, milk, whole-grain products, cheese, poultry, leafy green vegetables	Cheilosis (cracks at corners of mouth), lesions of eye	None reported
Niacin	Helps body transform carbohydrate, protein, and fat into energy	Nuts, poultry, fish, whole-grain products, dried fruit, leafy green vegetables, beans; can be formed in the body from tryptophan, an essential amino acid found in protein	Pellagra (skin and gut lesions; nervous and mental disorders)	Flushing, burning, and tingling around neck, face, and hands
Vitamin B_6	Aids in the use of fats and amino acids; aids in the formation of protein	Sunflower seeds, beans, poultry, nuts, leafy green vegetables, bananas, dried fruit	Irritability, convulsions, kidney stones	None reported
Folic acid	Aids in the formation of hemoglobin in red blood cells; aids in the formation of genetic material	Dark green leafy vegetables, nuts, beans, whole-grain products, fruit juices	Anemia, gastrointestinal disturbances, red tongue	None reported
Pantothenic acid	Aids in the formation of hormones and certain nerve-regulating substances; helps in the metabolism of carbohydrate, protein, and fat	Nuts, beans, seeds, dark green leafy vegetables, poultry, dried fruit, milk	Sleep disturbances, fatigue, impaired coordination, nausea (rare in humans)	None reported
Biotin	Aids in the formation of fatty acids; helps in the release of energy from carbohydrate	Occurs widely in foods, especially eggs. Made by bacteria in the human intestine	Fatigue, depression, nausea, dermatitis, muscular pains	None reported
Vitamin B_{12}	Aids in the formation of red blood cells and genetic material; helps the functioning of the nervous system	Milk, yogurt, cheese, fish, poultry, and eggs; not found in plant foods unless fortified (such as in some breakfast cereals)	Pernicious anemia, neurological disorders	None reported

*From David C. Nieman et al., *Nutrition.* Copyright © 1992 Wm. C. Brown Communications, Inc., Dubuque, Iowa. All Rights Reserved. Reprinted by permission.
[†]From Nevin S. Scrimshaw and Vernon R. Young, "The Requirements of Human Nutrition." Copyright © 1976 Scientific American Inc. All rights reserved.

Studies have shown, however, that calcium supplements cannot prevent osteoporosis after menopause even when the dosage is 3,000 mg a day. In postmenopausal women, bone-eating cells called osteoclasts are known to be more active than bone-forming cells called osteoblasts (p. 665). Until now, the most effective defense against osteoporosis in older women has been exercise, which encourages the work of osteoblasts, and estrogen replacement. Recently, however, studies have shown that the drug etidronate disodium, which inhibits osteoclast activity, is effective in osteoporotic women when administered at the proper dosage.

Young women can guard against the possibility of developing osteoporosis when they grow older by forming strong, dense bones before menopause. Eighteen-year-old women are apt to get only 679 mg of calcium a day when the RDA is 800 mg. They should consume more calcium-rich foods, such as milk and dairy products. Calcium supplements may not be as effective. A cup of milk supplies 270 mg of calcium, while a 500 mg tablet of calcium carbonate provides only 200 mg. The rest is just not taken up by the body; it is not in a form that is *bioavailable*. An excess of bioavailable calcium can lead to kidney stones.

> Dietary calcium and exercise, plus estrogen therapy if needed, are the best safeguards against osteoporosis.

Sodium

The recommended amount of sodium intake per day is 400–3,300 mg, and the average American takes in 4,000–4,700 mg, mostly as salt (sodium chloride). In recent years, this imbalance has caused concern because high-sodium intake has been linked to hypertension in some people. About one-third of the sodium we consume occurs naturally in foods; another third is added during commercial processing; and we add the last third either during home cooking or at the table in the form of table salt.

Clearly it is possible for us to cut down on the amount of sodium in the diet by reducing the amount of salt we eat. Table 36.3 gives recommendations for doing so.

Summary

1. All mammals have teeth. Herbivores need teeth that can clip off plant material and grind it up. Also, the herbivore's stomach contains bacteria that can digest cellulose.
2. Carnivores need teeth that can tear and rip meat into pieces. Meat is easier to assimilate, so the digestive system of carnivores has less specialization of parts and a shorter intestine than that of herbivores.
3. Some animals are continuous feeders (e.g., clams, which are sessile filter feeders); others are discontinuous feeders (e.g., squid). Discontinuous feeders need a storage area for food.
4. Some animals (e.g., planarians) have an incomplete digestive tract, which shows little specialization of parts. Other animals (e.g., earthworms) have a complete digestive tract, which does show specialization of parts.
5. In the human digestive tract, food is chewed and manipulated in the mouth, where salivary glands secrete saliva. Saliva contains salivary amylase, which begins carbohydrate digestion.
6. Food then passes down the esophagus to the stomach. The stomach stores and mixes food with mucus and gastric juice to produce chyme. Pepsin begins protein digestion here.
7. Chyme gradually enters the duodenum where bile, pancreatic juice, and the intestinal secretions are found. Enzymes in the small intestine hydrolyze all of the organic nutrients. Table 36.2 summarizes the enzymes involved in digesting food.
8. Most nutrient absorption takes place in the small intestine, but some water and minerals are absorbed in the colon. Digestive wastes leave the colon by way of the anus.
9. The liver produces bile, which is stored in the gallbladder. The liver is also involved in the processing of absorbed nutrient molecules and in maintaining the blood concentration of nutrient molecules, such as glucose. The liver converts ammonia to urea and breaks down toxins.
10. The pancreas produces digestive enzymes and hormones (insulin and glucagon) involved in the control of carbohydrate metabolism in the body.
11. There are 3 hormones that regulate digestive-tract secretions: gastrin, which stimulates acid- and enzyme-secreting cells in the stomach; secretin, which stimulates the pancreas to release sodium bicarbonate; and CCK-PZ, which stimulates the gallbladder to release bile and the pancreas to release digestive enzymes.
12. A balanced diet is required for good health. Food should provide us with all the necessary vitamins, minerals, amino acids, fatty acids, and an adequate amount of energy.

Writing Across the Curriculum

In order to practice writing skills, students should write out the answers to any or all of the study questions and the critical thinking questions. The study questions are sequenced in the same order as the text. Suggested answers to the critical thinking questions are in appendix D.

Study Questions

1. Contrast the incomplete with the complete gut using the planarian and earthworm as examples.
2. Contrast a continuous with a discontinuous feeder using the clam and squid as examples.
3. Contrast the dentition of the mammalian herbivore with the mammalian carnivore using the horse and lion as examples.

4. List the parts of the human digestive tract, anatomically describe them, and state the contribution of each to the digestive process.
5. State the location and describe the functions of both the liver and pancreas.
6. Assume that you have just eaten a ham sandwich. Discuss the digestion of the contents of the sandwich.
7. Discuss the absorption of the products of digestion into the circulatory system.
8. What are gastrin, secretin, and CCK-PZ? Where are they produced and what are their functions?
9. Give reasons why carbohydrates, fats, proteins, vitamins, and minerals are all necessary to good nutrition.

Objective Questions

1. Animals that feed discontinuously
   a. must have digestive tracts that permit storage.
   b. are able to avoid predators by limiting their feeding time.
   c. exhibit extremely rapid digestion.
   d. Both a and b.
2. In which of the following types of animals would you expect the digestive system to be more complex?
   a. those with a single opening for the entrance of food and exit of wastes
   b. those with 2 openings, one serving as an entrance and the other as an exit
   c. only those complex animals that also have a respiratory system
   d. Both b and c.
3. The typhlosole within the gut of an earthworm compares best to which of these organs in humans?
   a. teeth in the mouth
   b. esophagus in the thoracic cavity
   c. folds in the stomach
   d. villi in the small intestine
4. Which of these animals is a continuous feeder with a complete gut?
   a. planarian
   b. clam
   c. squid
   d. lion
5. The most common food digested in the human stomach is
   a. carbohydrate.
   b. fat.
   c. protein.
   d. nucleic acid.
6. Assuming normal body temperature, which of these combinations is most likely to result in complete digestion?
   a. fat, bile, sodium bicarbonate, lipase
   b. fat, bile, sodium bicarbonate, pepsin
   c. fat, HCl, maltase
   d. fat, bile, HCl, sodium bicarbonate, trypsin
7. Which association is incorrect?
   a. protein—trypsin
   b. fat—lipase
   c. maltose—pepsin
   d. starch—amylase
8. Most of the absorption of the products of digestion takes place in humans across the
   a. squamous epithelium of the esophagus.
   b. convoluted walls of the stomach.
   c. fingerlike villi of the small intestine.
   d. smooth wall of the large intestine.
9. The hepatic portal vein is located between the
   a. hepatic vein and the vena cava.
   b. mouth and the stomach.
   c. pancreas and the small intestine.
   d. small intestine and the liver.
10. Bile in humans
    a. is an important enzyme for the digestion of fats.
    b. is made by the gallbladder.
    c. emulsifies fat.
    d. All of these.
11. Which of these is not a function of the human liver?
    a. produce bile
    b. store glucose
    c. produce urea
    d. make red blood cells
12. The large intestine in humans
    a. digests all types of food.
    b. is the longest part of the intestinal tract.
    c. absorbs water.
    d. is connected to the stomach.
13. Label this diagram of the digestive tract.

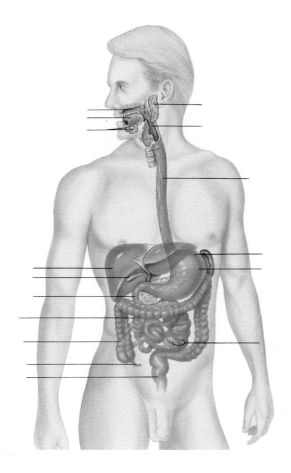

Animal Structure and Function

14. Predict and explain the digestive results per test tube for this experiment.

Test tube 1:

Test tube 2:

Test tube 3:

Test tube 4:

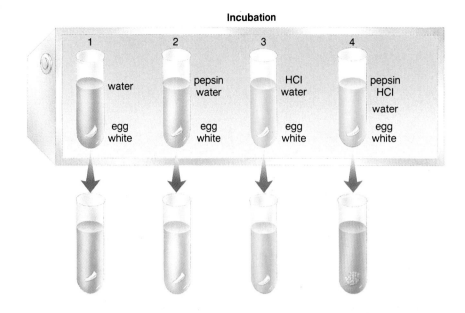

Incubation

Concepts and Critical Thinking

1. *All systems of the animal's body contribute to homeostasis.*

Tell several ways the human digestive system contributes to homeostasis.

2. *Organisms are adapted to obtaining a share of available resources.*

Animals vary according to how they obtain and process food. Choose either the grasshopper, earthworm, clam, or squid to support this concept.

3. *Forms of life depend on each other for materials and energy.*

Review figure 1.5, and explain how heterotrophs (e.g., animals) are dependent on autotrophs (e.g., plants) and how autotrophs in turn are dependent on heterotrophs.

Selected Key Terms

omnivore (om'nĭ-vōr) 588
herbivore (her'bĭ-vor) 588
carnivore (kar'nĭ-vōr) 588
epiglottis (ep"ĭ-glot'is) 593

esophagus (ĕ-sof'ah-gus) 593
peristalsis (per"ĭ-stal'sis) 593
duodenum (du"o-de'num) 593
villus (vil'lus) 594
bile (bīl) 594

gallbladder (gawl'blad-der) 594
lacteal (lak'te-al) 596
essential amino acid (ĕ-sen'shal ah-me'no as'id) 598
vitamin (vi'tah-min) 600

37

Respiratory System

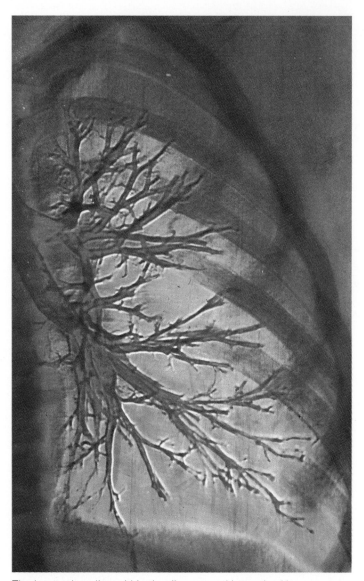

The human lung lies within the rib cage and is serviced by a branching pulmonary artery. The rib cage not only protects the lungs, it also assists breathing. When the rib cage moves up and out, the lungs expand, and when it moves down and in, the lungs recoil.

Your study of this chapter will be complete when you can

1. compare the oxygen-containing capacities of water and air;
2. describe the mechanism of gas exchange in a hydra and a planarian;
3. compare the respiratory organs of aquatic and terrestrial animals;
4. compare the various methods for ventilating the lungs among terrestrial vertebrates;
5. compare the incomplete method of ventilation used by amphibians, reptiles, and mammals with the complete method used by birds;
6. describe the path of air in humans, and describe in general the structure and function of all organs mentioned;
7. describe how the breathing rate is controlled in humans;
8. give the equations applicable to the exchange of gases in the lungs and tissues, and tell how the gases are transported in the blood;
9. name 3 types of hemoglobin found in humans;
10. tell how the respiratory tract protects itself from infection, and relate emphysema and lung cancer to cigarette smoking.

nimals do not have a storage area for gases and must continually acquire oxygen and rid the body of carbon dioxide. Without a constant supply of oxygen an animal dies. This was first observed in 1772 by the English scientist Joseph Priestley, who collected oxygen by heating red mercuric oxide. His contemporary, the Frenchman Antoine Lavoisier, correctly deduced that both combustion by a lit candle and respiration by an animal remove oxygen from the air. This was one of the first times that a physiological process was explained with reference to a nonliving mechanism.

We know today that oxygen is the final acceptor for electrons during cellular respiration, a metabolic process that supplies ATP energy for repair, growth, and movement.

Respiratory Gas-Exchange Surfaces

Gas exchange takes place by the physical process of diffusion. For diffusion to be effective, the gas-exchange region must be (1) moist, (2) thin, and (3) large in relation to the size of the body. Some animals are small and shaped in a way that allows the surface of the animal to be the gas-exchange surface. Other animals are complex and have a specialized gas-exchange surface. The effectiveness of diffusion is enhanced by vascularization, and delivery to cells is promoted when the blood contains a respiratory pigment such as hemoglobin.

Aquatic Environments

It is more difficult for animals to obtain oxygen from water than from air. Water fully saturated with air contains about 5% the amount of oxygen than does the same volume of air. Also, water is more dense than air. Therefore, aquatic animals expend more energy for breathing than do terrestrial animals. Fishes use up to 25% of their energy output, while terrestrial mammals use only 1%–2% of their energy output for breathing.

Hydras and planarians have a large surface area in comparison to their size (fig. 37.1). This makes it possible for most of their cells to exchange gases directly with the environment. In hydras, the outer layer of cells is in contact with the external environment and the inner layer can exchange gases with the water within the gastrovascular cavity. In planarians, the flattened body permits cells to exchange gases with the external environment.

A tubular shape also provides a surface area adequate enough for gas-exchange purposes. In addition to a tubular shape, polychaete worms (p. 435) have extensions of the body wall called parapodia, which are vascularized. Quite often aquatic animals have **gills**, which are finely divided and vascularized outgrowths of either the outer or inner body surface. Among mollusks, such as clams and many snails, water is drawn into the mantle cavity, where it passes through the gills. In decapod crustaceans the gills are located in brachial chambers covered by the exoskeleton. Water is kept moving by the action of specialized appendages located near the mouth.

Figure 37.1

Animal shapes and gas exchange. Some small aquatic animals use the body surface for gas exchange. This works because the body surface is large compared to the size of the animal. *a.* In hydras every cell is near a source of oxygen. The inner layer of cells exchanges gases with the water in the gastrovascular cavity, and the outer layer of cells exchanges gases with the external environment. *b.* In planarians the body is flattened and most cells can carry out gas exchange with the external environment. Planarians do not have a circulatory system.

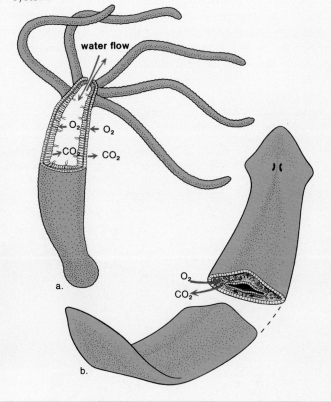

Among vertebrates, the gills of fishes are outward extensions of the pharynx (fig. 37.2). Ventilation is brought about by the combined action of the mouth and gill covers, or opercula (operculum, sing.). When the mouth is open, the opercula are closed and water is drawn in. Then the mouth closes and the opercula open, drawing the water from the pharynx through the gill slits located between the gill arches. On the outside of the gill arches, the gills are composed of *filaments* that are folded into platelike *lamellae*. In the capillaries of each lamella, the blood flows in a direction opposite to the movement of water across the gills. This *countercurrent flow* increases the amount of oxygen that can be taken up; as the blood in each lamella gains oxygen, it encounters water having a still higher oxygen content. The countercurrent mechanism allows about 80%–90% of the initial dissolved oxygen in water to be extracted.

Small aquatic animals sometimes use the body surface for gas exchange, but many larger ones have localized gas-exchange surfaces known as gills.

Terrestrial Environments

Air is a rich source of oxygen compared to water; however, it does have a drying effect on respiratory surfaces. A human loses about 350 ml of water per 24 hours when the air has a relative humidity of 50%.

The earthworm (see fig. 27.8) is an example of an invertebrate terrestrial animal that uses its body surface for respiration. An earthworm expends much energy to keep its body surface moist by secreting mucus and by releasing fluids from excretory pores. Further, the worm is behaviorally adapted to remain in damp soil during the day, when the air is driest.

Insects and certain other terrestrial arthropods have a respiratory system known as a tracheal system (fig. 37.3). Oxygen enters the tracheae at spiracles, which are valvelike openings on each side of the body. The tracheae branch and then rebranch, ending in tiny channels, the tracheoles, that are in direct contact with the body cells. Larger insects have a ventilation system to keep the air moving in and out of the tracheae. Many have air sacs located near major muscles; contraction of these muscles causes the air sacs to empty, and relaxation causes the air sacs to expand and draw air in. The tracheal system of insects is effective in delivering oxygen to the cells so that the insect circulatory system plays no role in gas transport.

Terrestrial vertebrates, in particular, have evolved vascularized outgrowths from the lower pharyngeal region known as **lungs.** The lungs of amphibians are simple, saclike structures

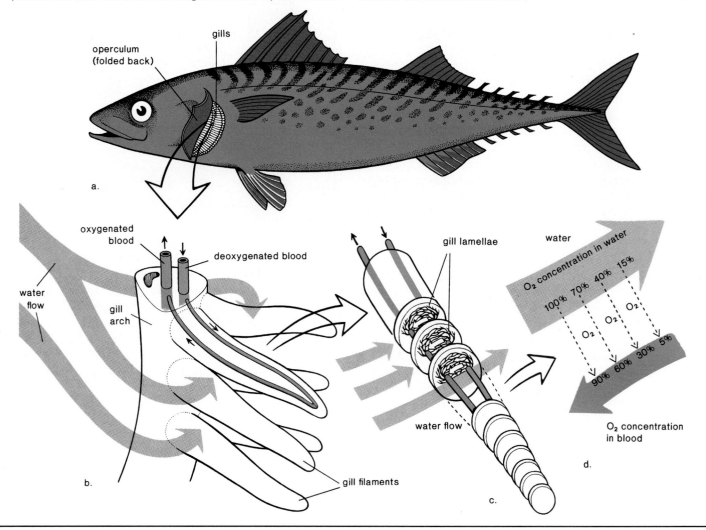

Figure 37.2
Anatomy of gills in detail. *a.* The operculum (folded back) covers and protects several layers of delicate gills. *b.* Each layer of gills has 2 rows of gill filaments. *c.* Each filament has many thin, platelike lamellae. Gases are exchanged between capillaries inside the lamellae and the water that flows between the lamellae. *d.* Blood in the capillaries flows in the direction opposite to that of the water. Blood takes up 90% of the oxygen in the water as a result of this countercurrent flow.

Animal Structure and Function

Figure 37.3

Tracheal system of insects. *a.* A system of air tubes extends throughout the body of an insect and they, rather than blood, carry oxygen to the cells. *b.* Air enters the tracheae at openings called spiracles. From here, it moves to the smaller tracheoles, which take it to the cells, where gas exchange takes place.

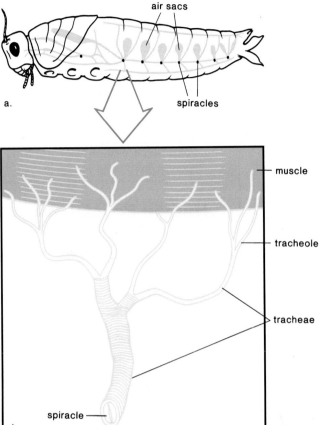

air sacs

spiracles

a.

muscle

tracheole

tracheae

spiracle

b.

(fig. 37.4*a*). Many amphibians possess a short trachea, which divides into 2 bronchi that open into the lungs. Most amphibians respire to some extent through the skin, which is kept moist by the presence of mucus produced by numerous glands on the surface of the body. During the winter in temperate climates, amphibians burrow in the mud, and all gas exchange occurs by way of the skin.

The inner lining of the lungs is more finely divided in reptiles than in amphibians (fig. 37.4*b*). The lungs of birds and mammals are even more elaborately subdivided into small passageways and spaces. It has even been estimated that human lungs have a total surface area that is at least 50 times the skin's surface area. To keep the lungs from drying out, air is moistened as it moves along the tubes leading to the lungs.

Terrestrial vertebrates ventilate the lungs by moving air in and out of the respiratory tract. Frogs use positive pressure to force air into the respiratory tract. With the nostrils firmly shut, the floor of the mouth rises and pushes the air into the lungs. Reptiles, birds, and mammals use negative pressure (fig. 37.5). The lungs first expand, and then air comes rushing in. Reptiles have jointed ribs that can be raised to expand the lungs. Mammals have a rib cage that is lifted up and out and a muscular **diaphragm** that is flattened. Both of these actions increase the volume of the lungs, so that air is thereby drawn in, during the process called **inhalation,** or inspiration. After inhalation, exhalation occurs. During **exhalation** (also called expiration), air is pushed out of the lungs. In reptiles, lowering the ribs exerts a pressure that forces air out. In mammals, the rib cage is lowered, and the diaphragm rises, forcing the air out of the lungs.

Figure 37.4

Respiration in amphibians compared to reptiles. *a.* The amphibian lung is small and saclike—the moist, thin skin is often used as an auxiliary gas-exchange surface. Amphibians use positive pressure to push air into the lungs. *b.* The reptilian lung has more convolutions—the tough, scaly skin protects the animal from drying out. Reptiles use negative pressure to draw air into the lungs.

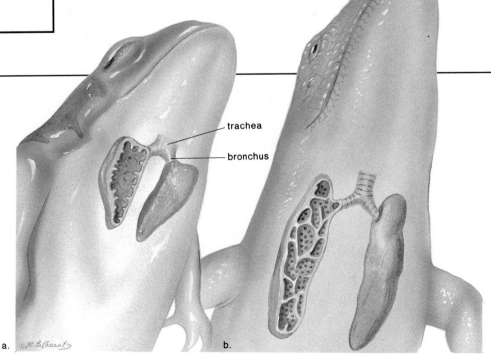

trachea

bronchus

a.

b.

The lungs of amphibians, reptiles, and mammals are not completely emptied and refilled during each breathing cycle. Because of this *incomplete ventilation* method, the air coming in mixes with used air still in the lungs. While this does help conserve water, it also decreases gas-exchange efficiency. The high oxygen requirement of flying birds, however, necessitates a method of *complete ventilation* (fig. 37.6). Incoming air is carried past the lungs by a bronchus that takes it to a set of posterior air sacs. The air then passes forward through the lungs into a set of anterior air sacs. From here, it is finally expelled. Fresh, oxygen-rich air passes through the lungs in a one-way direction only and does not mix with used air.

A few terrestrial animals use the body surface for gas exchange, but most utilize a specialized region. Insects have a tracheal system, while terrestrial vertebrates depend on lungs, which are ventilated variously.

Human Respiratory System

Path of Air

The human respiratory system includes all structures that conduct air to and from the lungs (fig. 37.7; table 37.1). The lungs lie deep within the **thoracic cavity**, where they are protected from drying out. As air moves through the nose, pharynx, trachea, and bronchi to the lungs, it is fil-

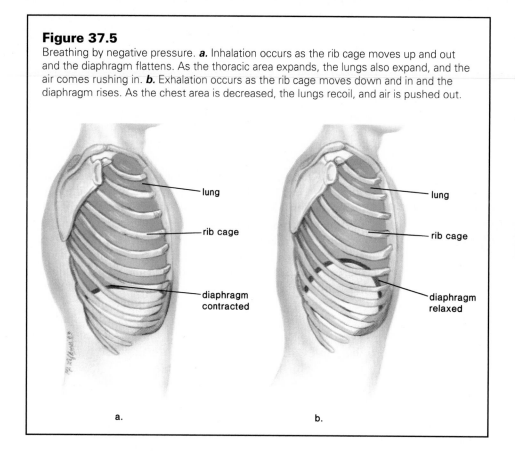

Figure 37.5

Breathing by negative pressure. *a.* Inhalation occurs as the rib cage moves up and out and the diaphragm flattens. As the thoracic area expands, the lungs also expand, and the air comes rushing in. *b.* Exhalation occurs as the rib cage moves down and in and the diaphragm rises. As the chest area is decreased, the lungs recoil, and air is pushed out.

lung

rib cage

diaphragm contracted

a.

lung

rib cage

diaphragm relaxed

b.

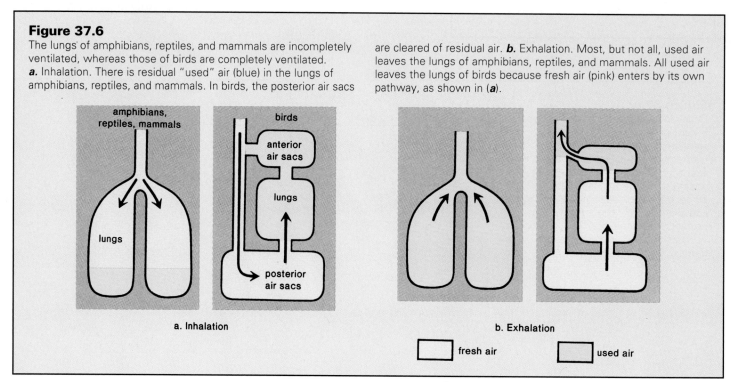

Figure 37.6

The lungs of amphibians, reptiles, and mammals are incompletely ventilated, whereas those of birds are completely ventilated. *a.* Inhalation. There is residual "used" air (blue) in the lungs of amphibians, reptiles, and mammals. In birds, the posterior air sacs are cleared of residual air. *b.* Exhalation. Most, but not all, used air leaves the lungs of amphibians, reptiles, and mammals. All used air leaves the lungs of birds because fresh air (pink) enters by its own pathway, as shown in (*a*).

amphibians, reptiles, mammals

birds

anterior air sacs

lungs

lungs

posterior air sacs

a. Inhalation

b. Exhalation

fresh air   used air

Figure 37.7

Human respiratory system. The path of air can be traced from the nose and mouth to the alveoli. The enlargement shows that alveoli are air sacs surrounded by capillaries. When the concentration of oxygen is higher in the sacs than in the blood, oxygen diffuses into the blood. When the concentration of carbon dioxide is higher in the blood than in the sacs, carbon dioxide diffuses out of the blood.

Table 37.1
Path of Air

Structure	Function
Nasal cavities	Filter, warm, and moisten
Pharynx (throat)	Connection to larynx
Glottis	Permits passage of air
Larynx (voice box)	Sound production
Trachea (windpipe)	Passage of air to thoracic cavity
Bronchi	Passage of air to each lung
Bronchioles	Passage of air to each alveolus
Alveoli	Air sacs for gas exchange

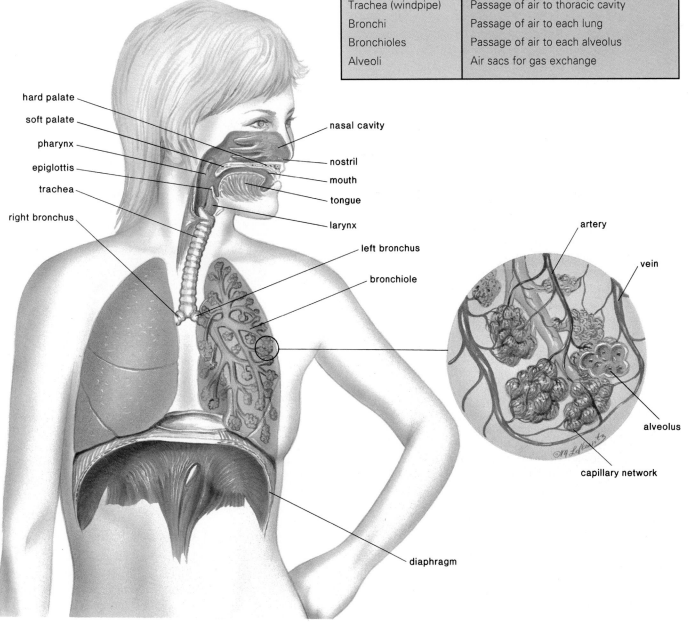

tered so that it is free of debris, warmed, and humidified. By the time the air reaches the lungs, it is at body temperature and is saturated with water. In the nose, hairs and cilia act as a screening device. In the trachea and the bronchi, cilia beat upward, carrying mucus, dust, and occasional bits of food that "went down the wrong way" into the throat, where the accumulation may be swallowed or expectorated.

The hard and soft palates separate the nasal cavities from the mouth, but the air and food passages cross in the **pharynx.** This may seem inefficient, and there is danger of choking if food accidentally enters the trachea, but this arrangement does have the advantage of letting you breathe through the mouth in case your nose is plugged up. In addition, it permits greater intake of air during heavy exercise, when greater gas exchange is required.

Air passes from the pharynx through the **glottis,** an opening into the **larynx,** or voice box. At the edges of the glottis, embedded in mucous membrane, are the *vocal cords.* These elastic ligaments vibrate and produce sound when air is expelled past them through the glottis from the larynx.

The larynx and the **trachea** are permanently held open to receive air. The larynx is held open by the complex of cartilages that form the Adam's apple. The trachea is held open by a series of C-shaped, cartilaginous rings that do not completely meet in the rear. When food is being swallowed, the larynx rises, and the glottis is closed by a flap of tissue called the **epiglottis.** A backward movement of the soft palate covers the entrance of the nasal passages into the pharynx. The food then enters the esophagus, which lies behind the larynx (see fig. 36.7).

The trachea divides into 2 **bronchi** (bronchus, sing.), which enter the right and left lungs; each then branches into a great number of smaller passages called **bronchioles.** The 2 bronchi resemble the trachea in structure, but as the bronchial tubes divide and subdivide, their walls become thinner and rings of cartilage are no longer present. Each bronchiole terminates in an elongated space enclosed by a multitude of air pockets, or sacs, called **alveoli** (alveolus, sing.), which make up the lungs.

> In humans, air moves through the nose, pharynx, and larynx to the trachea, which is held open by cartilaginous rings. Two bronchi divide into bronchioles, which finally take air into the lungs. Each bronchiole terminates in a large number of alveoli, which are covered by an extensive network of capillaries.

Breathing

Humans breathe using the same mechanism employed by all other mammals. The volume of the thoracic cavity and lungs is increased by muscle contractions that lower the diaphragm and raise the ribs (fig. 37.5). These movements create a negative pressure in the thoracic cavity and lungs, and air then flows into the lungs. When rib and diaphragm muscles relax, air is exhaled as a result of increased pressure in the thoracic cavity and lungs.

Increased carbon dioxide (CO_2) and hydrogen ion (H^+) concentrations in the blood are the primary stimuli that increase breathing rate. The chemical content of the blood is monitored by

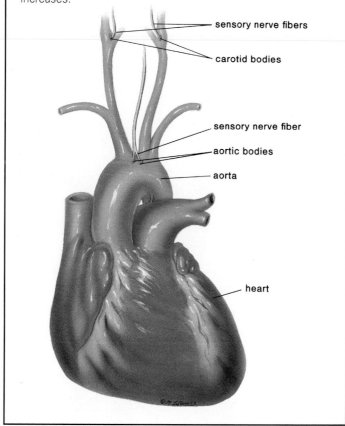

Figure 37.8
When chemoreceptors within the aortic and carotid bodies are stimulated by a rise in the H^+ concentration of the blood, caused by an increased amount of CO_2, nerve impulses travel to the respiratory center of the brain and the breathing rate increases.

sensory nerve fibers

carotid bodies

sensory nerve fiber

aortic bodies

aorta

heart

chemoreceptors called the *aortic* and *carotid bodies,* which are specialized structures located in the walls of the aorta and the carotid arteries (fig. 37.8). These receptors are very sensitive to changes in CO_2 and H^+ concentrations, but they are only minimally sensitive to a lower O_2 concentration. Information from the chemoreceptors goes to the respiratory center in the medulla oblongata of the brain, which then increases the breathing rate when CO_2 or H^+ concentration increases. This respiratory center is itself sensitive to the chemical content of the blood reaching the brain.

Gas Exchange and Transport

Diffusion primarily accounts for the exchange of gases between the air in the alveoli and the blood in the pulmonary capillaries (fig. 37.9). Atmospheric air contains little CO_2, but blood flowing into the pulmonary capillaries is almost saturated with the gas. Therefore, CO_2 diffuses out of the blood and into the alveoli. The pattern is the reverse for oxygen: blood coming into the pulmonary capillaries is oxygen poor and the alveolar air is oxygen rich; therefore, O_2 diffuses into the capillaries.

Animal Structure and Function

Figure 37.9

Gas exchanges at the lungs and the tissues. In the lungs, CO_2 leaves the blood and O_2 enters the blood. At the tissues, O_2 leaves the blood and CO_2 enters the blood. Steps necessary for gas exchange are shown for the lungs (*insert above*) and for the tissues (*insert below*).

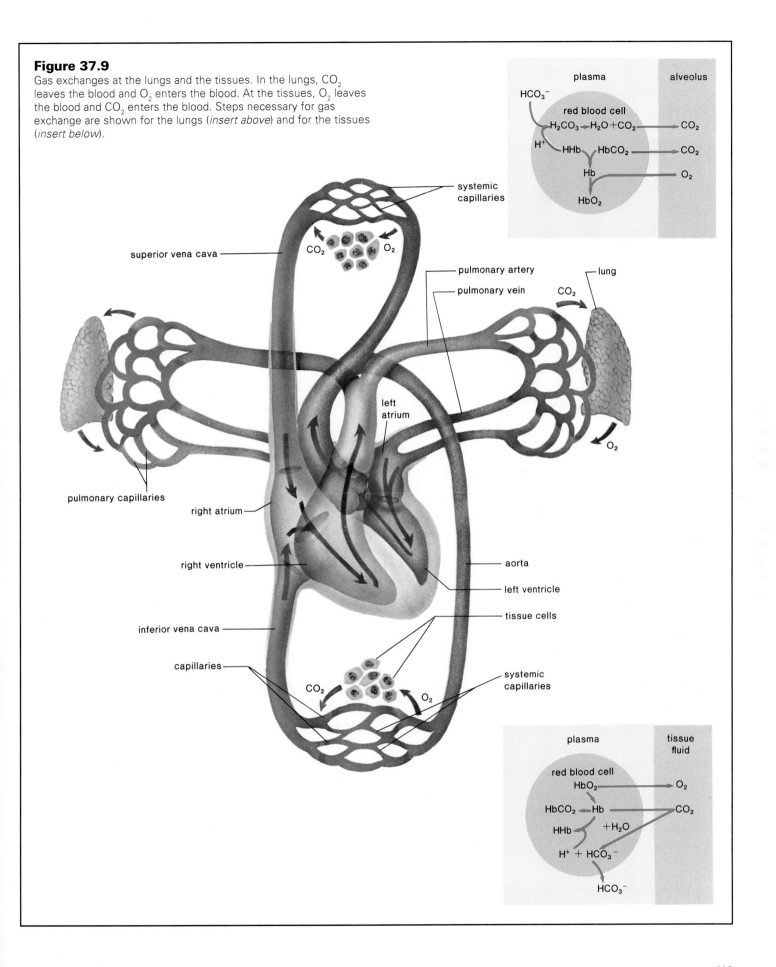

Figure 37.10

Hemoglobin, the respiratory pigment. *a.* Red blood cells move single file through capillaries. *b.* Each one is a biconcave disk containing many molecules of hemoglobin. *c.* Hemoglobin contains 4 polypeptide chains, 2 of which are alpha (α) chains and 2 of which are beta (ß) chains. The plane in the center of each chain represents an iron-containing heme group. Oxygen combines loosely with iron when hemoglobin is oxygenated.

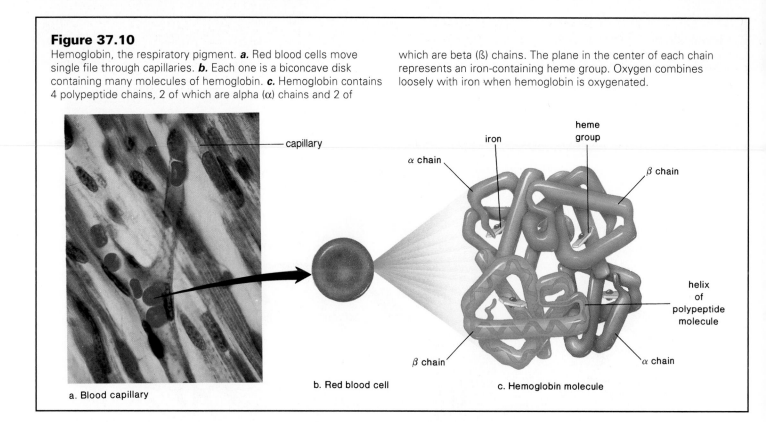

a. Blood capillary

b. Red blood cell

c. Hemoglobin molecule

Transport of O_2 and CO_2

Most oxygen entering the blood combines with **hemoglobin** (Hb) in red blood cells to form oxyhemoglobin (HbO_2):

$$Hb + O_2 \rightarrow HbO_2$$
$$\text{oxyhemoglobin}$$

Each hemoglobin molecule contains 4 polypeptide chains, and each chain is folded around an iron-containing group called heme (fig. 37.10). It is actually the iron that forms a loose association with oxygen. Since there are about 280 million hemoglobin molecules in each red blood cell, each cell is capable of carrying more than one billion molecules of oxygen.

Carbon monoxide, present in automobile exhaust, combines with hemoglobin more readily than oxygen, and it stays combined for several hours regardless of the environmental conditions. Accidental death or suicide from carbon monoxide poisoning occurs because the hemoglobin of the blood is not available for oxygen transport.

Oxygen-binding characteristics of hemoglobin can be studied by examining oxyhemoglobin dissociation curves (fig. 37.11). These curves show the percentage of oxygen-binding sites of hemoglobin that are carrying oxygen at various oxygen partial pressures (PO_2). A partial pressure of a gas is simply the amount of pressure exerted by that gas among all the gases present. At the normal partial pressures of O_2 in the lungs, hemoglobin becomes practically saturated with O_2, but at the partial pressures in the tissues, oxyhemoglobin quickly gives up much of its oxygen:

$$HbO_2 \longrightarrow Hb + O_2$$

The acid pH and warmer temperature of the tissues also promote this dissociation (breakdown).

In the tissues, some hemoglobin combines with carbon dioxide to form *carbaminohemoglobin*. Most of the carbon dioxide, however, is transported in the form of the **bicarbonate ion.** First carbon dioxide combines with water, forming *carbonic acid*, and then this dissociates to a hydrogen ion and the bicarbonate ion:

$$CO_2 + H_2O \longrightarrow H_2CO_3 \longrightarrow H^+ + HCO_3^-$$
$$\text{carbonic} \qquad \text{bicarbonate}$$
$$\text{acid} \qquad \text{ion}$$

Carbonic anhydrase, an enzyme in red blood cells, speeds up this reaction. The released hydrogen ions, which could drastically change the pH of the blood, are absorbed by the globin portions of hemoglobin, and the bicarbonate ions diffuse out of the red blood cells to be carried in the plasma. Hemoglobin, which combines with a hydrogen ion, is called reduced hemoglobin and can be symbolized as HHb. HHb plays a vital role in maintaining the pH of the blood.

Animal Structure and Function

The Risks of Smoking and Benefits of Quitting

Based on available statistics, the American Cancer Society informs us that smoking carries a high risk. Among the risks of smoking are the following:

Shortened life expectancy A 25-year-old who smoke 2 packs of cigarettes a day has a life expectancy 8.3 yrs shorter than a nonsmoker. The greater the number of packs smoked, the shorter the life expectancy.

Lung cancer The first event appears to be a thickening of the cells that line the bronchi. Then there is a loss of cilia so that it is impossible to prevent dust and dirt from settling in the lungs. Following this, cells with atypical nuclei appear in the thickened lining. A disordered collection of cells with atypical nuclei may be considered to be cancer in situ (at one location). The final step occurs when some cells break loose and penetrate the other tissues, a process called metastasis. This is true cancer (fig. 37.A*b*).

Cancer of the larynx, mouth, esophagus, bladder, and pancreas The chances of developing these cancers are from 2 to 17 times higher in cigarette smokers than in nonsmokers.

Emphysema Cigarette smokers have 4 to 25 times greater risk of developing emphysema. Damage is seen in the lungs of even young smokers. Smoking causes the lining of the bronchioles to thicken. If a large part of the lungs is involved, the lungs are permanently inflated and the chest balloons out due to this trapped air. The victim is breathless and has a cough. Since the surface area for gas exchange is reduced, not enough oxygen reaches the heart and brain. The heart works furiously to force more blood through the lungs, which may lead to a heart condition. Lack of oxygen for the brain may make the person feel depressed, sluggish, and irritable.

Coronary heart disease Cigarette smoking is the major factor in 120,000 additional U.S. deaths from coronary heart disease each year.

Reproductive effects Smoking mothers have more stillbirths and low-birthweight babies who are more vulnerable to disease and death. Children of smoking mothers are smaller and underdeveloped physically and socially even seven years after birth.

In the same manner, the American Cancer Society informs smokers of the benefits of quitting. These benefits include the following:

Risk of premature death is reduced Do not smoke for 10 to 15 years, and the risk of death due to any one of the cancers mentioned approaches that of the non-smoker.

Health of respiratory system improves The cough and excess sputum disappear during the first few weeks after quitting. As long as cancer has not yet developed, all the ill effects mentioned can reverse themselves and the lungs can become healthy again. In patients with emphysema, the rate of alveoli destruction is reduced and lung function may improve.

Coronary heart disease risk sharply decreases After only one year the risk factor is greatly reduced, and after 10 years an exsmoker's risk is the same as that of those who never smoked.

The increased risk of having stillborn children and underdeveloped children disappears Even for women who do not stop smoking until the fourth month of pregnancy, such risks to infants are decreased.

People who smoke must ask themselves if the benefits of quitting outweigh the risks of smoking.

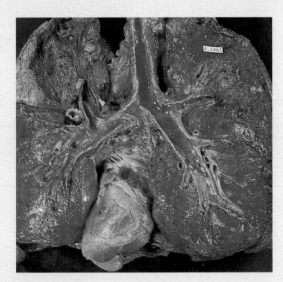

a.   b.

Figure 37.A
a. Normal lungs with heart in place. Notice the healthy red color. *b.* Lungs of a heavy smoker. Notice how black the lungs are except where cancerous tumors are located.

As blood enters the pulmonary capillaries, most of the carbon dioxide is present in plasma as the bicarbonate ion. The little free carbon dioxide remaining begins to diffuse out, and the following reaction is driven to the right:

$$H^+ + HCO_3^- \longrightarrow H_2CO_3 \longrightarrow H_2O + CO_2$$

Carbonic anhydrase also speeds up this reaction, during which hemoglobin gives up the hydrogen ions it has been carrying, HHb becoming Hb.

> Gas exchange in the lungs and the tissues is dependent upon the process of diffusion. Oxygen is transported in the blood by hemoglobin, and carbon dioxide is carried as the bicarbonate ion.

Types of Hemoglobin

Muscle cells contain an oxygen-binding pigment called *myoglobin*, which has a higher affinity for oxygen than does hemoglobin. Myoglobin tends to hold oxygen until the PO_2 falls very low. Myoglobin provides an excellent reserve source of oxygen for muscle cells when they are contracting and metabolizing rapidly.

Fetal hemoglobin has a higher affinity for oxygen than adult hemoglobin at the PO_2 levels found in the placenta. This facilitates transfer of oxygen from mother's blood to fetal blood. After birth, fetal hemoglobin is gradually replaced with the adult type of hemoglobin.

Respiration and Health

The full length of the respiratory tract is lined with a warm, moist mucous membrane that is constantly exposed to environmental air. The quality of this air, determined by the microorganisms and pollutants it contains, can affect our health.

Common Respiratory Infections

Droplets from one single sneeze can be loaded with billions of bacteria or viruses. The pseudostratified, ciliated, columnar epithelium of the trachea (p. 540) is protected by the production of mucus and by the constant beating of the cilia. If the number of infective agents is large and/or our resistance is reduced, however, a respiratory infection such as a cold or the flu can result.

Lung Disorders

Pneumonia and tuberculosis are 2 serious infections of the lungs that ordinarily can be controlled by antibiotics. Two other illnesses, emphysema and lung cancer, are not due to infections; in most instances, they are due to cigarette smoking, as discussed in this chapter's reading.

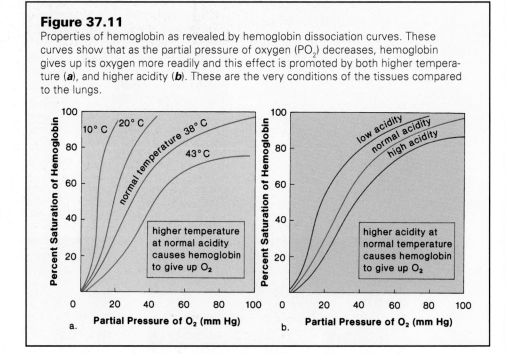

Figure 37.11
Properties of hemoglobin as revealed by hemoglobin dissociation curves. These curves show that as the partial pressure of oxygen (PO_2) decreases, hemoglobin gives up its oxygen more readily and this effect is promoted by both higher temperature (*a*), and higher acidity (*b*). These are the very conditions of the tissues compared to the lungs.

Summary

1. Some aquatic animals like hydras and planarians use their entire body surface for gas exchange.
2. Most animals have a localized, special gas-exchange area. Most aquatic animals pass water over gills. On land, insects utilize tracheal systems and vertebrates have lungs.
3. Lungs are found inside the body, where water loss is reduced. To ventilate the lungs, some vertebrates use positive pressure, but most inhale, using muscular contraction to produce a negative pressure that causes air to rush into the lungs. When the breathing muscles relax, air is exhaled.
4. Birds have a series of air sacs that allow a one-way flow of air over the gas-exchange area. This is called a complete ventilation system.
5. Table 37.1 lists the structures found in the human respiratory system.
6. Humans breathe by negative pressure, as do other mammals. During inhalation, the rib cage goes up and out, and the diaphragm lowers. The lungs expand and air comes rushing in.
7. During exhalation, the rib cage goes down and in, and the diaphragm rises. Therefore, air goes out.

8. The rate of breathing is dependent upon the amount of carbon dioxide in the blood, as detected by chemoreceptors such as the aortic and carotid bodies.
9. Gas exchange in the lungs and tissues is brought about by diffusion. Hemoglobin transports oxygen in the blood; carbon dioxide is mainly transported in plasma as the bicarbonate ion. The enzyme carbonic anhydrase found in red blood cells speeds up the formation of the bicarbonate ion.
10. The respiratory tract is subject to infections. Two disorders of the lungs, emphysema and cancer, are usually due to cigarette smoking.

Writing Across the Curriculum

In order to practice writing skills, students should write out the answers to any or all of the study questions and the critical thinking questions. The study questions are sequenced in the same order as the text. Suggested answers to the critical thinking questions are in appendix D.

Study Questions

1. Compare the respiratory organs of aquatic animals to those of terrestrial animals.
2. How does the countercurrent flow of blood within gill capillaries and water passing across the gills assist respiration in fishes?
3. Why don't insects require a blood respiratory pigment, and why is it beneficial for the body wall of earthworms to be moist?
4. Explain the phrase, "breathing by using negative pressure."
5. Contrast incomplete ventilation in humans with complete ventilation in birds.
6. Name the parts of the human respiratory system and list a function for each part.
7. The concentration of what gas in the blood controls the breathing rate in humans? Explain.

Objective Questions

1. One problem faced by terrestrial animals with lungs, but not by freshwater aquatic animals with gills, is
   a. gas exchange involves water loss.
   b. breathing requires considerable energy.
   c. oxygen diffuses very slowly in air.
   d. All of these.
2. Which of these exemplifies that not all active animals require that the circulatory system transport gases?
   a. mouse
   b. dragonfly
   c. trout
   d. sparrow
3. Birds have a more efficient lung than humans because the flow of air is
   a. the same during both inhalation and exhalation.
   b. in only one direction through the lungs.
   c. never backed up like in human lungs.
   d. not hindered by a larynx.

4. Which animal breathes by positive pressure?
   a. fish
   b. human
   c. bird
   d. frog
5. Which of these is a true statement?
   a. In lung capillaries, carbon dioxide combines with water to give carbonic acid.
   b. In tissue capillaries, carbonic acid breaks down to carbon dioxide and water.
   c. In lung capillaries, carbonic acid breaks down to carbon dioxide and water.
   d. In tissue capillaries, carbonic acid combines with hydrogen ions to form the carbonate ion.
6. Air comes into the human lungs because
   a. atmospheric pressure is less than the pressure inside the lungs.
   b. atmospheric pressure is greater than the pressure inside the lungs.
   c. although the pressures are the same inside and outside, the partial pressure of oxygen is lower within the lungs.

   d. the residual air in the lungs causes the partial pressure of oxygen to be less than outside.
7. If the digestive and respiratory tracts were completely separate in humans, there would be no need for
   a. swallowing.
   b. external nares.
   c. an epiglottis.
   d. a diaphragm.
8. To trace the path of air in humans you would place the trachea
   a. directly after the nose.
   b. directly before the bronchi.
   c. before the pharynx.
   d. Both a and c.
9. In humans, the respiratory center
   a. is stimulated by carbon dioxide.
   b. is located in the medulla oblongata.
   c. controls the rate of breathing.
   d. All of these.
10. Carbon dioxide is carried in the plasma
   a. in combination with hemoglobin.
   b. as the bicarbonate ion.
   c. combined with carbonic anhydrase.
   d. All of these.

11. Label this diagram of the human respiratory system:

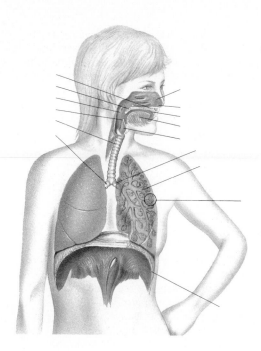

Concepts and Critical Thinking

1. *All systems of the animal's body contribute to homeostasis.*

Tell 2 ways the human respiratory system contributes to homeostasis.

2. *Organisms have localized boundaries for exchanging materials with the environment.*

Compare the similarities and differences between gills and lungs and the obvious localized boundaries with the environment.

3. *The structure of an organ suits its function.*

How does the structure of the lungs suit their function?

Selected Key Terms

gills (gilz) 607
tracheal system (tra'ke-al sis'tem) 608
lung (lung) 608
diaphragm (di'ah-fram) 609
inhalation (in"hah-la'shun) 609
exhalation (eks"hah-la'shun) 609

pharynx (far'inks) 612
glottis (glot'is) 612
larynx (lar'inks) 612
trachea (tra'ke-ah) 612
epiglottis (ep"ĭ-glot'is) 612
bronchus (brong'kus) 612

bronchiole (brong'ke-ōl) 612
alveolus (al-ve'o-lus) 612
hemoglobin (he"mo-glo-bin) 614
bicarbonate ion (bi-kar'bo-nāt i'on) 614
carbonic anhydrase (kar-bon'ik an-hi'drās) 614

38

Excretory System

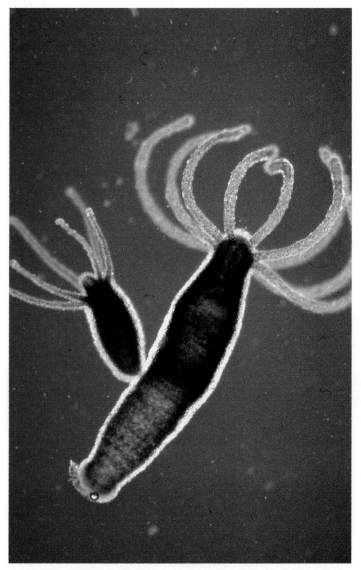

Some aquatic animals, like the hydra pictured here, don't have an excretory system. The body wall has only 2 layers of cells, and water enters and exits a large, central, fluid-filled cavity. Each cell excretes metabolic wastes directly into water, which washes them away. Magnification, X8.

Your study of this chapter will be complete when you can

1. relate the excretion of ammonia, urea, and uric acid to the animal's environment;
2. give examples of how various animals regulate the water and salt balance of the body;
3. contrast the manner in which marine bony fishes and freshwater bony fishes regulate the water and salt content of the blood;
4. compare the operation of planarian flame cells, earthworm nephridia, insect Malpighian tubules, and human kidneys;
5. trace the path of urine in humans, and describe in general the structure and function of each organ mentioned;
6. list the parts of the kidney nephron, and relate these to the macroscopic anatomy of the kidney;
7. describe the 3 steps in urine formation, and relate these to the parts of the nephron;
8. describe how water excretion is regulated and how the pH of the blood is adjusted by the kidneys;
9. tell, in general, how an artificial kidney machine works.

like the digestive and respiratory systems, the excretory system plays a major role in maintaining homeostasis. Excretion is the elimination of molecules that have taken part in metabolic reactions and should not be confused with defecation, which is the elimination of nondigested material from the gut.

Physiology of Excretion

The excretory system is largely responsible for ridding the body of nitrogenous and other waste products (fig. 38.1). It also helps regulate the amount of water and ions in body fluids.

Nitrogenous Wastes

The breakdown of various molecules, including nucleic acids and amino acids, results in nitrogenous wastes. For simplicity's sake, however, we will limit our discussion to amino acid metabolism. Amino acids, derived from protein in food, can be used by cells for synthesis of new body protein or other nitrogen-containing molecules. The amino acids not used for synthesis are oxidized to generate energy or are converted to fats or carbohydrates that can be stored. In either case, the *amino groups* (–NH$_2$) must be removed because they are not needed for any of these purposes (fig. 38.2). Once the amino groups have been removed from amino acids, they may be excreted from the body in the form of ammonia, urea, or uric acid, depending on the species (table 38.1). Removal of amino groups from amino acids requires a fairly constant amount of energy. The energy requirement for the conversion of amino groups to either ammonia, urea, or uric acid, however, differs, as indicated in figure 38.2.

Ammonia

Amino groups removed from amino acids immediately form *ammonia* (NH$_3$) by addition of a third hydrogen ion. Therefore, little or no energy is required to convert an amino group to ammonia. Ammonia is quite toxic and can only be used as a nitrogenous excretory product if a good deal of water is available to wash it from the body. The high solubility of ammonia permits this means of excretion in bony fishes, aquatic invertebrates, and amphibians whose gills and skin surfaces are in direct contact with the water of the environment.

Figure 38.1

The internal environment of cells (blood and tissue fluid) in humans and other animals stays relatively constant because the blood is continually refreshed by certain organs of the body.

a. These organs are involved in excretion—ridding the body of metabolic wastes. In particular, the gut (large intestine) excretes heavy metals, the lungs excrete carbon dioxide, and the kidneys and skin excrete nitrogenous wastes. The liver removes toxic substances from the blood and converts them to molecules that are excreted by the kidneys. *b.* Diagrammatic representation of exchanges that occur in the body and with the external environment. The arrows that point toward the external environment represent pathways of excretion.

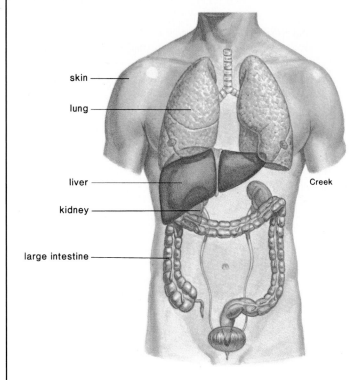

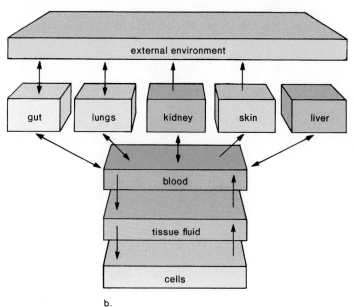

Animal Structure and Function

Figure 38.2
Nitrogenous wastes from protein breakdown. Proteins are hydrolyzed to amino acids, whose breakdown results in a carbon skeleton and amino groups. The carbon skeleton can be used as an energy source, but the amino groups must be excreted as either ammonia, urea, or uric acid. The energy and water requirements for these vary as indicated.

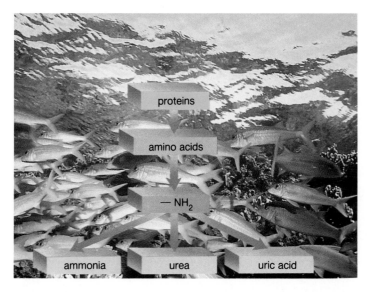

Table 38.1
Nitrogenous Waste Excretion

Product	Habitat	Animals
Ammonia	Water	Aquatic invertebrates Bony fishes Amphibian larvae
Urea	Land	Adult amphibians Mammals
Uric acid	Land	Insects Birds Reptiles

Urea

Terrestrial amphibians and mammals usually excrete **urea** as their main nitrogenous waste. Urea is much less toxic than ammonia and can be excreted in a moderately concentrated solution. This allows body water to be conserved, an important advantage for terrestrial animals with limited access to water.

Production of urea, however, requires the expenditure of energy. Urea is produced in the liver by a set of energy-requiring enzymatic reactions known as the *urea cycle*. In the cycle, carrier molecules take up carbon dioxide and 2 molecules of ammonia finally releasing urea.

Uric Acid

Uric acid is excreted by insects, reptiles, birds, and some dogs (e.g., dalmatians). Uric acid is not very toxic and is poorly soluble in water. Poor solubility is an advantage for water conservation because uric acid can be concentrated even more readily than can urea. In reptiles and birds, a dilute solution of uric acid passes from the kidneys to the *cloaca*, a common reservoir for the products of the digestive, urinary, and reproductive systems. After water is absorbed by the cloaca, the uric acid passes out with the feces.

Embryos of reptiles and birds develop inside shelled eggs that are completely enclosed. The production of insoluble, relatively nontoxic uric acid is advantageous for shelled embryos because all nitrogenous wastes are stored inside the shell until hatching takes place.

Uric acid is synthesized by a long, complex series of enzymatic reactions that require expenditure of even more ATP than does urea synthesis. Here again, there seems to be a trade-off between the advantage of water conservation and the disadvantage of energy expenditure for synthesis of an excretory molecule.

> Animals excrete nitrogenous wastes derived from protein breakdown as either ammonia, urea, or uric acid. Ammonia requires the most amount of water to excrete; uric acid requires the most amount of energy to produce. Habitat plays a primary role in determining which type of molecule is excreted.

Osmotic Regulation

In addition to their role in excretion of nitrogenous wastes, excretory organs also have the important function of regulating the water and salt balance of the body. Figure 38.3 shows that among animals, only marine invertebrates and cartilaginous fishes, such as sharks and rays, have body fluids that are nearly isotonic to seawater. These organisms have little difficulty maintaining their normal salt and water balance. Surprising, though, is the observation that while they are isotonic, the body fluids of cartilaginous fishes do not contain the same amount of salt as seawater. The answer to this paradox is that their blood contains a concentration of urea high enough to match the tonicity of the sea! For some unknown reason, this amount of urea is not toxic to them.

The body fluids of all bony fishes have only a moderate amount of salt. Apparently, their common ancestor evolved in fresh water, and only later did some groups invade the sea. Marine bony fishes (fig. 38.4*a*) are therefore prone to water loss and could become dehydrated. To counteract this, they drink seawater almost constantly. On the average, marine bony fishes swallow an amount of water estimated to be equal to 1% of their body weight every hour. This is equivalent to a human drinking about 700 ml of water every hour around the clock. While they get water by drinking, this habit also causes these fishes to acquire salt. Instead of forming a hypertonic urine, however, they actively transport sodium (Na$^+$) and chloride (Cl$^-$) ions into the surrounding seawater at the gills.

It is easy to see that the osmotic problems of freshwater bony fishes are exactly opposite to those of marine bony fishes (fig. 38.4b). The body fluids of freshwater bony fishes are hypertonic to fresh water, and they are prone to gain water. These fishes never drink water but instead eliminate excess water through production of large quantities of dilute (hypotonic) urine. They discharge a quantity of urine equal to one-third their body weight each day. Because they tend to lose salts, they actively transport salts into the blood across the membranes of their gills.

Excretory organs help regulate the water and salt balance of the body. For example, marine bony fishes drink water constantly and excrete salt from the gills. Freshwater bony fishes never drink water and excrete a dilute urine.

The difference in adaptation between marine and freshwater bony fishes makes it remarkable that some fishes actually can move between the 2 environments during their life cycle. Salmon, for example, begin their lives in freshwater streams and rivers, move to the ocean for a period of time, and finally return to fresh water to breed. These fishes alter their behavior and their gill and kidney functions in response to the osmotic changes they encounter when moving from one environment to the other.

Like marine bony fishes, some animals that evolved on land are also able to drink seawater despite its high tonicity. Birds

Figure 38.3

Comparison of relative ion concentration of animal body fluids and seawater (blue horizontal plane above) and fresh water (blue horizontal plane below). For example, marine invertebrates are the only animals with fluids with the ionic concentration of seawater, and freshwater invertebrates are the only animals with fluids that approach the ionic concentration of fresh water.

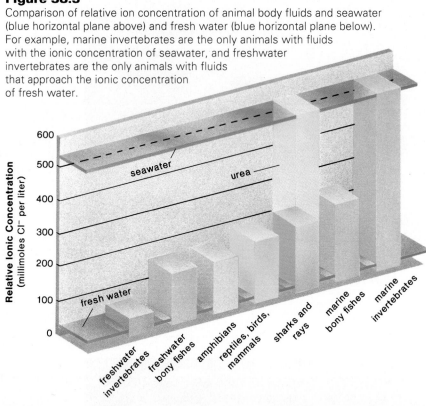

Figure 38.4

Water and salt balances in bony fishes. The black arrows represent passive transport from the environment, and the colored arrows represent active transport by the fishes to counteract environmental pressures. *a.* Marine bony fish. *b.* Freshwater bony fish.

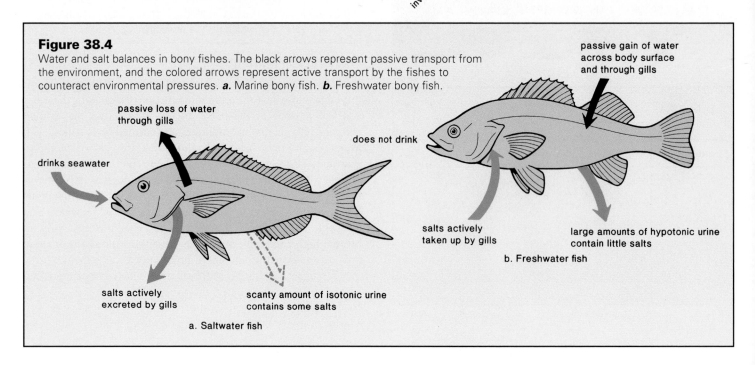

Figure 38.5

a. The flame cell excretory system in planarians. Two or more tracts of branching tubules run the length of the body and open to the outside by pores. At the ends of side branches there are small flame cells, bulblike cells whose beating cilia causes fluid to enter the tubules that remove excess fluid from the body.
b. The earthworm nephridium. The nephridium has a ciliated opening, the nephridiostome, that leads to a coiled tubule surrounded by a capillary network. Urine can be temporarily stored in the bladder before being released to the outside via a pore termed a nephridiopore. Most segments contain a pair of nephridia, one on each side.

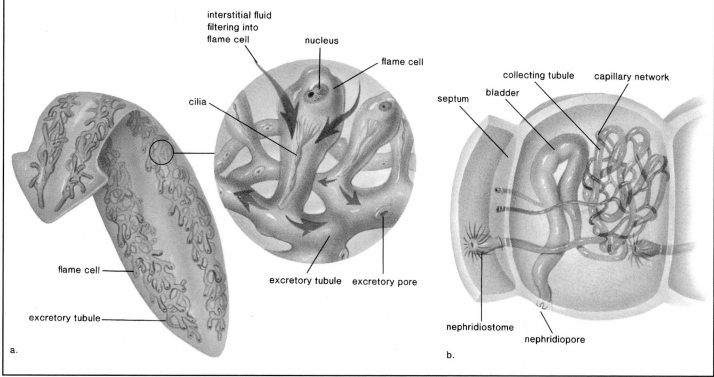

and reptiles that live near the sea have a nasal salt gland that can excrete large volumes of concentrated salt solution. Mammals that live at sea, like whales, porpoises, and seals, most likely can concentrate their urine enough to drink salt water. Humans cannot do this, and they die if they drink only seawater.

Most terrestrial animals need to drink water occasionally, but the kangaroo rat manages to get along without drinking water at all. It forms a very concentrated urine, and its fecal material is almost completely dry. These abilities allow it to survive using metabolic water derived from the breakdown of nutrient molecules alone.

Organs of Excretion

Most animals have tubular organs that function in the excretion of nitrogenous wastes and osmotic regulation. Sometimes the primary function of these organs is osmotic regulation.

Flame Cells in Planarians

Planarians have 2 strands of branching excretory tubules that open to the outside of the body through excretory pores (fig. 38.5a). Located along the tubules are bulblike *flame cells*, each of which contains a cluster of beating cilia that looks like a flickering flame under the microscope. The beating of flame-cell cilia propels fluid through the excretory canals and out of the body. The system is believed to function in osmotic regulation and also in excreting wastes.

Nephridia in Earthworms

The earthworm's body is divided into segments, and nearly every body segment has a pair of excretory structures called **nephridia** (nephridium, sing.). Each nephridium is a tubule with a ciliated opening (the nephridiostome) and an excretory pore (the nephridiopore) (fig. 38.5b). As fluid from the body cavity is propelled through the tubule by beating cilia, certain substances are reabsorbed and carried away by a network of capillaries surrounding the tubule. This process results in the formation of urine that contains only metabolic wastes, salts, and water.

Each day, an earthworm excretes a lot of water and may produce a volume of urine equal to 60% of its body weight. Its excretion of ammonia is consistent with this finding.

Malpighian Tubules in Insects

Insects have a unique excretory system consisting of long, thin tubules,(i.e., **Malpighian tubules**) attached to the gut (fig. 38.6). Water and uric acid simply flow from the surrounding hemolymph

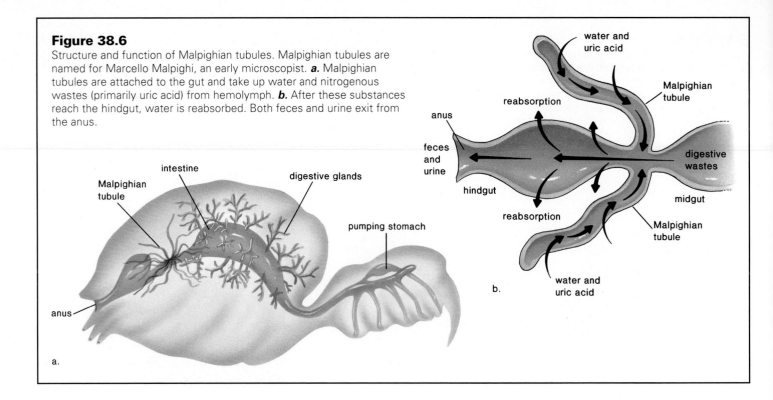

Figure 38.6
Structure and function of Malpighian tubules. Malpighian tubules are named for Marcello Malpighi, an early microscopist. **a.** Malpighian tubules are attached to the gut and take up water and nitrogenous wastes (primarily uric acid) from hemolymph. **b.** After these substances reach the hindgut, water is reabsorbed. Both feces and urine exit from the anus.

into these tubules before moving into the gut. Here, water and other useful substances are reabsorbed, but the uric acid eventually passes out of the gut. Insects that live in water or that eat large quantities of moist food reabsorb little water. But insects in dry environments reabsorb most of the water and excrete a dry, semisolid mass of uric acid.

> Most animals have a primary excretory organ for excreting nitrogenous wastes and regulating water and salt balance. Examples are flame cells in planaria, nephridia in earthworms, and Malpighian tubules in insects and spiders.

Human Kidney

The human **kidneys** are bean-shaped, reddish brown organs, each about the size of a fist. They are located one on either side of the vertebral column just below the diaphragm, where they are partially protected by the lower rib cage. They are a part of the human urinary system (fig. 38.7). This system is composed not only of the kidneys, which make **urine,** but also of organs that conduct urine from the body. Each kidney is connected to a **ureter,** a duct that carries urine from the kidney to the **urinary bladder,** where it is stored until it is voided from the body through the single **urethra.** In males, the urethra passes through the penis, and in females, it opens ventral to the opening of the vagina.

Structure

If a kidney is sectioned longitudinally, 3 major parts can be distinguished (fig. 38.8). The outer region is the *cortex,* which has a somewhat granular appearance. The *medulla* lies on the inner side

Figure 38.7
Human excretory system. Urine is formed and excreted by the kidneys and passes to the bladder by way of the ureters. After storage in the bladder, it exits when convenient by way of the urethra. These are the only organs in the body that ever contain urine.

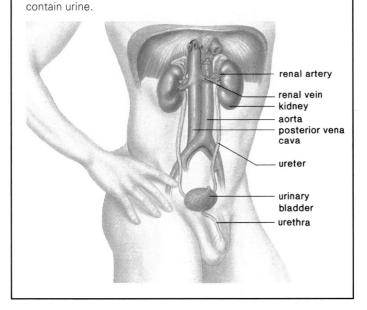

of the cortex and is arranged in a group of pyramid-shaped regions, each of which has a striped appearance. The innermost part of the kidney is a hollow chamber called the *renal pelvis.* Urine formed in the kidney collects in the renal pelvis before entering the ureter.

Animal Structure and Function

Figure 38.8

Macroscopic and microscopic anatomy of the kidney. **a.** Longitudinal section of kidney showing location of cortex, medulla, and renal pelvis. **b.** An enlargement of one renal pyramid showing the placement of nephrons.

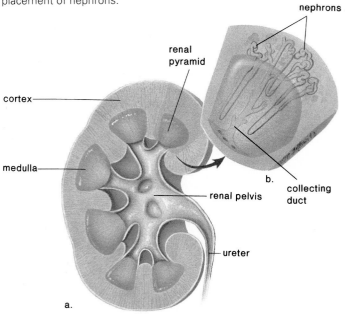

Nephrons

Microscopically, each kidney is composed of about one million tiny tubules called **nephrons.** Some nephrons are located primarily in the cortex, but others dip down into the medulla. Each nephron is made of several parts (fig. 38.9).

The blind end of a nephron is pushed in on itself to form a cuplike structure called the **glomerular capsule** (formerly called Bowman's capsule). Next, there is a region known as the **proximal** (near the glomerular capsule) **convoluted tubule,** which leads to a narrow U-turn known as the **loop of the nephron** (formerly called loop of Henle). This is followed by the **distal** (far from the glomerular capsule) **convoluted tubule.** Several distal convoluted tubules enter one **collecting duct.** The collecting duct transports urine down through the medulla and delivers it to the pelvis. The loop of the nephron and the collecting duct give the pyramids of the medulla their striped appearance (fig. 38.8).

Each nephron has its own blood supply (fig. 38.9). After the renal artery leaves the aorta to enter the kidney, it branches into numerous small arteries. These small arteries pass to all parts of the kidney and give off tiny arterioles, one for each nephron. Each arteriole, called an *afferent arteriole,* divides to form a capillary tuft, the **glomerulus,** which is surrounded by the glomerular capsule. The glomerular capillaries drain into an *efferent arteriole,* which subsequently branches into a second capillary network around the tubular parts of the nephron. These capillaries, called *peritubular capillaries,* lead to venules that join the renal vein, a vessel that enters the inferior vena cava.

Urine Formation

Human nephrons function to produce urine somewhat like earthworm nephridia in that they exchange molecules with the blood (table 38.2). Urine production requires 3 distinct processes (fig. 38.10):

1. Pressure filtration at the glomerular capsule
2. Reabsorption, including selective reabsorption, at the proximal convoluted tubule in particular
3. Tubular secretion at the distal convoluted tubule in particular

Pressure Filtration

When blood enters the glomerulus, blood pressure is sufficient to cause small molecules, such as nutrients, water, salts, and wastes, to move from the glomerulus to the inside of the glomerular capsule, especially since the glomerular walls are 100 times more permeable than the walls of most capillaries elsewhere in the body. The molecules that leave the blood and enter the glomerular capsule are called the *glomerular filtrate.* Blood proteins and blood cells are too large to be part of this filtrate, so they remain in the blood as it flows into the efferent arteriole. Glomerular filtrate has the same composition as tissue fluid, and if this composition were not altered in other parts of the nephron, death from loss of water (dehydration), loss of nutrients (starvation), and lowered blood pressure would quickly follow. *Selective reabsorption,* however, prevents this from happening.

Reabsorption

Reabsorption from the nephron to the blood takes place through the walls of the proximal convoluted tubule. Nutrients, water, and even some waste molecules diffuse passively back into the peritubular capillary network. Osmotic pressure further influences the movement of water. The nonfilterable proteins remain in the blood and exert an osmotic pressure. Also, after sodium (Na^+) is actively reabsorbed, chlorine (Cl^-) follows along, and these 2 ions together increase the osmotic pressure of the blood.

Selective reabsorption of nutrient molecules is brought about by active transport. The cells of the proximal convoluted tubule have numerous microvilli, which increase the surface area, and numerous mitochondria, which supply the energy needed for reabsorption. Reabsorption is selective since only molecules recognized by carrier molecules are actively transported through the tubule into the interstitial spaces. From there these molecules diffuse into the peritubular capillary network.

Glucose is an example of a molecule that ordinarily is reabsorbed completely because there is a plentiful supply of carrier molecules for it. However, every substance has a maximum rate of transport, and after all its carriers are in use, any excess in the filtrate will appear in the urine. For example, as reabsorbed levels of glucose approach 400mg/100ml plasma, the rest will appear in the urine. In diabetes mellitus, there is an excess of glucose in the filtrate because the liver fails to store glucose as glycogen.

Tubular Secretion

Tubular secretion is the means by which other nonfilterable wastes are added to the fluid as it passes through the distal convoluted

Figure 38.9

The human nephron. Each kidney contains over one million nephrons. The term *proximal* means nearer; *distal* means farther. In this case, the proximal convoluted tubule is proximal (closer) to the glomerular capsule and the distal convoluted tubule is farther from the glomerular capsule. Arrows indicate flow of blood to and away from the glomerulus.

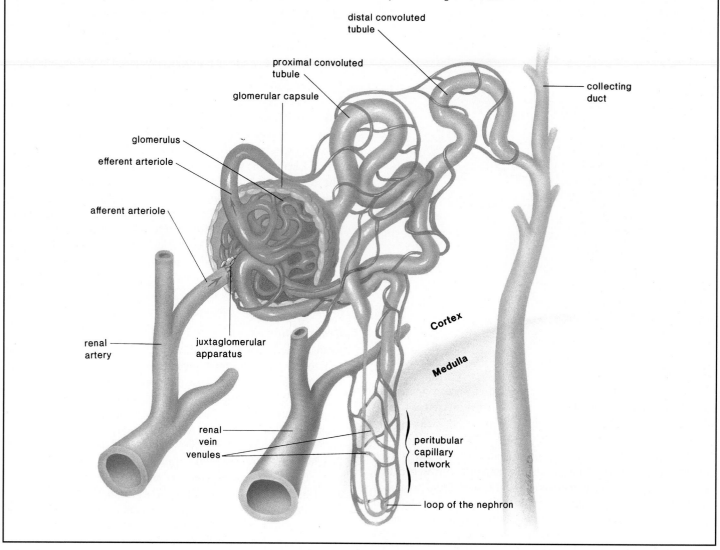

tubule. Toxic substances, such as foreign acids and bases that have been absorbed in the gut, are eliminated by tubular secretion. Penicillin and a number of other substances also are excreted in this way. But the process of tubular secretion is not as important to urine formation as are the first 2 steps studied.

Each step in urine formation is associated with a particular part of the nephron. Pressure filtration occurs at the glomerular capsule; reabsorption occurs at the proximal convoluted tubule; tubular secretion occurs at the distal convoluted tubule. The loop of the nephron plays the primary role in the reabsorption of water so that the urine of humans is hypertonic.

Control of Water Excretion

Reptiles and birds rely primarily on the gut to reabsorb water, but mammals rely on the kidneys. A *countercurrent mechanism* enables mammals to excrete a hypertonic urine. The arrangement of the loop of the nephron in relation to the collecting duct is essential to the process. A concentration gradient is established in the medulla by the extrusion of salt from the ascending limb of the loop of the nephron and by diffusion of urea from the collecting duct. The contents of the descending limb of the loop of the nephron are exposed to an increasing osmotic concentration that draws water from the descending limb into the medulla. From there the water enters the blood capillaries.

Animal Structure and Function

Figure 38.10

Steps in urine formation (simplified). *a.* The steps are noted at the location of the nephron where they occur. *b.* The molecules involved in the steps and the processes are listed.

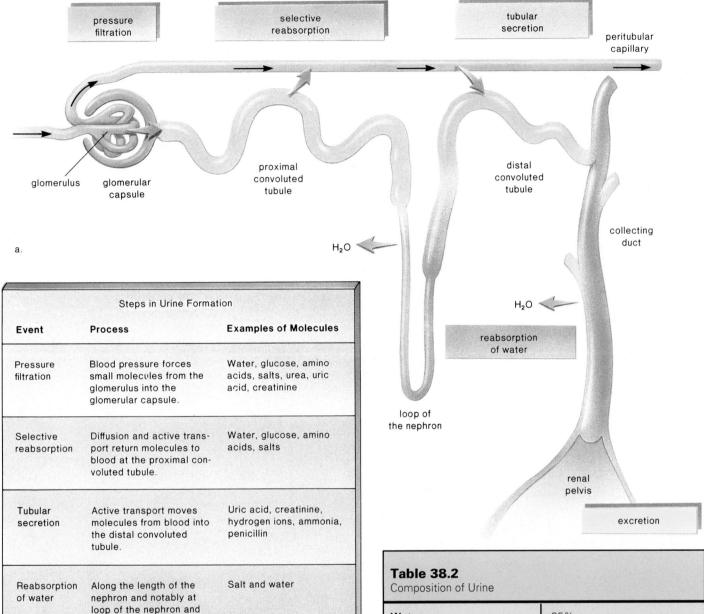

a.

Steps in Urine Formation

Event	Process	Examples of Molecules
Pressure filtration	Blood pressure forces small molecules from the glomerulus into the glomerular capsule.	Water, glucose, amino acids, salts, urea, uric acid, creatinine
Selective reabsorption	Diffusion and active transport return molecules to blood at the proximal convoluted tubule.	Water, glucose, amino acids, salts
Tubular secretion	Active transport moves molecules from blood into the distal convoluted tubule.	Uric acid, creatinine, hydrogen ions, ammonia, penicillin
Reabsorption of water	Along the length of the nephron and notably at loop of the nephron and collecting duct, water returns by osmosis following active reabsorption of salt.	Salt and water
Excretion	Urine formation rids body of metabolic wastes.	Water, salts, urea, uric acid, ammonia, creatinine

b.

Table 38.2
Composition of Urine

Water	95%
Solids	5%
Organic Wastes	(per 1,500 ml of urine)
Urea	30 g
Creatinine	1 g-2 g
Ammonia	1 g-2 g
Uric acid	1 g
Ions (Salts)	25 g

Positive Ions	*Negative Ions*
Sodium	Chlorides
Potassium	Sulfates
Magnesium	Phosphates
Calcium	

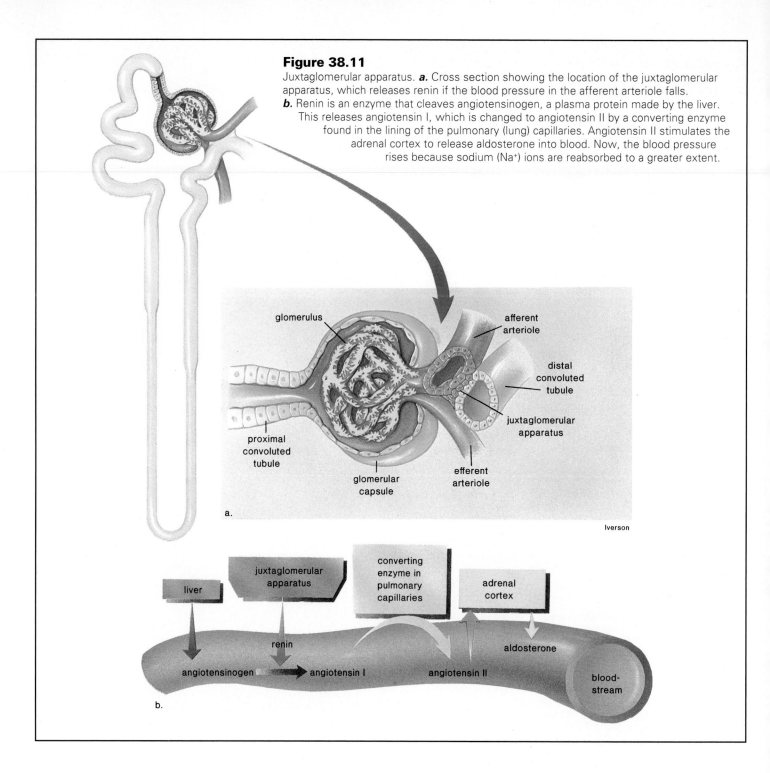

Figure 38.11

Juxtaglomerular apparatus. **a.** Cross section showing the location of the juxtaglomerular apparatus, which releases renin if the blood pressure in the afferent arteriole falls. **b.** Renin is an enzyme that cleaves angiotensinogen, a plasma protein made by the liver. This releases angiotensin I, which is changed to angiotensin II by a converting enzyme found in the lining of the pulmonary (lung) capillaries. Angiotensin II stimulates the adrenal cortex to release aldosterone into blood. Now, the blood pressure rises because sodium (Na⁺) ions are reabsorbed to a greater extent.

glomerulus

afferent arteriole

distal convoluted tubule

juxtaglomerular apparatus

proximal convoluted tubule

glomerular capsule

efferent arteriole

a.

Iverson

liver

juxtaglomerular apparatus

converting enzyme in pulmonary capillaries

adrenal cortex

renin

aldosterone

angiotensinogen

angiotensin I

angiotensin II

blood-stream

b.

The descending limb automatically loses water, but the amount of water that leaves the collecting duct is regulated by hormonal action. The hormone **ADH (antidiuretic hormone),** released by the posterior lobe of the pituitary, increases the permeability of the collecting duct so more water leaves it and is reabsorbed into the blood. If the osmotic pressure of the blood increases, the posterior pituitary releases ADH, more water is reabsorbed, and consequently there is less urine. On the other hand, if the osmotic pressure of the blood decreases, the posterior pituitary does not release ADH. The resulting impermeability of the collecting duct causes more water to be excreted and more urine to be formed. Drinking alcohol causes diuresis (increased urine flow) because it inhibits the secretion of ADH. Beer drinking also causes diuresis because of increased fluid intake. Drugs called diuretics are often prescribed for high blood pressure. These drugs cause increased urinary excretion and thus reduce blood volume and blood pressure.

How Do You Do Hemodialysis?

Hemodialysis is a way to remove nitrogenous wastes and regulate the pH of the blood when the kidneys are unable to perform these functions due to disease or injury. Dialysis occurs when small molecules pass through a membrane that will not allow the passage of large molecules. This is just the process that occurs when a patient is hooked up to a kidney machine (fig. 38.A). After the administration of an anticoagulant, blood from an artery moves through a tube to a cellulose acetate membrane that is continually washed by a solution called the dialysate. Later the blood is returned to the patient by way of a vein.

During dialysis, large substances like proteins and red blood cells cannot pass through the membrane, but small molecules like urea, glucose, and ions can pass through the dialyzing membrane. The composition of the dialysate is maintained at normal levels, so that if the blood lacks glucose and ions they will pass from the dialysate to the blood. On the other hand, nitrogenous wastes, toxic chemicals and drugs pass from the blood to dialysate. In the course of a 6-hour dialysis, from 50 g–250 g of urea can be removed from a patient, an amount that greatly exceeds the urea clearance rate of normal kidneys. Most patients undergo treatment only about twice a week.

There are drawbacks to hemodialysis, and they include the need to use an anticoagulant, possible damage to red blood cells, risk of infection, limitation on fluid and protein intake, and the time it takes.

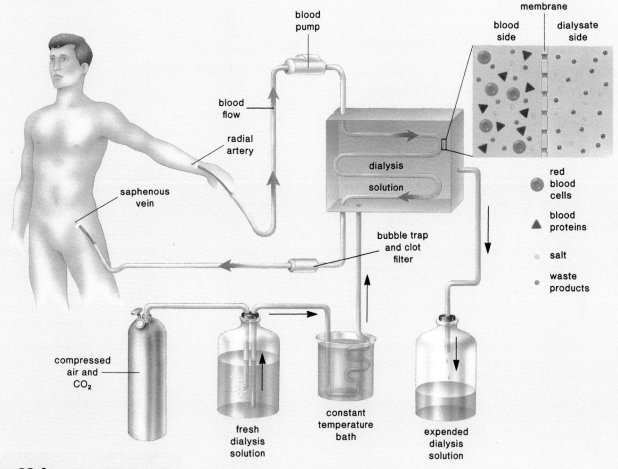

Figure 38.A
Diagram of an artificial kidney. As the patient's blood circulates through dialysis tubing, it is exposed to a dialysis solution. Wastes exit from blood into the solution because of a preestablished concentration gradient. In this way, blood is not only cleansed, the pH also can be adjusted.

The hormone **aldosterone,** which is secreted by the adrenal cortex, primarily maintains the sodium (Na^+) and potassium (K^+) balance of the blood. It causes the distal convoluted tubule to reabsorb Na^+ and to excrete K^+. The increase of Na^+ in the blood causes water to be reabsorbed, leading to an increase in blood volume and blood pressure.

Blood pressure is constantly monitored within the *juxtaglomerular apparatus,* a region of contact between the afferent arteriole and the distal convoluted tubule (fig. 38.11). The afferent arteriole cells in the apparatus secrete the enzyme *renin* when the blood pressure is insufficient to promote efficient filtration in the glomerulus. Renin converts a large plasma protein called angiotensinogen to *angiotensin.* Angiotensin then stimulates the adrenal cortex to release *aldosterone,* which allows Na^+ to be reabsorbed and blood volume and pressure to rise. This sequence of events is called the renin-angiotensin-aldosterone system (fig. 42.10). Notice that both ADH and aldosterone act to increase blood volume and raise the blood pressure. When blood pressure rises, the heart produces a peptide hormone called *atrial natriuretic factor (ANF)* that inhibits the secretion of renin by the juxtaglomerular apparatus and the release of ADH by the posterior pituitary. This is an example of how hormones serve as checks and balances to one another.

Adjustment of Blood pH

The kidneys help maintain the pH level of the blood within a narrow range, and the whole nephron takes part in this process. The excretion of hydrogen ions and ammonia, together with the reabsorption of sodium and bicarbonate ions, is adjusted to keep the pH within normal bounds. If the blood is acidic, hydrogen ions are excreted in combination with ammonia, while sodium and bicarbonate ions are reabsorbed. This will restore the pH because sodium ions promote the formation of hydroxyl ions, while bicarbonate takes up hydrogen ions when carbonic acid is formed:

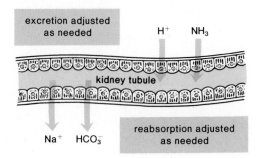

If the blood is basic, fewer hydrogen ions are excreted, and fewer sodium and bicarbonate ions are reabsorbed.

Reabsorption and/or excretion of ions (salts) by the kidneys illustrates their homeostatic ability: they maintain not only the pH of the blood but also its osmolarity.

Summary

1. Animals excrete nitrogenous wastes, which differ as to the amount of water and energy required to excrete them. Aquatic animals usually excrete ammonia, and land animals excrete either urea or uric acid.

2. Osmotic regulation is important to animals. Most have to balance their water and salt intake and excretion to maintain the normal concentration in the body fluids. Marine fishes constantly drink water, excrete salts at the gills, and pass an isotonic urine. Freshwater fishes never drink water, take in salts at the gills, and pass a hypotonic urine.

3. Animals often have an excretory organ. Earthworm nephridia exchange molecules with the blood in a manner similar to vertebrate kidneys. Malpighian tubules in insects take wastes and water from the hemocoel to the gut, where only water is reabsorbed.

4. Kidneys are a part of the human urinary system. Microscopically, each kidney is made up of nephrons, each of which has several parts and its own blood supply.

5. Urine formation requires 3 steps: pressure filtration, when nutrients, water, and wastes enter the glomerular capsule; selective reabsorption, when nutrients and some water are reabsorbed into the proximal convoluted tubule; and tubular secretion, when additional wastes are added to the distal convoluted tubule.

6. Humans excrete a hypertonic urine. The ascending limb of the loop of the nephron actively extrudes salt so that the medulla is hypertonic relative to the contents of the descending limb and the collecting duct. Therefore, water has a tendency to diffuse out of these.

7. Two hormones are involved in maintaining the water content of the blood. The hormone ADH, which makes the collecting duct more permeable, is secreted by the posterior pituitary in response to an increase in the osmotic pressure of the blood.

8. The hormone aldosterone is secreted by the adrenal cortex after the low Na^+ content of the blood and the resultant low blood pressure have caused the kidneys to release renin. The presence of renin leads to the formation of angiotensin, which causes the adrenal cortex to release aldosterone. Aldosterone causes the kidneys to retain Na^+.

9. The kidneys adjust the pH of the blood by excreting or conserving H^+, NH_3, HCO_3^-, and Na^+ as appropriate.

10. During hemodialysis, waste molecules diffuse out of the blood into the dialysis fluid. Other substances can be added to the blood.

Writing Across the Curriculum

In order to practice writing skills, students should write out the answers to any or all of the study questions and the critical thinking questions. The study questions are sequenced in the same order as the text. Suggested answers to the critical thinking questions are in appendix D.

Animal Structure and Function

Study Questions

1. Relate the 3 primary nitrogenous wastes to the habitat of animals.
2. Contrast the osmotic regulation of a marine bony fish with that of a freshwater bony fish.
3. Give examples of how other types of animals regulate their water and salt balance.
4. Describe how the excretory organs of the earthworm and the insect function.
5. Give the path of urine, and give a function for each structure mentioned.
6. Describe the macroscopic anatomy of a human kidney, and relate it to the placement of nephrons.
7. List the parts of a nephron, and give a function for each structure mentioned.
8. Describe how urine is made by telling what happens at each part of the nephron.
9. What role do ADH and aldosterone play in regulating the tonicity of urine? How does this effect blood pressure?
10. How does the nephron regulate the pH of the blood?

Objective Questions

1. Which of these is mismatched?
   a. insects—excrete uric acid
   b. humans—excrete urea
   c. fishes—excrete ammonia
   d. birds—excrete ammonia
2. One advantage of urea excretion over uric acid excretion is that urea
   a. requires less energy to form.
   b. can be concentrated to a greater extent.
   c. is not a toxic substance.
   d. requires less water to excrete.
3. Freshwater bony fishes maintain water balance by
   a. excreting salt across their gills.
   b. periodically drinking small amounts of water.
   c. excreting a hypotonic urine.
   d. excreting wastes in the form of uric acid.
4. Animals with which of these are most likely to excrete a semisolid, nitrogenous waste?
   a. nephridia
   b. Malpighian tubules
   c. human kidneys
   d. All of these.
5. In which of these human structures are you least apt to find urine?
   a. large intestine
   b. urethra
   c. ureter
   d. bladder

6. Excretion of a hypertonic urine in humans is associated with
   a. the glomerular capsule.
   b. the proximal convoluted tubule.
   c. the loop of the nephron.
   d. the distal convoluted tubule.
7. The presence of ADH causes an individual to excrete
   a. sugars.
   b. less water.
   c. more water.
   d. Both a and c.
8. In humans, water is
   a. found in the glomerular filtrate.
   b. reabsorbed from the nephron.
   c. in the urine.
   d. All of these.

9. Pressure filtration is associated with
   a. the glomerular capsule.
   b. the distal convoluted tubule.
   c. the collecting duct.
   d. All of these.
10. In humans, glucose
   a. is in the filtrate and urine.
   b. is in the filtrate and not in urine.
   c. undergoes tubular secretion and is in urine.
   d. undergoes tubular secretion and is not in urine.
11. Label this diagram of a nephron, and give the steps for urine formation in the boxes:

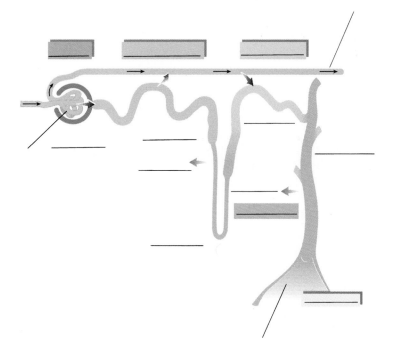

Concepts and Critical Thinking

1. *The excretory system plays a primary role in homeostasis.*

   Relate the need of excretion to figure 7.10 of the text.

2. *Animals utilize countercurrent mechanisms to increase blood concentrations of substances.*

   Tell how the countercurrent mechanism in the gills of fishes (p. 607) and in the kidneys of mammals (p. 626) helps adapt them to their environments.

3. *Structure suits the function.*

   Why would you expect the proximal convoluted tubule to be lined with cells, which have many mitochondria and microvilli?

Selected Key Terms

urea (u-re′ah) 621
uric acid (u′rik as′id) 621
nephridium (nĕ-frid′e-um) 623
ureter (u-re′ter) 624
urinary bladder (u′rĭ-ner″e blad′der) 624
urethra (u-re′thrah) 624
nephron (nef′ron) 625

glomerular capsule (glo-mer′u-lar kap′sūl) 625
proximal convoluted tubule (prok′sĭ-mal kon′vo-lūt-ed tu′būl) 625
loop of the nephron (lo͞op uv the nef′ron) 625

distal convoluted tubule (dis′tal kon′vo-lūt-ed tu′būl) 625
collecting duct (kŏ-lekt′ing dukt) 625
glomerulus (glo-mer′u-lus) 625
antidiuretic hormone (ADH) (an″tĭ-di″u-ret′ik hor′mōn) 628
aldosterone (al″do-ster′ōn) 630

Animal Structure and Function

39

Nervous System

A dramatic representation of neurons with many processes that reach out in all directions. When neurons communicate with one another, they release transmitter substances that are represented here as bursts of color.

Your study of this chapter will be complete when you can

1. state, in general, the overall function of the nervous system;
2. compare the organization and complexity of various invertebrate nervous systems with that of the human nervous system;
3. describe, in general, the structure of a neuron; name 3 types of neurons, and state their specific functions;
4. describe the nerve impulse as an electrochemical change that is recorded as the action potential by the oscilloscope;
5. describe the structure and function of a synapse, including transmission across a synapse;
6. describe the structure and function of the peripheral nervous system;
7. draw a diagram depicting a spinal reflex, and explain the function of all parts included;
8. state the parts of the autonomic nervous system; cite similarities and differences in structure and function of its 2 divisions;
9. describe, in general, the structure and function of the central nervous system;
10. list the major parts of the brain, and give a function for each part;
11. discuss, in general, current research into learning and memory;
12. list several well-known excitatory and inhibitory neurotransmitters in the peripheral and central nervous systems;
13. discuss and give examples of the effects of drugs at synapses, and list 5 criteria for drug abuse.

he ability to respond to stimuli is a characteristic of all living things. In complex animals, such as vertebrates, this ability depends on the nervous, endocrine (hormonal), sensory, and musculo-skeletal systems. Working together, the nervous and endocrine systems coordinate the actions of all the other systems of the body to produce effective behavior and to keep the internal environment within safe limits for life. Because hormones are transported in the blood, it may require seconds, minutes, hours, or even longer for these chemical messengers to produce their effects. The nervous system, on the other hand, communicates rapidly, requiring only thousandths of a second. The nervous system receives and processes information before sending out signals to the muscles and glands for an appropriate response. In this way, the nervous system integrates and controls the other systems of the body.

Evolution of Nervous Systems

A comparative study of animal nervous systems indicates the steps that may have led to the centralized nervous system found in vertebrates (fig. 39.1). Hydras have a simple nervous system that looks like a net of threads extending throughout the radially symmetrical body. The net is actually composed of neurons located in the mesoglea that contact one another and muscle fibers within epidermal cells. Experiments with sea anemones and jellyfishes suggest that cnidarians may have 2 nerve nets. A fast-acting one allows major responses, particularly in times of danger, and the other coordinates slower and more delicate movements.

Planarians are bilaterally symmetrical, and they have a more complex nervous system that resembles a ladder. Cephalization is present—there is a brain, a concentration of neurons at the anterior end of the animal that receives sensory information from photoreceptors in the eyespots and sensory cells in the auricles. Two longitudinal nerve cords allow a rapid transfer of information from anterior to posterior. Transverse nerves between the nerve cords keep the movement of the 2 sides coordinated. Therefore, planarians have the rudiments of a central nervous system (brain and nerve cords) and a peripheral nervous system (transverse nerves).

The annelids and arthropods have what is usually considered the typical invertebrate nervous system. The **central nervous system** consists of a brain and a single, ventral solid nerve cord; the **peripheral nervous system** consists of nerves that contain both sensory and motor fibers. The ventral solid nerve cord has a ganglion in each segment, which apparently controls the muscles of that segment. Even so, the brain, which normally receives sensory information, controls the activity of the ganglia so that the entire animal is coordinated.

Vertebrates also have a central nervous system and a peripheral nervous system (fig. 39.1d). There is a vast increase, however, in the number of neurons. For example, an insect's entire

Figure 39.1

Evolution of the nervous system. **a.** The nerve net of a hydra, a cnidarian. **b.** The paired nerve cords with cross connections of a planarian, a flatworm, have a ladder appearance. **c.** The earthworm, a segmented worm, has a central nervous system consisting of the brain and a ventral solid nerve cord. It also has a peripheral nervous system consisting of nerves. **d.** A rabbit, like other vertebrates, has a dorsal, hollow nerve cord in its central nervous system.

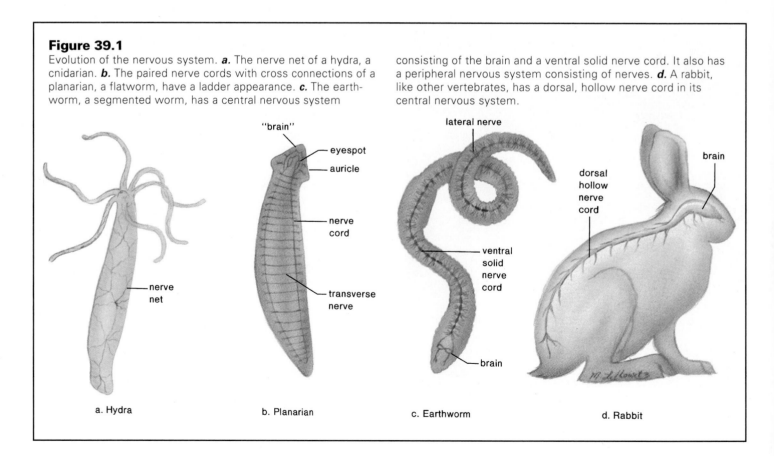

a. Hydra

b. Planarian

c. Earthworm

d. Rabbit

Animal Structure and Function

Figure 39.2
Overall organization of the nervous system in human beings. *a.* Pictorial representation. The central nervous system (brain and spinal cord) and some of the nerves of the peripheral nervous system are shown. *b.* The central nervous system is at the top of the diagram, and the peripheral nervous system is below. The nerves of the peripheral nervous system belong to either the somatic nervous system or the autonomic nervous system. The autonomic nervous system has 2 portions, the sympathetic and parasympathetic systems.

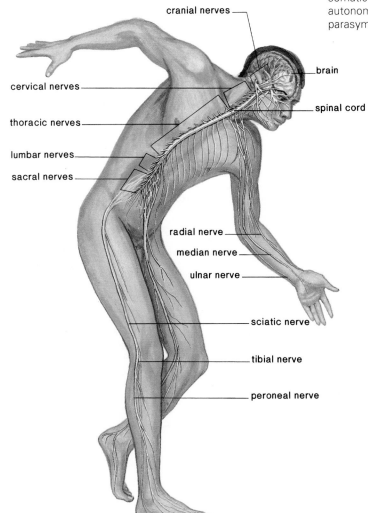

cranial nerves

brain

cervical nerves

spinal cord

thoracic nerves

lumbar nerves

sacral nerves

radial nerve

median nerve

ulnar nerve

sciatic nerve

tibial nerve

peroneal nerve

a.

Waldrop

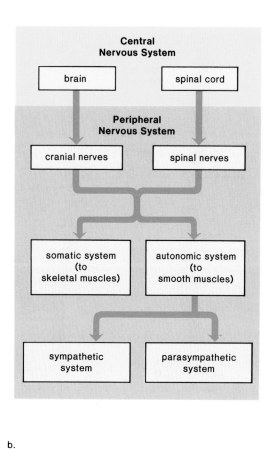

b.

nervous system may contain a total of about one million neurons, while a vertebrate nervous system may contain many thousand to several billion times that number.

In complex animals the nervous system has 2 parts: the central nervous system and the peripheral nervous system.

Human Nervous System

In the human nervous system, the *central nervous system* includes the brain and spinal cord (dorsal nerve cord), which lie in the midline of the body, where the skull protects the brain and the vertebrae protect the spinal cord (fig. 39.2). The *peripheral nervous system* contains both cranial nerves, which originate in the brain, and spinal nerves, which project from either side of the spinal cord. The peripheral nervous system is further divided into the somatic system and the autonomic (or visceral) system. The somatic system contains nerves that control skeletal muscles, skin, and joints. The autonomic system contains nerves that control the glands and smooth muscles of the internal organs. Before we continue our discussion of the nervous system, we must first examine the anatomy and functions of nerve cells, or neurons.

Neurons

Neurons are cells that vary in size and shape, but they all have 3 parts: the dendrite(s), the cell body, and the axon (fig. 39.3). The **dendrites** receive information from other neurons and generally conduct nerve impulses *toward* the cell body. The **axon,** on the

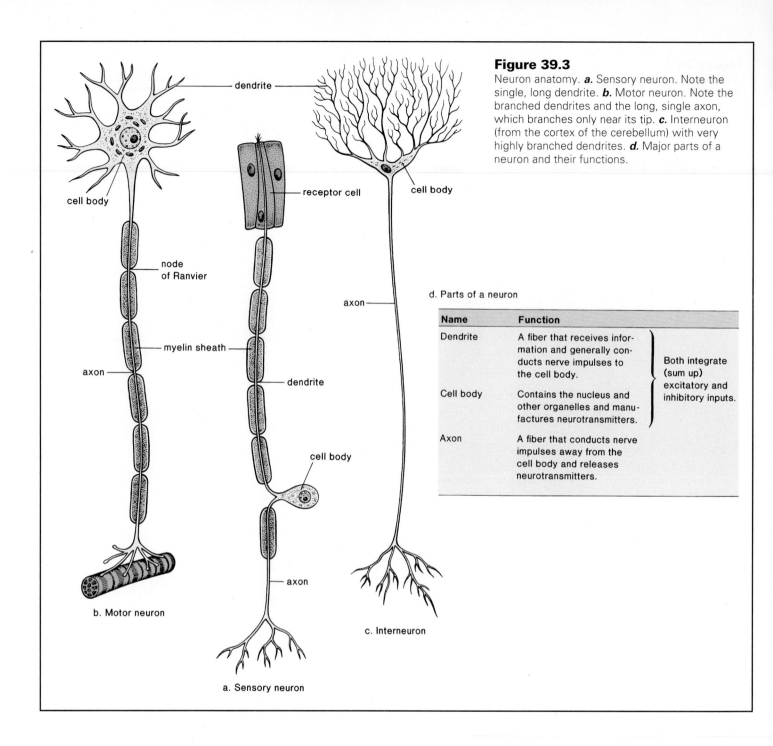

Figure 39.3
Neuron anatomy. ***a.*** Sensory neuron. Note the single, long dendrite. ***b.*** Motor neuron. Note the branched dendrites and the long, single axon, which branches only near its tip. ***c.*** Interneuron (from the cortex of the cerebellum) with very highly branched dendrites. ***d.*** Major parts of a neuron and their functions.

dendrite

cell body

node of Ranvier

myelin sheath

axon

b. Motor neuron

receptor cell

dendrite

cell body

axon

a. Sensory neuron

cell body

axon

c. Interneuron

d. Parts of a neuron

Name	Function	
Dendrite	A fiber that receives information and generally conducts nerve impulses to the cell body.	Both integrate (sum up) excitatory and inhibitory inputs.
Cell body	Contains the nucleus and other organelles and manufactures neurotransmitters.	
Axon	A fiber that conducts nerve impulses away from the cell body and releases neurotransmitters.	

other hand, conducts nerve impulses *away* from the cell body. The cell body contains the nucleus and other organelles typically found in cells. One of the main functions of the cell body is to manufacture neurotransmitters, which are chemicals stored in secretory vesicles at the ends of axons. When neurotransmitters are released, they influence the excitability of nearby neurons.

Three types of neurons are shown in figure 39.3. *Sensory neurons,* each with a long dendrite and a short axon, take messages from sense organs to the central nervous system. *Motor neurons,* each with a long axon and short dendrites, take messages from the central nervous system to muscle fibers or glands. Because motor

neurons cause muscle fibers or glands to react, they are said to innervate these structures. The third type of neuron, called an **interneuron**, is only found within the central nervous system. It conveys messages between various parts of the central nervous system, such as from one side of the brain or spinal cord to the other or from the brain to the cord, and vice versa. An interneuron has short dendrites and either a long axon or a short axon.

The dendrites and axons of these neurons are sometimes called fibers, or processes. Most long fibers, whether dendrite or axon, are covered by a white **myelin sheath** formed from the membranes of the tightly spiraled Schwann cells surrounding these fibers.

Animal Structure and Function

Schwann cells are one of the several types of glial cells in the nervous system. Neuroglial cells service the neurons and have supportive and nutritive functions.

Nerve Impulse

Italian investigator Luigi Galvani discovered in 1786 that a nerve can be stimulated by an electric current. But it was realized later that the speed of the nerve impulse is too slow to be due simply to the movement of electrons or current within a nerve fiber. In the early 1900s, Julius Bernstein, at the University of Halle, Germany, suggested that the nerve impulse is an electrochemical phenomenon involving the movement of unequally distributed ions on either side of a neuron plasma membrane. It was not until 1939, however, that the investigators developed a technique that enabled them to substantiate this hypothesis. A. L. Hodgkin and A. F. Huxley, English neurophysiologists, received the Nobel Prize in 1963 for their work in this field. They and a group of researchers, headed by K. S. Cole and J. J. Curtis, at Woods Hole, Massachusetts, managed to insert a very tiny electrode into the giant axon of the squid *Loligo* (fig. 39.4). This internal electrode was then connected to a voltmeter and an *oscilloscope,* an instrument with a screen that shows a trace or pattern indicating a change in voltage with time (fig. 39.4*d*). *Voltage* is a measure of the electrical potential difference between 2 points, which in this case is the difference between 2 electrodes, one placed inside and another placed outside the axon. (An electrical potential difference across a membrane is called the membrane potential.) When a potential difference exists, we can say that a plus and minus pole exist; therefore, an oscilloscope indicates the existence of polarity and records polarity changes.

Resting Potential

When the axon is not conducting an impulse, the oscilloscope records a membrane potential equal to about –65 mV (millivolts), indicating that the inside of the neuron is more negative than the outside (fig. 39.4*b*). This is called the **resting potential** because the axon is not conducting an impulse.

The existence of this polarity can be correlated with a difference in ion distribution on either side of the axomembrane (plasma membrane of axon). As figure 39.5*a* shows, there is a higher concentration of sodium ions (Na^+) outside than inside the axon and a higher concentration of potassium ions (K^+) inside than outside the axon. The unequal distribution of these ions is due to the action of the sodium-potassium pump. This pump is an active transport system in the plasma membrane that pumps sodium ions out of and potassium ions into the axon. The work of the pump maintains the unequal distribution of sodium and potassium ions across the axomembrane.

The pump is always working because the axomembrane is somewhat permeable to these ions and they tend to diffuse toward their lesser concentration. Since the axomembrane is more permeable to potassium ions than to sodium ions there are always more positive ions outside the axomembrane than inside and this accounts for the polarity recorded by the oscilloscope. There are also large, negatively charged proteins in the axoplasm

Figure 39.4

The original nerve impulse studies utilized giant squid axons and a voltage recording device known as an oscilloscope. *a.* The squid axons shown produce rapid muscle contraction so that the squid can move quickly. *b.* These axons are so large (about 1 mm in diameter) that a microelectrode can be inserted inside them. When the axon is not conducting a nerve impulse, the electrode registers and the oscilloscope records a resting potential of –65 mV. *c.* When the axon is conducting a nerve impulse, the *threshold* for an action potential has been achieved, and there is a rapid change in potential from –65 mV to +40 mV (called depolarization), followed by a return to –65 mV (called repolarization). *d.* Enlargement of action potential of nerve impulse.

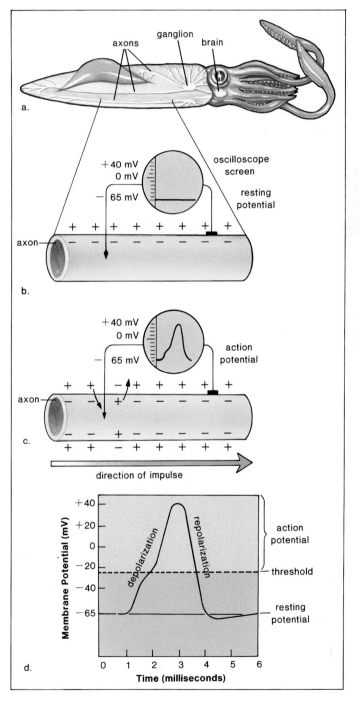

Figure 39.5

Action potential and resting potential. The action potential is the result of an exchange of sodium (Na^+) ions and potassium (K^+) ions, and it is recorded as a change in polarity by an oscilloscope (as shown on the right). So few ions are exchanged for each action potential that it is possible for a nerve fiber to conduct nerve impulses repeatedly. Whenever the fiber rests, the sodium-potassium pump restores the original distribution of ions.

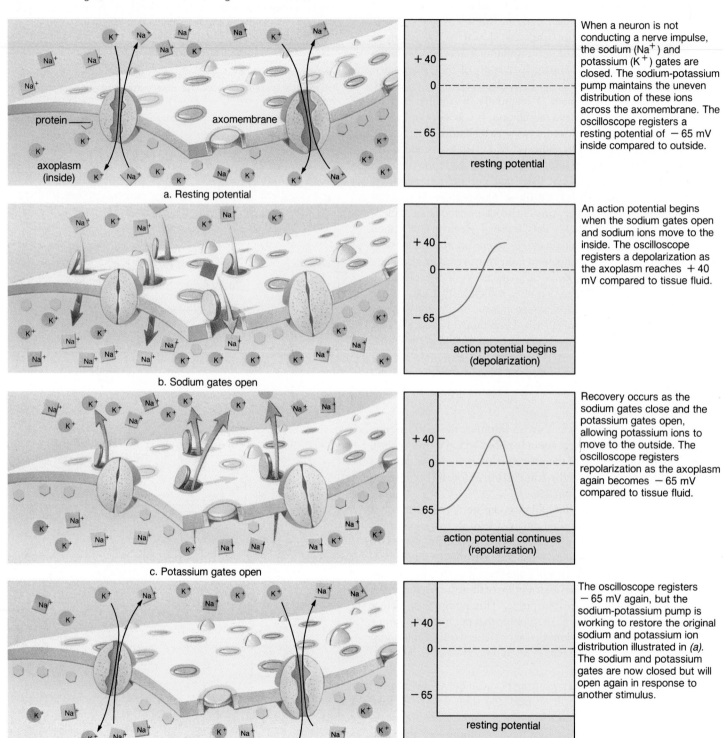

a. Resting potential

When a neuron is not conducting a nerve impulse, the sodium (Na^+) and potassium (K^+) gates are closed. The sodium-potassium pump maintains the uneven distribution of these ions across the axomembrane. The oscilloscope registers a resting potential of -65 mV inside compared to outside.

b. Sodium gates open

An action potential begins when the sodium gates open and sodium ions move to the inside. The oscilloscope registers a depolarization as the axoplasm reaches $+40$ mV compared to tissue fluid.

c. Potassium gates open

Recovery occurs as the sodium gates close and the potassium gates open, allowing potassium ions to move to the outside. The oscilloscope registers repolarization as the axoplasm again becomes -65 mV compared to tissue fluid.

d. Sodium-potassium pump

The oscilloscope registers -65 mV again, but the sodium-potassium pump is working to restore the original sodium and potassium ion distribution illustrated in (a). The sodium and potassium gates are now closed but will open again in response to another stimulus.

Animal Structure and Function

(cytoplasm of axon); altogether, then, the oscilloscope records that the axoplasm is −65 mV compared to tissue fluid. This is the resting potential.

Action Potential

If the axon is stimulated to conduct a nerve impulse by either an electric shock, a sudden change in pH, or a pinch, there is a rapid change in the polarity recorded as a trace on the oscilloscope screen. This change in polarity is called the **action potential** (fig. 39.5). First, the trace goes from −65 mV to +40 mV (called *depolarization*), indicating that the axoplasm is now more positive than tissue fluid. Then the trace returns to −65 mV again (called *repolarization*), indicating that the inside of the axon is negative again.

The action potential is due to special protein-lined channels in the axomembrane, which can open to allow either sodium or potassium ions to pass through (fig. 39.5*b* and *c*). These channels have gates called "sodium gates" and "potassium gates." During the depolarization phase of the action potential, the sodium gates open and sodium rushes into the axon. Once this phase is complete, repolarization occurs. During repolarization, the potassium gates open and potassium rushes out of the axon.

Notice that at the completion of an action potential, the original ion distribution has been altered somewhat (fig. 39.5*d*). There are now more sodium ions inside the axon than before, and there are more potassium ions outside the axon than before. The sodium-potassium pump is able to restore the former distribution, however.

The oscilloscope records changes at only one location in a fiber, but actually the action potential travels along the length of a fiber. It is self-propagating because the ion channels are prompted to open whenever the membrane potential decreases in an adjacent area. In invertebrates, some fibers are larger than others and a few axons are called giant axons because they are so large. The nerve impulse travels a thousand times faster in giant axons compared to thin fibers (25 m/s compared to 0.025 m/s). In vertebrates the speed of travel is improved not by increase in size but because most long fibers have a myelin sheath (fig. 39.3). There are gaps in the myelin sheath called nodes of Ranvier, where one Schwann cell ends and the next begins. The action potential simply jumps from one node of Ranvier to the next and may reach speeds of 200 m per sec. This is called **saltatory** (saltatory—jumping) **conduction.**

All neurons transmit the same type of nerve impulse—an electrochemical change that is self-propagating along the fiber(s).

Transmission across a Synapse

In 1897, the English scientist Sir Charles Sherrington and others noted 2 important aspects of nerve impulse transmission between neurons. First, an impulse passing from one vertebrate nerve cell to another always moves in only one direction. Second, there is a very short delay in transmission of the nerve impulse from one neuron to another. This latter observation led to the hypothesis that there is a minute space between neurons. Sherrington called the region where the impulse moves from one neuron to another a **synapse,** meaning "to clasp." Synapses occur between the end of an axon and a dendrite or between the end of an axon and a cell body (fig. 39.6).

A synapse has 3 components: a presynaptic membrane, a gap now called the synaptic cleft, and a postsynaptic membrane. As mentioned previously, there are vesicles at the ends of axons where transmitter substances, called **neurotransmitters,** are stored. When nerve impulses reach a presynaptic membrane, the vesicles fuse with the membrane and discharge their contents into the synaptic cleft. The discharged neurotransmitter molecules diffuse across the synaptic cleft and bind in a lock-and-key manner to the postsynaptic membrane at *receptor sites.* This binding process alters the membrane potential of the postsynaptic membrane in the direction of either excitation or inhibition. If excitation occurs, the axoplasm becomes less negative, and if inhibition occurs, the axoplasm becomes more negative compared to tissue fluid. For example, if a neurotransmitter causes only potassium gates to open, potassium will exit and the axoplasm will become even more negative.

Summation and Integration

A dendrite or cell body is on the receiving end of many synapses. Whether the neuron fires (initiates a nerve impulse) or not depends on the summary, or net, effect of all the excitatory and inhibitory neurotransmitters it receives. If the amount of excitatory neurotransmitter received is sufficient to raise the membrane potential above the threshold level (fig. 39.4*d*), the neuron fires. If the amount of excitatory neurotransmitter received is insufficient, only local excitation occurs. For this reason, the dendrites and the cell body are the parts of a neuron where *integration* (a summing up) occurs (fig. 39.3). This is the way that the nervous system fine tunes its response to the environment.

Neurotransmitters

Acetylcholine (ACh) and **norepinephrine** (NE) are the primary excitatory neurotransmitters and are active in both the peripheral and the central nervous systems. Examples of inhibitory substances, so far discovered only in the central nervous system, are given on page 647.

Once neurotransmitters have been released into a synaptic cleft, they remain active for only a short time. In some neurons, the cleft contains enzymes that rapidly inactivate the neurotransmitter. For example, the enzyme acetylcholinesterase (AChE), or simply cholinesterase, breaks down acetylcholine. A single molecule of AChE catalyzes the breakdown of 25,000 molecules of acetylcholine per second. The breakdown products are then taken up into the presynaptic neuron and used in the synthesis of new acetylcholine molecules. The enzyme monoamine oxidase breaks down norepinephrine after it is absorbed.

Transmission of a nerve impulse across a synapse is dependent on neurotransmitter substances that alter the potential difference across the postsynaptic membrane.

Figure 39.6

Diagrammatic representation of a synapse at 3 different magnifications. *a.* Typically, a synapse is located wherever an axon is close to a dendrite or cell body. Drawing based on a photomicrograph shows that there are several synaptic endings per axon because of terminal branching of the axon. *b.* Drawing based on low-power electron micrographs shows that an axon ending has numerous synaptic vesicles, each filled with a neurotransmitter. This drawing makes it clear that a synapse contains a cleft (space) between the axon and the dendrite. *c.* Drawing based on high-power electron micrographs shows that a synapse consists of the presynaptic membrane, the synaptic cleft, and the postsynaptic membrane. When a nerve impulse reaches the synaptic vesicles, they move toward and fuse with the presynaptic membrane, discharging their contents. The neurotransmitter diffuses across the cleft and combines with a receptor. A nerve impulse may follow.

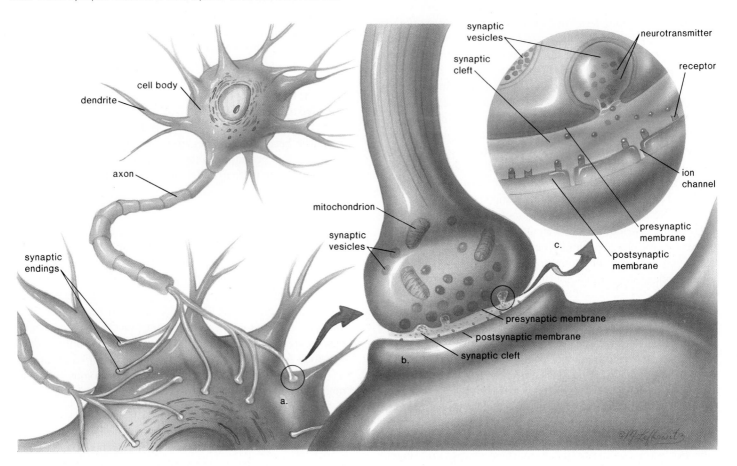

Peripheral Nervous System

The peripheral nervous system (PNS) contains **nerves** (fig. 39.2), structures that contain only long dendrites and/or long axons. Each of these fibers is surrounded by a myelin sheath (fig. 39.3), and therefore these nerves have a white glistening appearance. There are no cell bodies in nerves because cell bodies are found only in the central nervous system (CNS) or in the ganglia. **Ganglia** (ganglion, sing.) are collections of cell bodies within the PNS.

Humans have 12 pairs of cranial nerves and 31 pairs of spinal nerves. *Cranial nerves* are either sensory nerves (having long dendrites of sensory neurons only), motor nerves (having long axons of motor neurons only), or mixed nerves (having both long dendrites and long axons). All cranial nerves, except the vagus nerve, control the head, neck, and face. The *vagus nerve* controls the internal organs.

Spinal nerves are all mixed nerves that take impulses to and from the **spinal cord** (fig. 39.7). Their arrangement shows that humans are segmented animals: there is a pair of spinal nerves for each segment. Spinal nerves project from the spinal cord, which is a part of the central nervous system. The spinal cord is a thick, whitish nerve cord that extends longitudinally down the back, where it is protected by the vertebrae (vertebra, sing.). The cord contains a tiny, central canal filled with cerebrospinal fluid, gray matter consisting of cell bodies and short fibers, and white matter consisting of myelinated fibers.

In the PNS, cranial nerves take impulses to and/or from the brain and spinal nerves take impulses to and from the spinal cord.

Figure 39.7

The anatomy of the spinal cord. ***a.*** Cross section of the spine, showing spinal nerves. The human body has a total of 31 pairs of spinal nerves. ***b.*** This cross section of the spinal cord shows that a spinal nerve has a dorsal and a ventral root. Also, the cord is protected by 3 layers of tissue called the meninges. Spinal meningitis is an infection of these layers.

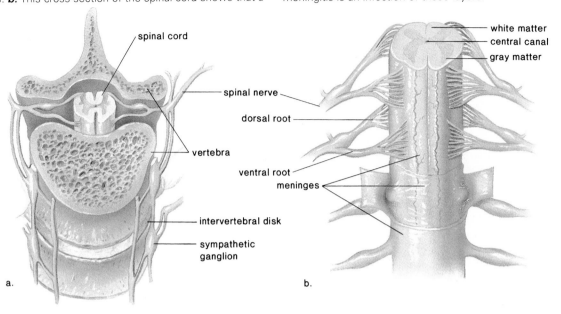

Somatic Nervous System

The *somatic nervous system* includes all nerves that serve the musculoskeletal system and the exterior sense organs, including those in the skin. Exterior sense organs are **receptors,** which receive environmental stimuli and then initiate nerve impulses. Muscle fibers are **effectors,** which bring about a reaction to the stimulus. Receptors are discussed in chapter 40, and muscle effectors are discussed in chapter 41.

Reflex Arc

Reflexes are automatic, involuntary responses to changes occurring inside or outside the body. In the somatic nervous system, outside stimuli often initiate a reflex action. Some reflexes, such as blinking the eye, involve the brain, but others, such as withdrawing the hand from a hot object, do not necessarily involve the brain. Figure 39.8 illustrates the path of the second type of reflex action involving the spinal cord and a spinal nerve, called a *spinal reflex,* or *reflex arc.* Various other reactions occur; the person may look in the direction of the object, jump back, and utter appropriate exclamations. This whole series of responses is explained by the fact that the sensory neuron stimulates several interneurons, which take impulses to all parts of the central nervous system, including the cerebrum, which in turn makes the person conscious of the stimulus and his or her reaction to it.

The reflex arc is a major functional unit of the nervous system. It allows us to react to internal and external stimuli.

Autonomic Nervous System

The **autonomic nervous system** is a part of the PNS (fig. 39.9). It is made up of motor neurons that control the internal organs automatically and usually without conscious intervention. There are 2 divisions to the autonomic nervous system (table 39.1): the **sympathetic nervous system** and the **parasympathetic nervous system.** Both of these (1) function automatically and usually subconsciously in an involuntary manner; (2) innervate all internal organs; and (3) use 2 motor neurons for each impulse. The cell body of the first motor neuron is located in the CNS, and the cell body of the second motor neuron is located in a ganglion. The axon that occurs before the ganglion is called the preganglionic fiber, and the axon that occurs after the ganglion is called the postganglionic fiber.

Sympathetic Nervous System

Preganglionic fibers of the sympathetic nervous system arise from the middle, or thoracic-lumbar, portion of the cord and almost immediately terminate in ganglia that lie near the cord (fig. 39.9). The sympathetic nervous system is especially important during emergency situations and may be associated with the "fight or flight response." For example, it inhibits the digestive tract but dilates the pupil of the eye, accelerates the heartbeat, and increases the breathing rate. It is not surprising, then, that the neurotransmitter released by the postganglionic axon is *norepinephrine* (NE), a chemical close in structure to epinephrine (adrenalin), a well-known heart stimulant.

Figure 39.8

Diagram of a reflex arc shows the detailed composition of a spinal nerve. When the receptors in the skin are stimulated, nerve impulses (see arrows) move along a sensory neuron to the spinal cord. (Note that the cell body of a sensory neuron is in a ganglion outside the cord.) The nerve impulses are picked up by an interneuron, which lies completely within the cord, and pass to the dendrites and cell body of a motor neuron, which lies ventrally within the cord. The nerve impulses then move along the axon of the motor neuron to an effector, such as a muscle fiber that contracts. The brain receives information concerning sensory stimuli by way of other interneurons having long fibers in tracts that run up and down the cord within the white matter.

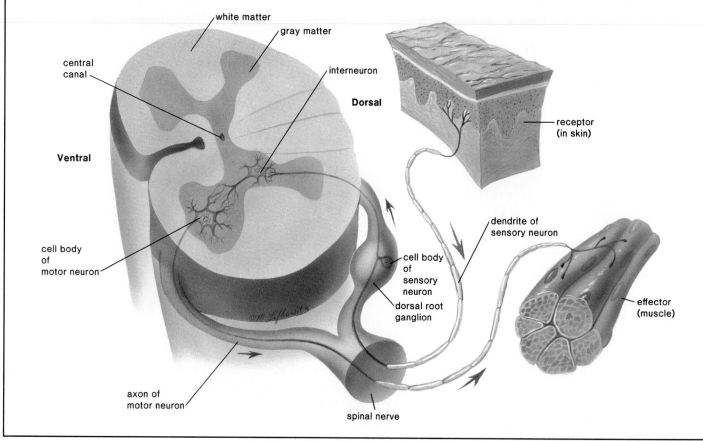

Parasympathetic Nervous System

In the parasympathetic nervous system, the preganglionic fibers arise from the brain and the bottom (sacral) portion of the cord system. This system is therefore often referred to as the craniosacral portion of the autonomic nervous system. The preganglionic fibers terminate in ganglia that lie near or within the organ (fig. 39.9). The parasympathetic system, sometimes called the "housekeeper system," promotes all the internal responses we associate with a relaxed state; for example, it causes the pupil of the eye to contract, promotes digestion of food, and retards the heartbeat. The neurotransmitter utilized by the parasympathetic system is *acetylcholine* (ACh).

The autonomic nervous system controls the functioning of internal organs without conscious control. The sympathetic nervous system brings about those responses we associate with "fight or flight." The parasympathetic nervous system brings about those responses we associate with normally restful activities.

Central Nervous System

The central nervous system (CNS) consists of the spinal cord and brain (fig. 39.10). The CNS is protected by bone: the brain is enclosed within the skull, and the spinal cord is surrounded by vertebrae. Also, both the brain and the spinal cord are wrapped in 3 layers of protective membranes known as the *meninges;* spinal meningitis is a well-known infection of these coverings. The interior of the brain contains spaces known as *ventricles.* Like the central canal of the spinal cord, these are filled with cerebrospinal fluid.

The spinal cord and brain contain gray matter and white matter. As mentioned, the *gray matter* is made up of cell bodies and short fibers. The **white matter** consists of the long, myelinated fibers of interneurons. Some of these are grouped together in bundles called *tracts,* which run between the brain and the spinal cord. The tracts cross so that the left side of the brain controls skeletal muscles on the right side of the body and vice versa.

Figure 39.9

Structure and function of the autonomic nervous system. The sympathetic fibers arise from the thoracic and lumbar portions of the cord; the parasympathetic fibers arise from the brain and sacral portion of the cord. Each system innervates the same organs but has a contrary effect. For example, the sympathetic system speeds up the beat of the heart, whereas the parasympathetic system slows it down.

Table 39.1
Sympathetic versus Parasympathetic System

Sympathetic	Parasympathetic
Fight or flight	Normal activity
Norepinephrine is neurotransmitter	Acetylcholine is neurotransmitter
Postganglionic fiber is longer than preganglionic	Preganglionic fiber is longer than postganglionic
Preganglionic fiber arises from middle portion of cord	Preganglionic fiber arises from brain and lower portion of cord

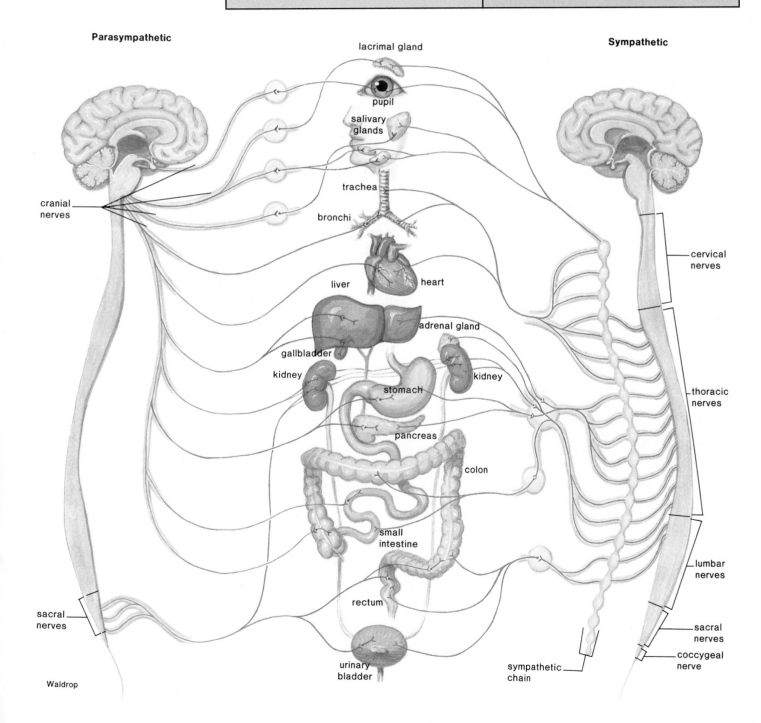

Parasympathetic

Sympathetic

lacrimal gland
pupil
salivary glands
trachea
bronchi
cranial nerves
liver
heart
adrenal gland
gallbladder
kidney
kidney
stomach
pancreas
colon
small intestine
rectum
sacral nerves
urinary bladder
sympathetic chain
cervical nerves
thoracic nerves
lumbar nerves
sacral nerves
coccygeal nerve

Waldrop

Figure 39.10
The human brain. Note how large the cerebrum is,
compared to the rest of the brain.

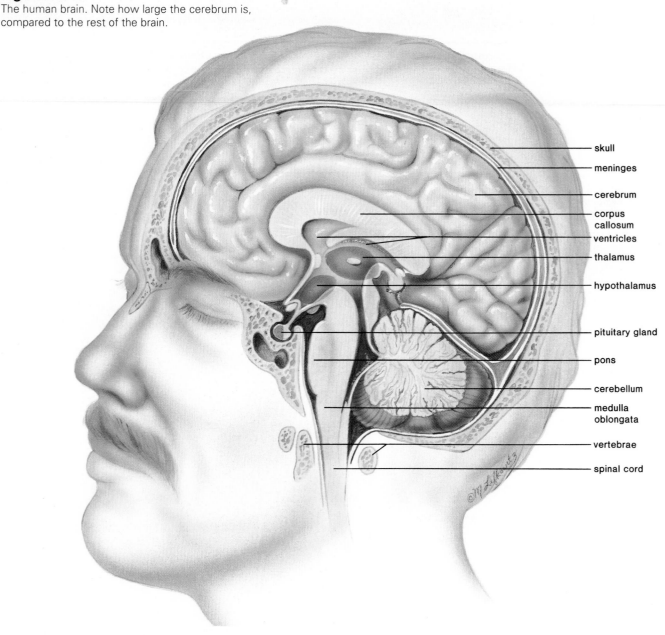

skull
meninges
cerebrum
corpus callosum
ventricles
thalamus
hypothalamus
pituitary gland
pons
cerebellum
medulla oblongata
vertebrae
spinal cord

The CNS consists of the brain and the spinal cord and is the place where sensory information is received and motor control is initiated.

The Brain

The largest and most prominent portion of the human brain is the **cerebrum** (fig. 39.10). Consciousness resides only in the cerebrum, which we will discuss later; the rest of the brain functions below the level of consciousness.

The Unconscious Brain

The unconscious brain has a number of different regions. The **medulla oblongata** is the part of the brain that lies closest to the spinal cord. It contains centers for regulating heartbeat, breathing, and vasoconstriction (blood pressure) and also reflex centers for vomiting, coughing, sneezing, hiccupping, and swallowing.

The **hypothalamus** forms the lower walls and floor of the third ventricle. This part of the brain is concerned with homeostasis, or the constancy of the internal environment, and contains centers for regulating hunger, sleep, thirst, body temperature, water balance, and blood pressure. The hypothalamus controls the pituitary gland and thereby serves as a link between the nervous and the endocrine systems.

The medulla oblongata and the hypothalamus are both concerned with control of the internal organs.

Figure 39.11

Comparison of brain sizes among vertebrates. Note the enlargement in the size of the cerebrum counterclockwise from bass to human.

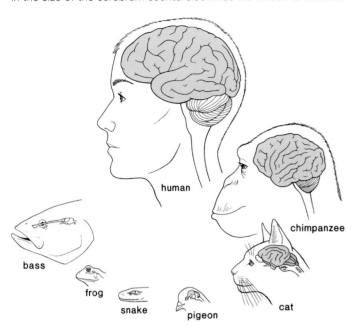

The *midbrain* and *pons* contain tracts that connect the cerebrum with other parts of the brain. The midbrain controls head and eyeball movements in response to visual and auditory stimuli. The pons helps the medulla regulate the breathing rate.

The **thalamus** is an egg-shaped structure in the third ventricle. It is the last portion of the brain for sensory input before the cerebrum. It serves as a central relay station for sensory impulses traveling upward from other parts of the spinal cord and the brain to the cerebrum. It receives all sensory impulses (except those associated with the sense of smell) and channels them to appropriate regions of the cortex for interpretation. The thalamus is sometimes called the gatekeeper to the cerebrum because it monitors the sensory data to be sent to the cerebrum. The thalamus allows us to ignore extraneous sensory information and pay attention instead to more important matters.

The **cerebellum,** a bilobed structure that resembles a butterfly, is the second largest portion of the brain (fig. 39.10). It is located dorsal to the pons and medulla oblongata. The cerebellum functions in muscle coordination by integrating impulses received from higher centers to ensure that all the skeletal muscles work together to produce smooth and graceful motions. The cerebellum is also responsible for maintaining normal muscle tone and transmitting impulses that maintain

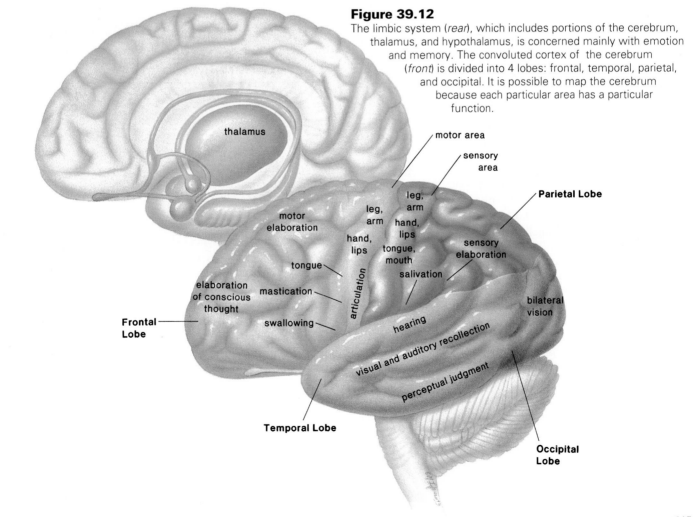

Figure 39.12

The limbic system (*rear*), which includes portions of the cerebrum, thalamus, and hypothalamus, is concerned mainly with emotion and memory. The convoluted cortex of the cerebrum (*front*) is divided into 4 lobes: frontal, temporal, parietal, and occipital. It is possible to map the cerebrum because each particular area has a particular function.

posture. It receives information from the inner ear indicating the position of the body and sends impulses to muscles that maintain or restore balance.

> The thalamus is concerned with sensory reception, and the cerebellum is concerned with muscle control.

The Conscious Brain

The cerebrum, the foremost part of the brain, is the largest part of the brain in humans. It consists of 2 large masses called cerebral hemispheres, which are connected by a bridge of nerve fibers called the corpus callosum. The outer portion of the cerebral hemispheres, the cerebral cortex, is highly convoluted and gray in color because it contains cell bodies and short fibers.

A comparative study of vertebrates indicates a progressive increase in the size of the cerebrum from fishes to humans, and the cerebral cortex is more convoluted in humans than other vertebrates (fig. 39.11). The function of the cerebrum has also changed. In fishes and amphibians, the cerebrum largely has an olfactory function, but in reptiles, birds, and mammals, the cerebrum receives information from other parts of the brain and coordinates sensory data and motor functions. Only the cerebrum is responsible for consciousness, and it is the portion of the brain that governs intelligence and reason. These qualities are particularly well developed in humans.

The cerebral cortex of each hemisphere contains 4 surface lobes: frontal, parietal, temporal, and occipital (fig. 39.12). Different functions are associated with each lobe. For example, the *frontal lobe* controls motor functions and permits us to control our muscles consciously. The *parietal lobe* receives information from receptors located in the skin, such as those for touch, pressure, and pain. The *occipital lobe* interprets visual input. The *temporal lobe* has sensory areas for hearing and smelling.

Certain areas of the cerebral cortex have been "mapped" in great detail. For example, we know which portions of the frontal lobe control various parts of the body and which portions of the parietal lobe receive sensory information from these same parts. Each of the 4 lobes of the cerebral cortex contains an association area, which receives information from the other lobes and integrates it into higher, more complex levels of consciousness. These areas are concerned with intellect, artistic and creative ability, learning, and memory.

> The cerebrum is the most highly developed region of the brain and it alone carries on conscious thought processes. The cerebrum interprets incoming sensory data and initiates voluntary muscle movements.

The Limbic System

The **limbic system** involves portions of both the unconscious and conscious brain. It lies just beneath the cortex and contains neural pathways that connect portions of the frontal lobes, temporal lobes, thalamus, and hypothalamus (fig. 39.12). Several masses of gray matter that lie deep within each hemisphere of the cerebrum, termed the *basal nuclei,* are also a part of the limbic system.

Figure 39.13
Individual nerve cells in a snail, *Hermissenda,* are being stimulated by microelectrodes, which produce the signals scientists previously recorded when a snail learns to avoid light. When this snail is freed, it automatically avoids the light and does not need to be taught like other snails. Normally, to teach snails to avoid light, they are placed on a table that rotates every time they venture toward light.

Stimulation of different areas of the limbic system causes the subject to experience rage, pain, pleasure, or sorrow. By causing pleasant or unpleasant feelings about experiences, the limbic system apparently guides the individual into behavior that is likely to increase the chance of survival.

Learning and Memory The limbic system is also involved in the processes of learning and memory. Learning requires memory, but just what permits memory development is not definitely known. Investigators have been working with invertebrates such as slugs and snails because their nervous system is very simple and yet they can be conditioned to perform a particular behavior. To study this simple type of learning, it has been possible to insert electrodes into individual cells and to alter or record the electrochemical responses of these cells (fig. 39.13). This type of research has shown that learning is accompanied by an increase in the number of synapses, while forgetting involves a decrease in the number of synapses. In other words, the nerve-circuit patterns are constantly changing as learning, remembering, and forgetting occur. Within the individual neuron, learning involves a change in gene regulation and nerve protein synthesis and an increased ability to secrete transmitter substances.

Animal Structure and Function

Figure 39.14
Some drug actions at synapses. **a.** Drug stimulates the release of the neurotransmitter. **b.** Drug blocks the release of the neurotransmitter. **c.** Drug combines with the neurotransmitter, preventing its breakdown. **d.** Drug mimics the action of the neurotransmitter. **e.** Drug blocks the receptor so that the neurotransmitter cannot be received. Generally, only one of these actions occurs at a given synapse.

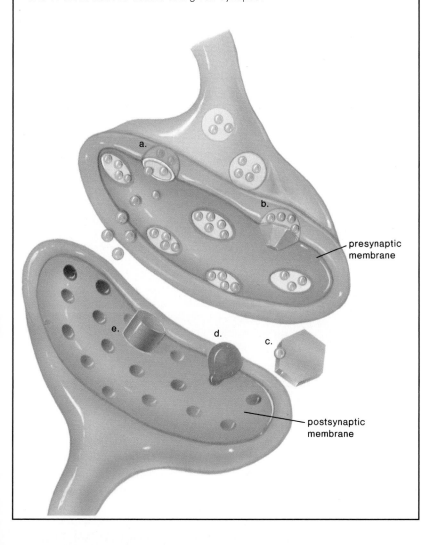

presynaptic membrane

postsynaptic membrane

Table 39.2
Drug Action

Drug Action	Neurotransmitter	Result
Blocks	Excitatory	Depression
Enhances	Excitatory	Stimulation
Blocks	Inhibitory	Stimulation
Enhances	Inhibitory	Depression

At the other end of the spectrum, some investigators have been studying learning and memory in monkeys. This work has led to the conclusion that the limbic system is absolutely essential to both short-term and long-term memory. An example of short-term memory in humans is the ability to recall a telephone number long enough to dial the number; an example of long-term memory is the ability to recall the events of the day. After nerve impulses circulate within the limbic system, the basal nuclei stimulate the sensory areas where memories are stored. The involvement of the limbic system certainly explains why emotionally charged events result in our most vivid memories. The fact that the limbic system communicates with the sensory areas for touch, smell, vision, and so forth accounts for the ability of any particular sensory stimulus to awaken a complex memory.

The limbic system is particularly involved in the emotions and in memory and learning.

Neurotransmitters in the Brain

As discussed previously, neurotransmitters released at the ends of axons affect the membrane potential of postsynaptic membranes. Some excitatory transmitters are amines, such as acetylcholine (ACh), norepinephrine (NE), serotonin, and dopamine. We have already had occasion to mention that ACh and NE are transmitters for the autonomic nervous system. *Serotonin* and *dopamine* are associated with behavior states, such as mood, sleep, attention, learning, and memory. Increasingly, it appears that feelings of pleasure are accompanied by the release of dopamine. The inhibitory transmitters include the amino acids gamma-aminobutyric acid (GABA) and glycine.

Both excitatory and inhibitory neurotransmitters are active in the brain.

In addition to the neurotransmitters already mentioned, a number of different types of peptides have been discovered in the CNS. The *endorphins,* molecules that are the body's natural opioids, are of particular interest. They are called natural opioids because the psychoactive drugs morphine and heroin, which are both derived from opium, attach to the receptors for endorphins in the CNS. When endorphins are released, they, like opium, produce a feeling of elation and reduce the sensation of pain. For example, if endorphins are present, neurons do not release substance P, a neurotransmitter that brings about the sensation of pain. Exercise has been associated with their presence, and this may account for the so-called runner's high.

Neurotransmitter Disorders It has been discovered that several neurological illnesses, such as *Parkinson disease* and *Huntington disease,* are due to an imbalance of neurotransmitters.

Alcohol

It is possible to drink alcoholic beverages in moderation, but they are often abused. Alcohol use becomes "abuse," or an illness, when alcohol ingestion impairs an individual's social relationships, health, job efficiency, or judgment. While it is general knowledge that alcoholics are prone to drink until they become intoxicated, there is much debate as to what causes alcoholism. Some believe that alcoholism is due to an underlying psychological disorder, while others maintain that the condition is due to an inherited physiological disorder.

Alcohol effects on the brain are biphasic. After consuming several drinks, blood alcohol concentration rises rapidly and the drinker reports feeling "high" and happy (euphoric); however, after about 90 minutes and lasting until about 5–6 hours after consumption, the drinker feels depressed and unhappy (dysphoric). On the other hand, if the drinker continues to drink to maintain a high blood level of alcohol, he or she experiences ever-increasing loss of control. Coma and death are even possible if a substantial amount of alcohol is consumed within a limited period.

Marijuana

The dried flowering tops, leaves, and stems of the Indian hemp plant *Cannabis sativa* contain and are covered by a resin that is rich in THC (tetrahydrocannabinol). The names *cannabis* and *marijuana* can apply either to the plant or to THC.

The effects of marijuana differ depending upon the strength and the amount consumed, the expertise of the user, and the setting in which it is taken. Usually, the user reports experiencing a mild euphoria, along with alterations in vision and judgment that result in distortions of space and time. The inability to concentrate or to speak coherently and motor incoordination may also be involved.

Intermittent use of low-potency marijuana is not generally associated with obvious symptoms of toxicity, but heavy use may produce chronic intoxication. Intoxication is recognized by the presence of hallucinations, anxiety, depression, rapid flow of ideas, body image distortions, paranoid reactions, and similar psychotic symptoms. The terms *cannabis psychosis* and *cannabis delirium* refer to such reactions.

The use of marijuana has ill effects, but it does not seem to produce physical dependence. There may be a psychological dependence on the euphoric and sedative effects. Craving or difficulty in stopping may also occur as a part of regular heavy use.

Marijuana has been called a *gateway* drug because adolescents who have used marijuana tend to also try other drugs. For example, in a study of 100 cocaine abusers, 60% had also smoked marijuana for more than 10 years.

Cocaine

Cocaine, currently the second most popular illegal drug after marijuana, is derived from the shrub *Erythroxylon coca.* Cocaine is sold in powder form and in the form of crack, a more potent extract. Users often use the word *rush* to describe the feeling of euphoria that follows intake of the drug. Snorting (inhaling through the nose) produces this effect in a few minutes; injection, within 30 seconds; and smoking in less than 10 seconds. Persons dependent upon the drug are, therefore, most likely to use the last method of intake. The rush only lasts a few seconds and is then replaced by a state of arousal that lasts from 5 minutes–30 minutes. Then the user begins to feel restless, irritable, and depressed. To overcome these symptoms the user is apt to take more of the drug, repeating the cycle until there is no more drug left. A binge of

---

Parkinson disease is a condition characterized by a wide-eyed, unblinking expression, an involuntary tremor of the fingers and thumbs, muscle rigidity, and a shuffling gait. All these symptoms are due to dopamine deficiencies. Huntington disease is characterized by a progressive deterioration of the individual's nervous system; this eventually leads to constant thrashing and writhing movements and finally to insanity and death. The problem is believed to be malfunction of the inhibitory neurotransmitter GABA. *Alzheimer disease*, a severe form of senility with marked memory loss in 5%–10% of all people over age 65, is recognized by the presence of innumerable diffuse beta-protein plaques in the brain. Many different types of neurons and neurotransmitters are adversely affected.

Some neurological illnesses are associated with the deficiency of a particular neurotransmitter in the brain.

Action of Neurological Drugs Drugs that alter the mood of a person either enhance or block the action of a particular neurotransmitter found particularly in the limbic system. There are a number of different ways drugs can influence the action of neurotransmitters, some of which are shown in figure 39.14. It is clear, as outlined in table 39.2 that stimulants can either enhance the action of an excitatory transmitter or block the action of an inhibitory transmitter. On the other hand, depressants can either enhance the action of an inhibitory transmitter or block the action of an excitatory transmitter.

One group of scientists believes that the neurotransmitter dopamine is involved in all forms of pleasure. Substantiating this position is the knowledge that cocaine interferes with the uptake of dopamine from synaptic clefts. Also, neurons producing dopamine make up a significant part of the limbic system. Many new medications developed to counter drug addiction and mental illness affect the release, reception, or breakdown of dopamine. There is some evidence to suggest that those with mental illness and those prone to drug addiction are afflicted with the same genetic defects; perhaps ones affecting dopamine transmission across the synapse.

A wide variety of drugs can be used to alter mood and/or emotional state. This chapter's reading discusses 5 of the most commonly abused drugs: alcohol, marijuana, cocaine, methamphetamine, and heroin. Drug abuse occurs when a person takes

---

Animal Structure and Function

this sort can go on for days, after which the individual suffers a *crash*. During the binge period, the user is hyperactive and has little desire for food or sleep but has an increased sex drive. During the crash period, the user is fatigued, depressed, and irritable and has memory and concentration problems and displays no interest in sex. Indeed, men are often impotent. Other drugs, such as marijuana, alcohol, or heroin, are often taken to ease the symptoms of the crash.

With continued cocaine use, the body begins to make less dopamine to compensate for a seemingly excess supply. The user, therefore, now experiences *tolerance* (always needing more of the drug for the same effect), *withdrawal* (symptoms such as the crash symptoms described previously when a drug is not taken), and an intense *craving* for cocaine. These are indications that the person is highly dependent upon the drug or, in other words, that cocaine is extremely addictive.

Methamphetamine (Ice)

Methamphetamine is related to amphetamine, a well-known stimulant. Both methamphetamine and amphetamine have been drugs of abuse for some time, but a new form of methamphetamine known as "ice"

is now being used as an alternative to cocaine. Ice is a pure, crystalline hydrochloride salt that has the appearance of sheet-like crystals. Unlike cocaine, ice can be produced in this country in illegal laboratories and does not need to be imported.

Ice, like crack, will vaporize in a pipe and can be smoked, avoiding the complications of intravenous injections. After rapid absorption into the bloodstream, the drug moves quickly to the brain. It has the same stimulatory effect as cocaine, and subjects report they cannot distinguish between the 2 after intravenous administration. Methamphetamine effects, however, persist for hours instead of a few seconds. Therefore, it is the preferred drug of abuse by many.

Heroin

Heroin is derived from morphine, a derivative of *opium*. Heroin is usually injected. After intravenous injection, the onset of action is noticeable within one minute and reaches its peak in 3 minutes–6 minutes. There is a feeling of euphoria along with relief of pain. Side effects can include nausea, vomiting, dysphoria, and respiratory and circulatory depression leading to death.

Heroin binds to receptors meant for the body's own opioids, the endorphins. As mentioned previously, the opioids are be-

lieved to alleviate pain by preventing the release of the neurotransmitter termed substance P from certain sensory neurons in the region of the spinal cord. When substance P is released, pain is felt, and when substance P is not released, pain is not felt. Evidence also indicates that there are opioid receptors in neurons that travel from the spinal cord to the limbic system and that stimulation of these can cause a feeling of pleasure. This explains why opium and heroin not only kill pain but also produce a feeling of tranquility.

Individuals who inject heroin become physically dependent on the drug. With time, the body's production of endorphins decreases; now *tolerance* develops so that the user needs to take more of the drug just to prevent *withdrawal* symptoms. The euphoria originally experienced upon injection is no longer felt.

The withdrawal symptoms include perspiration, dilation of pupils, tremors, restlessness, abdominal cramps, goose flesh, defecation, vomiting, and increase in systolic pressure and respiratory rate. Those who are excessively dependent may experience convulsions, respiratory failure, and death. Infants born to women who are physically dependent also experience these withdrawal symptoms.

---

excessive amounts of a drug under circumstances that increase the likelihood of a harmful effect. Individuals who are drug abusers are also apt to display a *physical dependence* on the drug (formerly called an addiction to the drug). Dependence is present when the person (1) spends much time thinking about the drug or arranging to get it; (2) often takes more of the drug than was intended; (3) is tolerant to the drug—that is, must increase the

amount of the drug to get the same effect; (4) has withdrawal symptoms when he or she stops taking the drug; and (5) has a repeated desire to cut down on use.

Neurological drugs interfere with normal neurotransmitter function in the brain, particularly in the limbic system. Drug abuse often results in physical dependence on the drug.

Summary

1. In humans, the nervous system, along with the endocrine system, regulates the other systems of the body and coordinates body functions.
2. A comparative study of the invertebrates shows a gradual increase in the complexity of nervous system. The human nervous system, like that of the earthworm, is divided into the central and peripheral nervous systems.

3. Neurons, cells that conduct nerve impulses, have 3 parts: dendrite(s), cell body, and axon. The dendrites and the cell body receive information from the environment, and if stimulation is sufficient, nerve impulses travel down the axon to where neurotransmitters are stored in vesicles.
4. The nerve impulse, which is recognized when the resting potential becomes an action potential, is an electrochemical phenomenon involving

the movement first of sodium and then of potassium ions across the axomembrane.

Resting potential: Sodium-potassium pump at work, inside of the neuron is negative (−65 mV) compared to the outside of the neuron.

Action potential: (1) Sodium ions move to the inside, making it positive compared to outside (+40 mV). (2) Potassium ions move to the outside, making the inside negative again compared to the outside (–65 mV).

5. The nerve impulse is self-propagating because the gated channels that allow sodium and potassium ions to flow down their concentration gradients are sensitive to a nearby decrease in the membrane potential.

6. Saltatory conduction occurs in myelinated fibers—the action potential jumps from node of Ranvier to node of Ranvier. This accounts for the great speed of impulses in these fibers.

7. Transmission across a synapse usually requires neurotransmitters because there is a small space, the synaptic cleft, that separates neuron from neuron. The neurotransmitters released at the ends of axons may be either excitatory or inhibitory.

8. Acetylcholine and norepinephrine are well-known excitatory transmitters. After its release, acetylcholine is destroyed by acetylcholinesterase in the synaptic cleft.

9. Within the human peripheral nervous system, the somatic system includes cranial and spinal nerves. Spinal nerves extend from the spinal cord and contain both sensory and motor fibers. It is possible to use them to trace the path of a reflex arc from receptor to effector, as described here.

Sensory neuron: receptor generates nerve impulses that travel in the dendrite to the cell body and then to the axon, which enters the cord.
Interneuron: transmits impulses from the dorsal to the ventral root of the cord.
Motor neuron: impulses begin in the dendrites and cell body and then pass out of the cord by way of the axon, which innervates an effector.

10. The autonomic nervous system controls internal organs and includes the sympathetic and parasympathetic nervous systems, which are contrasted in table 39.1.

11. The human central nervous system includes the spinal cord and the brain. In the brain, the medulla oblongata and hypothalamus regulate internal organs. The thalamus receives sensory input and passes it to the cerebrum. The cerebellum functions in muscle coordination.

12. A survey of vertebrates shows a continual evolutionary increase in the size of the cerebrum, with its greatest development in humans. The highly convoluted cerebral cortex is divided into lobes, each of which has specific functions. In general, the cerebrum is responsible for consciousness, sensory perception, motor control, and all the higher forms of thought.

13. Research in invertebrates indicates that learning is accompanied by an increase in the number of synapses; research in monkeys indicates that the limbic system is involved. There are short-term and long-term memories; the involvement of the limbic system explains why emotionally charged events result in vivid long-term memories.

14. Various illnesses such as Parkinson, Huntington, and Alzheimer diseases are due to an imbalance in a specific neurotransmitter in the brain. Drugs that people take to alter the mood and/or emotional state often interfere with normal neurotransmitter function in the brain, particularly the limbic system.

Writing Across the Curriculum

In order to practice writing skills, students should write out the answers to any or all of the study questions and the critical thinking questions. The study questions are sequenced in the same order as the text. Suggested answers to the critical thinking questions are in appendix D.

Study Questions

1. What is the overall function of the nervous and endocrine systems? How do they differ in regard to this function?

2. Trace the evolution of the nervous system by contrasting the organization of the nervous system in hydras, planarians, earthworms, and humans.

3. Describe the structure of a neuron and give a function for each part mentioned. Name 3 types of neurons, and give a function for each.

4. What are the major events of an action potential and the ion changes that are associated with each event?

5. Describe the mode of action of a neurotransmitter at a synapse, including how it is stored and how it is destroyed.

6. Contrast the structure and function of the peripheral and the central nervous systems.

7. Trace the path of a spinal reflex.

8. Contrast the sympathetic and parasympathetic divisions of the autonomic nervous system.

9. Name the major parts of the human brain and give a principal function for each part.

10. Define the limbic system and discuss its possible involvement in learning and memory.

11. Name several specific neurotransmitters and, in general, describe how various types of drugs can affect the action of neurotransmitters.

Objective Questions

1. Which is the most complete list of animals that have both a central and a peripheral nervous system?
   a. hydra, planaria, earthworm, rabbit, human
   b. planaria, earthworm, rabbit, human
   c. earthworm, rabbit, human
   d. rabbit, human

2. Which of these are the first element and the last element in a spinal reflex?
   a. axon and dendrite
   b. sense organ and muscle effector
   c. ventral horn and dorsal horn
   d. motor neuron and sensory neuron

Animal Structure and Function

3. Which phrase does not belong with the others?
   a. cerebrum
   b. cerebral cortex
   c. cerebral hemispheres
   d. cerebellum
4. A spinal nerve takes nerve impulses
   a. to the CNS.
   b. away from the CNS.
   c. both to and away from the CNS.
   d. only inside the CNS.
5. Which of these correctly describes the distribution of ions on either side of an axon when it is not conducting a nerve impulse?
   a. Na$^+$ outside and K$^+$ inside
   b. K$^+$ outside and Na$^+$ inside
   c. charged protein outside; Na$^+$ and K$^+$ inside
   d. Na$^+$ and K$^+$ outside and water only inside
6. When the action potential begins, sodium gates open, allowing sodium ions to cross the axomembrane. Now the polarity changes to
   a. negative outside and positive inside.
   b. positive outside and negative inside.
   c. no difference in charge between outside and inside.
   d. Any one of these.

7. Transmission of the nerve impulse across a synapse is accomplished by the
   a. movement of sodium and potassium ions.
   b. release of neurotransmitters.
   c. Both of these.
   d. Neither of these.
8. The autonomic nervous system has 2 divisions called the
   a. CNS and peripheral systems.
   b. somatic and skeletal systems.
   c. efferent and afferent systems.
   d. sympathetic and parasympathetic systems.

9. Synaptic vesicles are
   a. at the ends of dendrites and axons.
   b. at the ends of axons only.
   c. along the length of all long fibers.
   d. All of these.
10. Which of these is mismatched?
    a. cerebrum—consciousness
    b. thalamus—motor and sensory centers
    c. hypothalamus—internal environment regulator
    d. cerebellum—motor coordination
11. Label this diagram of a reflex arc. State a function for each structure labeled:

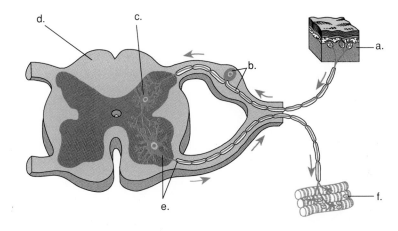

Concepts and Critical Thinking

1. *All systems of an animal's body contribute to homeostasis.*

   Tell several ways in which the nervous system contributes to homeostasis.

2. *Segmentation leads to specialization of parts.*

   How is segmentation along with specialization of parts reflected in the structure of the nervous system?

3. *Structure suits function.*

   How does the structure of a neuron suit its function?

Selected Key Terms

central nervous system (sen'tral ner'vus sis'tem) 634
peripheral nervous system (pĕ-rif'er-al ner'vus sis'tem) 634
neuron (nu'ron) 635
dendrite (den'drīt) 635
axon (ak'son) 635
myelin sheath (mi'ĕ-lin shĕth) 636
resting potential (rest'ing po-ten'shal) 637
action potential (ak'shun po-ten'shal) 639

saltatory conduction (sal'tah-to"re kon-duk'shun) 639
synapse (sin'aps) 639
neurotransmitter (nu"ro-trans-mit'er) 639
acetylcholine (as"ĕ-til-ko'lēn) 639
norepinephrine (nor"ep-i-nef'ron) 639
nerve (nerv) 640
ganglion (gang'gle-on) 640
reflex (re'fleks) 641
autonomic nervous system (aw"to-nom'ik ner'vus sis'tem) 641

sympathetic nervous system (sim"pah-thet'ik ner'vus sis'tem) 641
parasympathetic nervous system (par"ah-sim"pah-thet'ik ner'vus sis'tem) 641
medulla oblongata (mĕ-dul'ah ob"long-ga'tah) 644
hypothalamus (hi"po-thal'ah-mus) 644
thalamus (thal'ah-mus) 645
cerebellum (ser"ĕ-bel'um) 645
limbic system (lim'bic sis'tem) 646

40

Sense Organs

The eyes of an octopus and other cephalopods are highly developed and similar to those of humans. Light is brought to a focus on a single array of photo-receptor cells. Predators need good eyesight, and studies indicate that an octopus can see an object as small as 0.5 cm from a distance of 1 m.

Your study of this chapter will be complete when you can

1. state the function of sense organs;
2. describe the receptors for smell and taste in humans;
3. tell why the receptors for smell and taste are categorized as chemoreceptors;
4. in general, contrast the eye of arthropods with the eye of humans;
5. list the structures of the human eye, and give a function for each;
6. contrast the action of the sight receptors, the rods and cones;
7. give examples of various types of mechanoreceptors in animals;
8. list the structures of the human ear, and give a function for each one;
9. explain how the ear serves as an organ for static and dynamic equilibrium;
10. describe the organ of Corti and how it functions to permit hearing.

Sense organs receive external and internal stimuli; therefore, they are called **receptors.** Each type of receptor is designed to respond to a particular stimulus: for example, eyes respond only to light, and ears respond only to sound waves. Receptors do not interpret stimuli—they act merely as transducers that transform the energy of stimuli into nerve impulses. Interpretation is the function of the brain, which has a specific region for receiving nerve impulses from each of the sense organs. Impulses arriving at a particular sensory area of the brain can be interpreted in only one way; for example, those arriving at the visual area result in sight sensation, and those arriving at the olfactory area result in smell sensation. Usually the brain is able to discriminate the type of stimulus, the intensity of the stimulus, and the origin of the stimulus. On occasion, the brain can be fooled, as when a blow to the eyes causes you to "see stars."

Interoceptors located within the body monitor such conditions as blood pressure, expansion of lungs and bladder, and movement of limbs. *Exteroceptors* are located near the surface of the animal and respond to outer stimuli. Both types of receptors send information to the brain, and both are needed for homeostasis

and to promote appropriate behavior. Interoceptors will be discussed later in connection with the internal organs; now we will limit our coverage to exteroceptors.

Chemoreceptors

The receptors responsible for taste and smell are termed **chemoreceptors** because they are sensitive to certain chemical substances in food, liquids, and air. Chemoreception is found universally in animals and is therefore believed to be the most primitive sense. Chemoreceptors are present all over the body of planarians, but they are concentrated on the auricles at the sides of the head. In arthropods, chemoreceptors are found on the antennae and mouthparts, but in humans, unlike these animals, taste and smell have distinguishable receptors, as we will now discuss.

Smell

In humans, the receptors for the sense of smell are called the **olfactory cells** (*olfactus* means smell in Latin), located high in the roof of the nasal cavity (fig. 40.1). The olfactory cells are actually modified neurons that synapse with the nerve fibers making up the

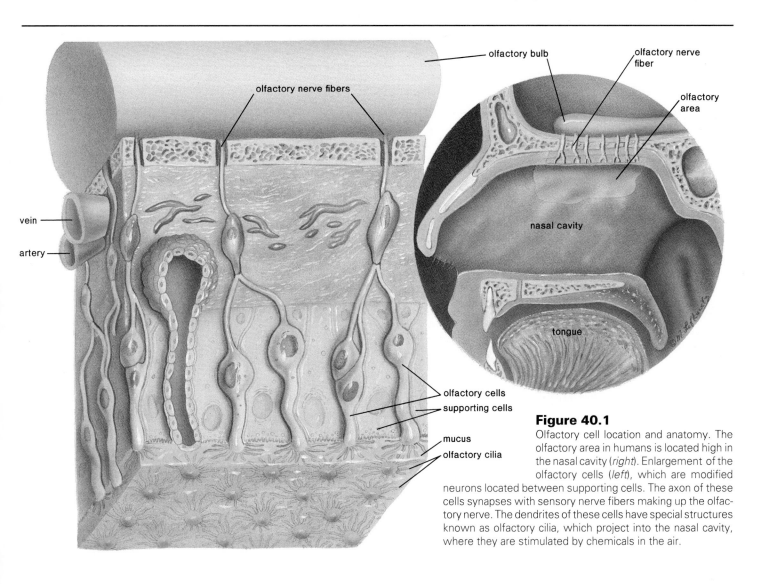

Figure 40.1

Olfactory cell location and anatomy. The olfactory area in humans is located high in the nasal cavity (*right*). Enlargement of the olfactory cells (*left*), which are modified neurons located between supporting cells. The axon of these cells synapses with sensory nerve fibers making up the olfactory nerve. The dendrites of these cells have special structures known as olfactory cilia, which project into the nasal cavity, where they are stimulated by chemicals in the air.

Figure 40.2

Taste buds. **a.** Elevations on the tongue indicate the presence of taste buds. The location of those containing taste buds responsive to sweet, sour, salt, and bitter is indicated. **b.** Enlargement of elevations, called papillae. **c.** The taste buds occur along the walls of the papillae. **d.** Drawing shows the various cells that make up a taste bud. Sensory cells in a bud end in microvilli that have receptors for the chemicals that exhibit the tastes noted in (**a**). When the chemicals combine with the receptors, nerve impulses are generated.

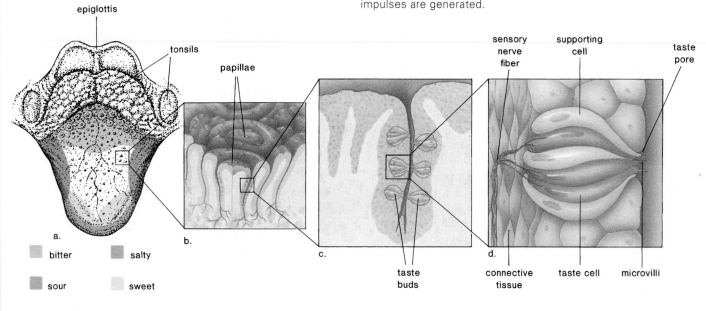

olfactory nerve. Each olfactory cell has cilia, which are stimulated by many chemicals in the air. Research resulting in the stereochemical theory of smell has shown that different smells may be related to the various shapes of the molecules rather than to the atoms that make up the molecules. These shapes fit specific olfactory sites on the olfactory cells' cilia.

The sense of smell is generally much more acute than the sense of taste. The human nose, for example, can detect one 25-millionth of 1 mg of mercaptan, the odoriferous chemical given off by a skunk. This averages out to approximately one molecule per sensory ending. Yet, humans have a weak sense of smell compared to other vertebrates, such as dogs.

Taste

The receptors for the sense of taste in mammals are the **taste buds.** Most of these are located on the tongue, but they are also present on the surface of the soft palate, pharynx, and epiglottis (fig. 40.2). Each taste bud contains a number of elongated cells. These cells have *microvilli,* which project through the taste pore, an opening in the taste bud. It is the microvilli that are stimulated by various chemicals in the environment. Humans are believed to have 4 types of taste buds, each type stimulated by chemicals that result in a bitter, a sour, a salty, or a sweet sensation.

The sense of taste and the sense of smell supplement each other, creating a combined effect when interpreted by the cerebral cortex. For example, when you have a cold, food seems to lose its taste, but actually the ability to sense its smell is temporarily absent.

This may work in reverse also. When we smell something, some of the molecules move from the nose down into the mouth and stimulate certain taste buds. Thus, part of what we refer to as smell is actually taste.

Photoreceptors

Many animals have light-sensitive receptors called **photoreceptors.** In its simplest form, a photoreceptor indicates only the presence of light and its intensity. The "eyespots" of planarians also allow the animal to determine the direction of light. More complex eyes have lenses that focus light on certain receptors. Arthropods have **compound eyes** composed of many independent visual units, each of which has its own lens and views a separate portion of the object (fig. 40.3). How well the brain combines this information to see the entire object is not known, but compound eyes seem especially well suited to detecting motion, as anyone who has tried to catch an insect knows.

Insects have color vision, but they make use of a slightly shorter range of the electromagnetic spectrum compared to humans. They can see the longest of the ultraviolet rays, and this enables them to be especially sensitive to the reproductive parts of flowers, which reflect particular ultraviolet patterns (fig. 40.4). Fishes, reptiles, and most birds are believed to have color vision, but among mammals, only humans and other primates have color vision. It would seem, then, that this trait was adaptive for life in trees, which accounts for its retention in these few mammals.

Animal Structure and Function

Figure 40.3

Each visual unit of a compound eye has a cornea and lens that focus light onto photoreceptor cells. The cells generate nerve impulses that are transmitted to the brain, where interpretation produces a mosaic image.

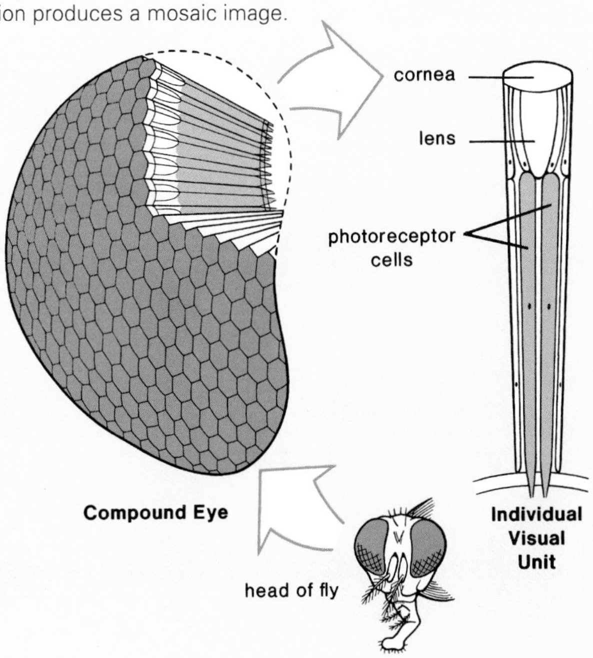

cornea

lens

photoreceptor cells

Compound Eye

head of fly

Individual Visual Unit

Figure 40.4

Marsh marigold as seen by humans (*left*) and insects (*right*). Humans see no markings, but insects see distinct blotches because their eyes respond to ultraviolet rays. These types of markings often highlight the reproductive parts of flowers where insects feed on nectar and pick up pollen at the same time.

Vertebrates and certain mollusks, like the squid, have a *camera type of eye*. A *single* lens focuses an image of the visual field on the photoreceptors, which are closely packed together. All of the photoreceptors taken together can be compared to a piece of film in a camera. The human eye is more complex than a camera, however, as we shall see.

Human Eye

The most important parts of the human eye and their functions are listed in table 40.1. The human eye, which is an elongated sphere about 2.5 cm in diameter, has 3 layers, or coats (fig. 40.5). The outer layer, the **sclera,** is an opaque, white, fibrous layer that surrounds the transparent cornea, the window of the eye. The middle, thin, dark-brown layer, the *choroid,* contains many blood vessels and pigment that absorbs stray light rays. Toward the front of the eye, the choroid thickens and forms the ring-shaped ciliary body and finally becomes a thin, circular, muscular diaphragm, the *iris,* which regulates the size of an opening called the *pupil.* The *lens,* which is attached to the ciliary body by ligaments, divides the cavity of the eye into 2 portions. A basic, watery solution called *aqueous humor* fills the anterior cavity between the cornea and the lens. A viscous, gelatinous material, the *vitreous humor,* fills the large posterior cavity behind the lens.

The inner layer of the eye, the **retina,** contains the receptors for sight: the **rods** and **cones** (fig. 40.6). Nerve impulses initiated by the rods and cones are passed to the bipolar cells, which in turn pass them to the ganglionic cells. The fibers of these cells pass in front of the retina forming the **optic nerve,** which carries the nerve impulses to the brain. Notice in figure 40.6*b* that there are many more rods and cones than nerve fibers leaving ganglionic cells. This means that there is considerable mixing of messages and a certain amount of integration before nerve impulses are sent to the brain. There are no rods or cones at the point where the optic nerve passes through the retina; therefore, this point is called the blind spot.

The center of the retina contains a special region called the fovea centralis, an oval, yellowish area with a depression where there are only cone cells (fig. 40.5). In the fovea centralis or fovea, color vision is most acute in daylight; at night, it is barely sensitive. At this time, the rods in the rest of the retina are active.

The human eye has 3 layers: the outer sclera, the middle choroid, and the inner retina. Only the retina contains receptors for sight.

Vision Process

Light rays entering the eye are bent (refracted) as they pass through the cornea, lens, and humors and brought to a focus on the retina. The lens is relatively flat when viewing distant objects but rounds up for close objects because light rays must be bent to a greater degree when viewing a close object. These changes of the lens shape are called accommodation (fig. 40.7). With

Figure 40.5

Anatomy of the human eye. Notice that the sclera becomes the cornea; the choroid becomes the ciliary body and iris. The ciliary body contains the ciliary muscle and ligaments, which hold and adjust the shape of the lens. The retina contains the receptors for sight, and vision is most acute in the fovea centralis, where there are only cones. A blind spot exists where the optic nerve leaves the retina and where there are no receptors for sight.

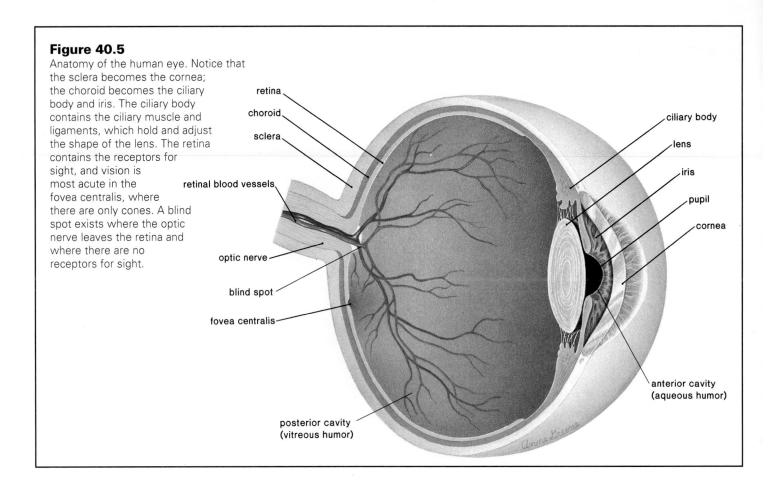

Figure 40.6

a. The retina is the inner layer of the eye. *b.* Rods and cones, which are located toward the back of the retina, are preceded by the bipolar cells and the ganglionic cells, whose fibers become the optic nerve. Notice that rods share bipolar cells but cones do not. Cones, therefore, distinguish more detail. *c.* The photosensitive pigment is located in the disks of the outer segment of rods and cones. A disk is apparently a modified cilium. *d.* Scanning electron micrograph of rods and cones. The cones are responsible for color vision, and the rods are responsible for night vision.

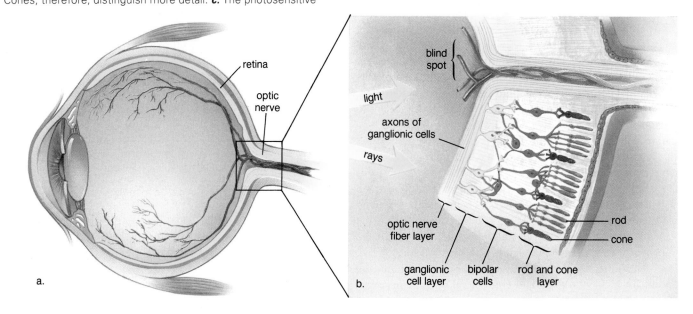

Animal Structure and Function

normal aging, the lens loses its ability to accommodate for close objects; therefore, persons frequently need reading glasses once they reach middle age.

Because of refraction, the image on the retina is rotated 180° from the actual, but it is believed that this image is righted in the brain. In one experiment, scientists wore glasses that inverted and reversed the field. At first, they had difficulty adjusting to the placement of objects, but they soon became accustomed to their inverted world. Experiments such as this suggest that if the retina sees the world "upside down," the brain has learned to see it right side up.

Table 40.1
Function of the Parts of the Eye

Part	Function
Lens	Refracts and focuses light
Iris	Regulates light entrance
Pupil	Admits light
Choroid	Absorbs stray light
Sclera	Protects
Cornea	Refracts light
Humors	Refracts light
Ciliary body	Holds lens in place
Retina	Contains receptors
Rods	Allow black-and-white vision
Cones	Allow color vision
Optic nerve	Transmits impulse
Fovea centralis	Region of cones in retina

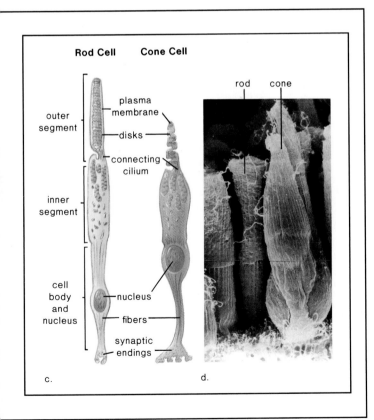

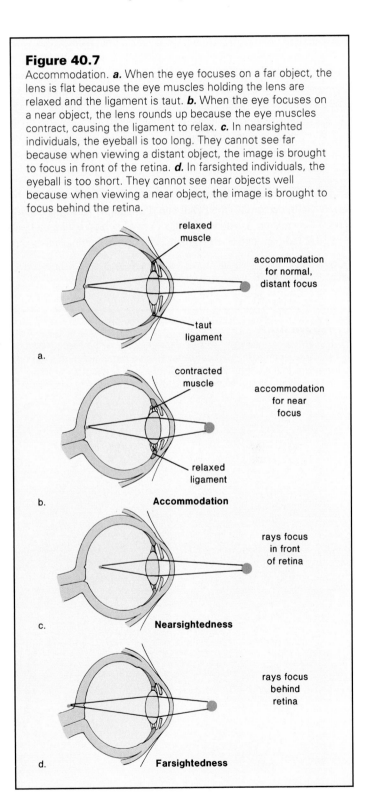

Figure 40.7

Accommodation. **a.** When the eye focuses on a far object, the lens is flat because the eye muscles holding the lens are relaxed and the ligament is taut. **b.** When the eye focuses on a near object, the lens rounds up because the eye muscles contract, causing the ligament to relax. **c.** In nearsighted individuals, the eyeball is too long. They cannot see far because when viewing a distant object, the image is brought to focus in front of the retina. **d.** In farsighted individuals, the eyeball is too short. They cannot see near objects well because when viewing a near object, the image is brought to focus behind the retina.

Rods Only dim light is required to stimulate rods; therefore, they are responsible for *night vision*. The numerous rods are also better at detecting motion than cones, but they cannot provide distinct and/or color vision. This causes objects to appear blurred and look gray in dim light. Many molecules of **rhodopsin** are located within the membrane of the disks (lamellae) found in the outer segment of the rods (fig. 40.6c). Rhodopsin is a complex molecule that contains protein (opsin) and a pigment molecule called *retinal,* which is a derivative of vitamin A. When light strikes retinal it changes shape, and opsin is activated. The reactions that follow eventually end after many molecules of GMP (guanosine monophosphate) have been converted to cyclic GMP.[1] Cyclic GMP, in turn, initiates nerve impulses in the rod, which pass through the retina to the optic nerve. Each stimulus generated lasts about one-tenth of a second. This is why we continue to see an image if we close our eyes immediately after looking at an object. It also allows us to see motion if still frames are presented at a rapid rate, as in "movies."

Cones The cones, located primarily in the fovea and activated by bright light, detect the fine detail and the color of an object. *Color vision* depends on 3 different kinds of cones, which contain a blue, green, or red pigment. Each pigment is made up of retinal and opsin,

---

[1]GMP is a nucleotide that contains the base guanine, the sugar ribose, and one phosphate group. In cyclic GMP, the single phosphate is attached to ribose in 2 places.

but there is a slight difference in the opsin structure of each, which accounts for their individual absorption patterns. Various combinations of cones are believed to be stimulated by in-between shades of color, and the combined nerve impulses are interpreted in the brain as a particular color.

> In the human eye, the receptors for sight are the rods and cones. The rods are responsible for vision in dim light, and the cones are responsible for vision in bright light and color vision. When a receptor is stimulated, nerve impulses are transmitted by the optic nerve to the brain.

Mechanoreceptors

Mechanoreceptors are sensitive to mechanical stimuli, such as pressure, sound waves, and gravity. Human skin contains various types of mechanoreceptors, such as *touch receptors* and pressure receptors (fig. 40.8). A pressure receptor, called the Pacinian corpuscle, is shaped like an onion and consists of a series of concentric layers of connective tissue wrapped around the end (dendrite) of a sensory neuron. In contrast, *pain receptors* are only the unmyelinated ("naked") ends (dendrites) of the fibers of sensory neurons. Some pain receptors are especially sensitive to mechanical stimuli; others are most sensitive to temperature or chemicals.

Figure 40.8

Receptors in human skin. The classical view is that each receptor has the function indicated; however, investigations in this century indicate that matters are not so clear-cut. For example, microscopic examination of the skin of the ear shows only free nerve endings (pain receptors), and yet the skin of the ear is sensitive to all sensations. Therefore, it appears that the receptors of the skin are somewhat but not completely specialized. Touch and pressure receptors are considered to be membrane receptors.

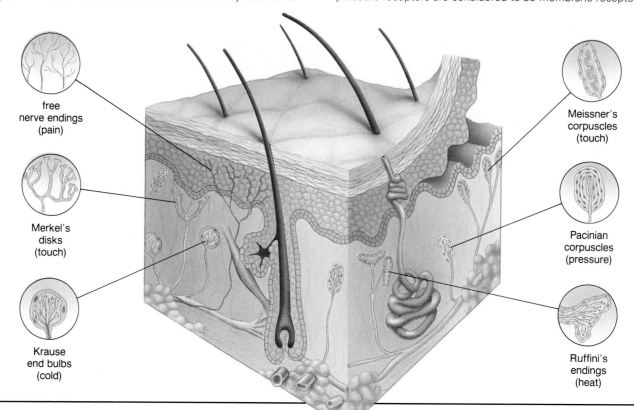

free nerve endings (pain)

Merkel's disks (touch)

Krause end bulbs (cold)

Meissner's corpuscles (touch)

Pacinian corpuscles (pressure)

Ruffini's endings (heat)

Animal Structure and Function

Hair cells, which are named for the cilia they bear, are often mechanoreceptors in various specialized vertebrate sense organs. Fishes and amphibians have a series of such receptors called the lateral line system, which detect water currents and pressure waves from nearby objects. In primitive fishes and aquatic amphibians, the receptors are located on the body surface, but in advanced fishes, they are located within a canal that has openings to the outside (fig. 40.9). A lateral line receptor is a collection of hair cells with cilia embedded in a mass of gelatinous material known as a cupula. When the cupula bends due to pressure waves, the hair cells initiate nerve impulses.

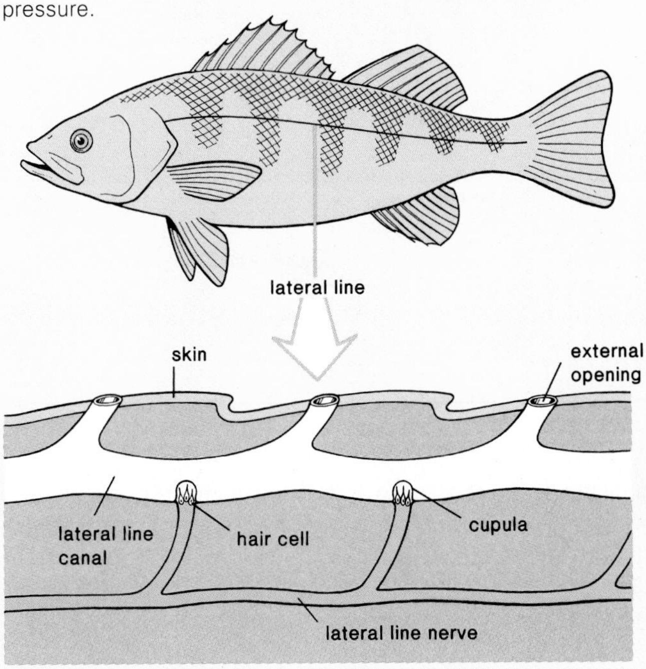

The ears of fishes, which function mainly as equilibrium organs, are derived from a portion of the lateral line system. The evolution of the inner ear of humans can also be traced back to the lateral line system of fishes. The inner ear of humans contains both equilibrium and second receptors.

Human Ear

Table 40.2 lists the most important parts of the human ear and their function.

As figure 40.10 shows, the human ear has an outer, a middle, and an inner portion. The *outer ear* consists of the pinna (external flap) and *auditory canal*. The auditory canal is lined with

Figure 40.9
Lateral line system of fishes. Location of the system (*upper*); longitudinal section of the system (*lower*). A main canal has openings to the exterior. Lining the canal are hair cells (embedded in cupulae) that act as sense receptors for pressure.

Table 40.2	
Function of the Parts of the Ear	
Part	**Function**
Outer Ear Pinna	Collects sound waves
Middle Ear Tympanic membrane and ossicles	Amplify sound waves
Inner Ear Oval window Cochlea Organ of Corti Utricle and saccule Semicircular canals	Initiates pressure waves Transmits pressure waves Receptor for hearing Static equilibrium Dynamic equilibrium

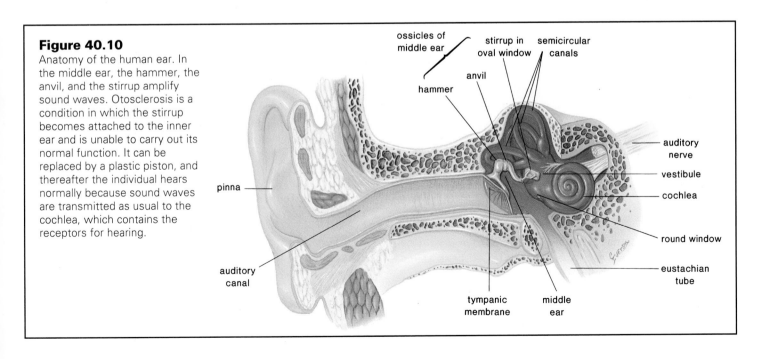

Figure 40.10
Anatomy of the human ear. In the middle ear, the hammer, the anvil, and the stirrup amplify sound waves. Otosclerosis is a condition in which the stirrup becomes attached to the inner ear and is unable to carry out its normal function. It can be replaced by a plastic piston, and thereafter the individual hears normally because sound waves are transmitted as usual to the cochlea, which contains the receptors for hearing.

Figure 40.11

Inner ear. *a.* The inner ear contains the semicircular canals, a vestibule, and the cochlea. The cochlea has been cut to show the location of the organ of Corti. *b.* There is an ampulla at the base of each semicircular canal that contains the receptors (hair cells) for dynamic equilibrium. *c.* In a vestibule are the utricle and the saccule, small sacs that contain the receptors (hair cells) for static equilibrium. *d.* The sense organ for hearing, the organ of Corti, is in the cochlea. The organ of Corti consists of hair cells with cilia that touch the tectorial membrane. Pressure waves cause the basilar membrane beneath the hair cells to vibrate. When the cilia touch the tectorial membrane, nerve impulses are initiated and taken up by sensory nerve fibers within the auditory nerve. The intensity of the sound is determined by how many cells are stimulated; the quality of the sound is determined by which particular cells along the entire organ of Corti are stimulated.

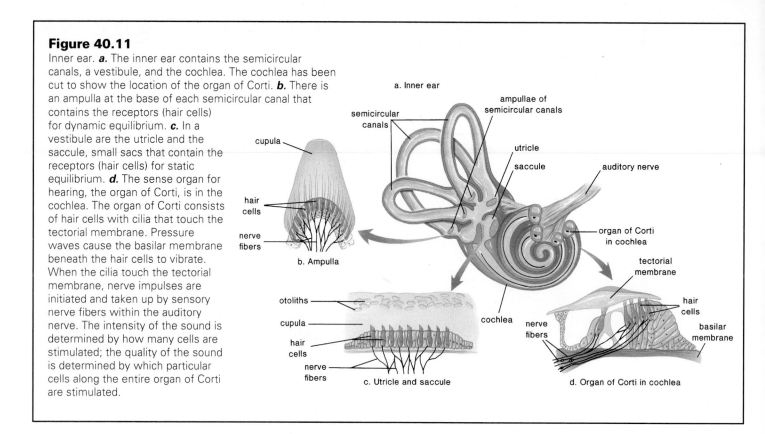

fine hairs, which filter the air. Modified sweat glands are located in the upper wall of the canal; these secrete earwax, which helps guard the ear against entrance of foreign materials.

The *middle ear* begins at the **tympanic membrane** (eardrum) and ends at a bony wall that has small openings covered by membranes, the *oval window* and the *round window*. Three small bones are located between the tympanic membrane and the oval window. Collectively called *ossicles,* individually they are the hammer (malleus), anvil (incus), and stirrup (stapes), named for their structural resemblance to these objects. The Eustachian tube extends from the middle ear to the pharynx and permits the equalization of air pressure between the inside and the outside of the ear. Chewing gum, yawning, and swallowing help move air through the Eustachian tubes during ascent and descent in airplanes or elevators.

The inner ear has 3 anatomic areas: the first 2 consist of a vestibule (chamber) and the semicircular canals and are concerned with balance; the third, the cochlea, is concerned with hearing. Whereas the outer ear and middle ear contain air, the inner ear is filled with fluid.

Balance

The **semicircular canals** are *dynamic equilibrium organs* because they initiate a sensation of movement (fig. 40.11*a*). At their bases are the ampullae, each one having touch-sensitive hair cells (fig. 40.11*b*). As the body moves about, there is first a slight lag, and then the fluid within the canals moves; this causes the cilia of the hair cells to bend. The 3 semicircular canals are each orien-

tated in a different plane to detect movement in any direction. The brain integrates the impulses it receives from each of the 3 canals and therefore can determine the direction and rate of movement. Very rapid or prolonged movements of the head may cause uncomfortable side effects, such as the dizziness and nausea of seasickness.

The vestibule between the semicircular canals and the cochlea contain 2 small sacs, the utricle and saccule (fig. 40.11*c*). Within each of these are groups of little hair cells; calcium carbonate granules called *otoliths* rest on their cilia. When the head tilts, the otoliths press on the cilia in a certain direction, and this initiates nerve impulses that inform the brain of the position of the head. Similar types of organs, called statocysts, are also found in cnidarians, mollusks, and crustacea. These organs only give information about the position of the head and do not result in a sensation of movement (fig. 40.12). They are therefore called *static equilibrium organs.*

The outer ear, middle ear, and cochlea are necessary for hearing in humans. The vestibule, containing the utricle and the saccule, and the semicircular canals are concerned with the sense of balance.

Hearing

The **cochlea** within the inner ear resembles the shell of a snail because it is spiral shaped (fig. 40.11*a*). A cross section of the cochlea shows that it contains 3 canals: the vestibular, cochlear, and tympanic canals. Along the length of the *basilar membrane,*

Picturing the Effects of Noise

We have an idea of what noise does to the ear," David Lipscomb [of the University of Tennessee Noise Laboratory] says. "There's a pretty clear cause-effect relationship." And these photomicrographs (see fig. 40.A) of the cochlea's tiny structures graphically document noise trauma to the inner ear.

Hair cells transmit the mechanical energy of sound waves into those neural impulses that the brain interprets as sound. Loud noise can damage or destroy hair cells as these scanning electron micrographs illustrate.

Hair cells come in 2 varieties: a single row of inner cells and a triple row of outer ones. "Outer cells degenerate before inner cells," notes Clifton Springs, N.Y.-otolaryngologist Stephen Falk. The most subtle change wrought by noise is a development of vesicles, or blisterlike protrusions along the walls of the hair cells' stereocilia. Continued assault by noise will lead to a rupturing of the vesicles and damage. In addition, the "cuticular plate"—base tissue supporting the stereocilia—may soften, followed by a swelling and ultimate degeneration of hair cells.

But sensory hair cells are not the only structures at risk. Adjacent inner-ear cells ... may undergo vacuolation—development

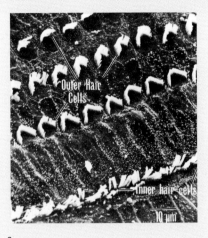
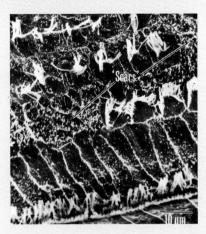

a.   b.

Figure 40.A
Electron micrographs showing damage to the organ of Corti due to loud noise. **a.** Normal organ of Corti. The tectorial membrane has been removed to reveal the hair cells. **b.** Organ of Corti after 24-hour exposure to noise level typical of rock music. Note scars where cilia on hair cells have worn away. Magnification, X1,400.

of degenerative empty spaces in cells. Even nerve fibers synapsing at the hair cells' roots may die. In the final phase of noise-induced cochlear damage, the organ of Corti—of which hair cells and supporting cells are a part—is completely denuded and covered by a layer of scar tissue.

From *Science News*, the weekly newsmagazine of science, copyright 1982 by Science Service, Inc., Washington, DC.

Figure 40.12
Generalized statocysts as found in mollusks and crustaceans. A small particle, the statolith, moves in response to a change in the animal's position. When the statolith stops moving, it stimulates the closest cilia of hair cells. These cilia transmit impulses, indicating the position of the body.

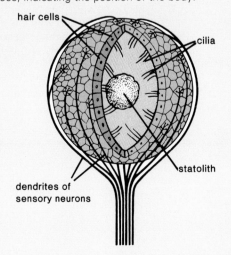

hair cells

cilia

statolith

dendrites of sensory neurons

which forms the floor of the cochlear canal, there are at least 24,000 ciliated hair cells. Just above them is another membrane called the *tectorial membrane* (fig. 40.11*d*). The hair cells plus the tectorial membrane form the **organ of Corti,** the sense organ for hearing.

Sound waves reach the organ of Corti by way of the outer and middle ear. Ordinarily, sound waves do not carry much energy, but when a large number of waves strike the eardrum, it moves back and forth (vibrates) very slightly. The hammer transfers the pressure from the inner surface of the eardrum to the anvil, and then to the stirrup (fig. 40.10). The pressure is multiplied about 20-fold as it moves from the eardrum to the stirrup. The stirrup vibrates the oval window, which transmits pressure waves to the fluid in the inner ear. These waves cause the basilar membrane to move up and down and the cilia of the hair cells to rub against the tectorial membrane. Bending of the cilia initiates nerve impulses, which pass by way of the auditory nerve to the brain, where the impulses are interpreted as a sound.

The organ of Corti is narrow at its base, but it widens as it approaches the tip of the cochlear canal. Various cells along it are sensitive to different wave frequencies, or pitches. Near its apex, the organ of Corti responds to low pitches, such as a bass drum, and near the base, it responds to high pitches, such as a bell or whistle. The nerve impulses from each region along the organ lead to slightly different areas in the brain. The pitch sensation we experience depends on which brain area is stimulated. Volume is a function of the *amplitude* of sound waves. Loud noises (measured in decibels, db) cause the fluid of the cochlea to oscillate to a greater degree, and this, in turn, causes the basilar membrane to move up and down to a greater extent. The resulting increased stimulation

is interpreted by the brain as loudness. It is believed that tone is interpreted by the brain according to the distribution of hair cells that are stimulated.

> The sense receptors for sound are ciliated hair cells on the basilar membrane (the organ of Corti). When the basilar membrane vibrates, the delicate cilia touch the tectorial membrane, producing nerve impulses, which are transmitted by the auditory nerve to the brain.

Summary

1. Sense organs are transducers; they transform the energy of a stimulus to the energy of nerve impulses. It is the brain, not the sense organ, that interprets the stimulus.
2. Human olfactory cells and taste buds are chemoreceptors. They are sensitive to chemicals in water and air.
3. The human eye is a photoreceptor. The compound eye of arthropods is made up of many individual units, whereas the human eye is a camera-type eye with a single lens. Table 40.1 lists the parts of the eye and the function of each part.
4. The receptors for sight in humans are the rods and cones, which are in the retina. The rods work in minimum light and detect motion, but they do not detect color. The cones require bright light and do detect color.
5. When light strikes rhodopsin, a molecule composed of opsin and retinal, retinal changes shape and opsin is activated. Chemical reactions that produce nerve impulses follow. These impulses are eventually picked up by the optic nerve.
6. There are 3 kinds of cones, containing blue, green, or red pigment. Each pigment is also made up of retinal and opsin, but opsin structure varies among the 3.
7. Mechanoreceptors include the touch receptors and pressure receptors in the human skin.
8. Many mechanoreceptors are hair cells with cilia, such as those found in the lateral line of fishes as well as the inner ear of humans. Table 40.2 lists the parts of the ear and the function of each part.
9. The inner ear contains the sense organs for balance. Just like the statocysts of invertebrates, portions of the human inner ear contain calcium carbonate granules resting on hair cells. The movement of these granules gives us a sense of static equilibrium. Movement of fluid past hair cells in the semicircular canals gives us a sense of dynamic equilibrium.
10. Hair cells on the basilar membrane (the organ of Corti) are responsible for hearing. Pressure waves, which begin at the oval window, cause the basilar membrane to vibrate so that the cilia of the hair cells touch the tectorial membrane. This causes the hair cells to initiate nerve impulses, which are carried by the auditory nerve to the brain.

Writing Across the Curriculum

In order to practice writing skills, students should write out the answers to any or all of the study questions and the critical thinking questions. The study questions are sequenced in the same order as the text. Suggested answers to the critical thinking questions are in appendix D.

Study Questions

1. In what ways are all receptors similar; in what ways are they different?
2. Discuss the structure and function of human chemoreceptors.
3. In general, how does the arthropod eye differ from that in humans? What types of animals have eyes that are constructed similarly to the human eye?
4. Name the parts of the eye, and give a function for each part.
5. Contrast the location and function of rods to that of cones.
6. What are the types of mechanoreceptors in human skin?
7. Describe how the lateral line system of fishes works and why it is considered to contain mechanoreceptors.
8. Describe the anatomy of the ear and how we hear.
9. Describe the role of the utricle, saccule, and semicircular canals in balance.

Objective Questions

1. A receptor
   a. is the first portion of a reflex arc.
   b. initiates nerve impulses.
   c. responds to only one type of stimulus.
   d. All of these.
2. Which of these gives the correct path for light rays entering the human eye?
   a. sclera, retina, choroid, lens, cornea
   b. fovea centralis, pupil, aqueous humor, lens
   c. cornea, pupil, lens, vitreous humor, retina
   d. optic nerve, sclera, choroid, retina, humors

Animal Structure and Function

3. Which gives an incorrect function for the structure?
   a. lens—focusing
   b. iris—regulation of amount of light
   c. choroid—location of cones
   d. sclera—protection
4. Which of these contain mechanoreceptors?
   a. human skin
   b. lateral line of fishes
   c. statocysts of arthropods
   d. All of these.
5. Which association is incorrect?
   a. lateral line—fishes
   b. compound eye—arthropods
   c. camera-type eye—squid
   d. statocysts—sea stars
6. Which one of these wouldn't you mention if you were tracing the path of sound vibrations?
   a. auditory canal
   b. tympanic membrane
   c. semicircular canals
   d. cochlea
7. Which one of these correctly describes the location of the organ of Corti?
   a. between the tympanic membrane and the oval window in the inner ear
   b. in the utricle and saccule within the vestibule
   c. between the tectorial membrane and the basilar membrane in the cochlear canal
   d. between the outer and inner ear within the semicircular canals

8. Which of these is mismatched?
   a. semicircular canals—inner ear
   b. utricle and saccule—outer ear
   c. auditory canal—outer ear
   d. ossicles—middle ear
9. Retinal is
   a. sensitive to light energy.
   b. a part of rhodopsin.
   c. found in both rods and cones.
   d. All of these.

10. Both olfactory receptors and sound receptors have cilia, and they both
    a. are chemoreceptors.
    b. are mechanoreceptors.
    c. initiate nerve impulses.
    d. All of these.
11. Label this diagram of the human eye. State a function for each structure labeled:

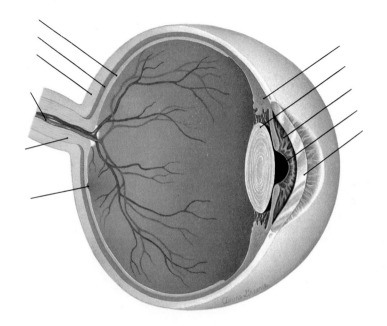

Concepts and Critical Thinking

1. *All organ systems contribute to homeostasis.*

   Tell several ways the sense organs contribute to homeostasis.

2. *In the whole animal, all systems work together and influence one another.*

   How do the nervous system and the sense organs work together?

3. *Animals are sensitive to only certain types of stimuli.*

   Why might animals be sensitive to only certain types of stimuli?

Selected Key Terms

receptor (re-sep'tor) 653
chemoreceptor (ke"mo-re-sep'tor) 653
olfactory cell (ol-fak'to-re sel) 653
taste bud (tāst bud) 654
photoreceptor (fo"to-re-sep'tor) 654
compound eye (kom'pownd i) 654

sclera (skle'rah) 655
retina (ret'ĭ-nah) 655
rod (rod) 655
cone (kōn) 655
rhodopsin (ro-dop'sin) 658
mechanoreceptor (mek"ah-no-re-sep'tor) 658

tympanic membrane (tim-pan'ik mem'brān) 660
semicircular canal (sem"e-ser'ku-lar kah-nal') 660
cochlea (kok'le-ah) 660
organ of Corti (or'gan uv kor'ti) 661

41

Musculoskeletal System

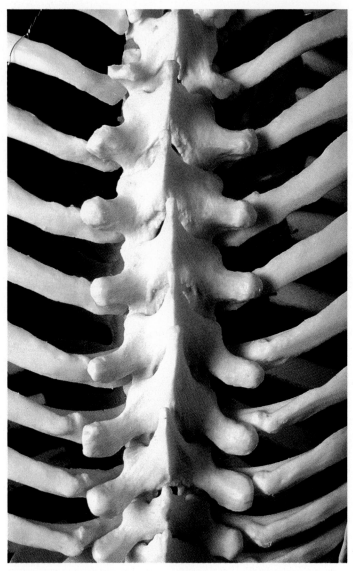

The human vertebral column illustrates that vertebrates have a strong but flexible skeleton. In humans, the vertebral column supports the head and trunk, yet it allows the body to bend forward, backward, and to the side. Only arthropods and vertebrates have a rigid skeleton that is jointed and segmented.

Your study of this chapter will be complete when you can

1. distinguish among the 3 types of skeletons in the animal kingdom, and give examples of animals that have each type;
2. give the functions of a skeletal system in animals;
3. name the 2 parts of the human skeleton, and list the location of the bones for both;
4. describe the anatomy of a long bone and the tissues found in a long bone;
5. explain how bone is continually being broken down and rebuilt;
6. name the different types of joints, and give an example of each;
7. describe the anatomy and location of the 3 types of muscles found in the vertebrate body;
8. describe how whole muscles work in antagonistic pairs;
9. describe the anatomy of a muscle fiber and sarcomere;
10. explain the sliding filament theory of muscle contraction;
11. list the sources of ATP energy for muscle contraction, and explain the occurrence of oxygen debt;
12. list the series of events that occur after a muscle fiber is innervated by nervous stimulation.

The ability to move by means of muscle fibers is one of the distinctive characteristics of animals. Being mobile helps animals obtain food, escape enemies, locate mates, and disperse the species. The title of this chapter—"Musculoskeletal System"—recognizes that the muscular system and the skeletal system work together to provide movement. Nevertheless, we will first consider the skeleton and then show how muscles cause the skeleton to move.

Skeletal System

The skeleton is the framework of the body. It helps protect the internal organs and assists in movement. To produce body movements, the force of muscle contractions must be specifically directed against other parts of the body.

Table 41.1
Classification of Skeletons

Type	Phylum	Example
Hydrostatic skeleton		
gastrovascular cavity	Cnidaria	Hydras, anemones
	Platyhelminthes	Flatworms
coelom*	Annelida	Earthworms
Exoskeleton	Arthropoda	Lobsters, grasshoppers
Endoskeleton	Chordata	Vertebrates

*See figure 26.9.

Figure 41.1
Arthropods have an external skeleton. In order to grow, they must periodically shed this exoskeleton, a process called molting. In this photo, a cicada is just emerging from its discarded skeleton. It will be more vulnerable than usual until its new exoskeleton has hardened.

discarded exoskeleton

Hydras and planarians use their fluid-filled gastrovascular cavity and annelids use their fluid-filled coelom as a hydrostatic skeleton. In these animals, muscular contraction is applied against the fluid-filled cavity (table 41.1). For example, in the earthworm, the contractions of first the circular muscle and then the longitudinal muscle of the body wall enable the worm alternately to extend and shorten. In this way, the earthworm can move forward.

Other members of the animal kingdom have an exoskeleton. The rigid calcium carbonate exoskeleton of a clam (the bivalve shell) is largely for protection. It grows as the animal grows. The chitinous exoskeleton of arthropods is jointed and movable. Chitin is a strong flexible nitrogenous polysaccharide. Arthropods molt to rid themselves of an exoskeleton that has become too small (fig. 41.1).

Vertebrates have an endoskeleton, composed of bone and cartilage, which grows with the animal. It is also jointed, as is the arthropod skeleton. A rigid but flexible skeleton helped the arthropods and vertebrates successfully colonize the terrestrial environment.

> Cnidarians, flatworms, and annelids have a hydrostatic skeleton. Arthropods and vertebrates have a rigid but flexible skeleton that allows complex body movements.

Human Skeleton

Bone and cartilage, the 2 types of tissue found in the human skeleton, were discussed in chapter 33. The human skeleton has many functions. It provides support and protection; it allows movement; it serves as a storage area for calcium and phosphorus salts; and it is the site of blood cell production in adults.

Growth and Development

Most of the bones of the skeleton are cartilaginous during prenatal development. Later, bone-forming cells known as *osteoblasts* replace cartilage with bone. At first, there is only a primary ossification center at the middle of a long bone (fig. 33.4*a*), but later, secondary centers form at the ends of the bones. A *cartilaginous disk* remains between the primary ossification center and each secondary center. The length of a bone is dependent on how long the cartilage cells within the disk continue to divide. Eventually, though, the disks disappear, and the bone stops increasing in size when the individual attains adult height.

In the adult, bone is *continually* being broken down and then built up again. Bone-absorbing cells called *osteoclasts* are derived from cells carried in the bloodstream. As they break down bone, they remove worn cells and deposit calcium in the blood. Apparently, after about 3 weeks, osteoclasts disappear. The destruction caused by the work of osteoclasts is repaired by osteoblasts. As they form new bone, they take calcium from the blood. Eventually some of these cells get caught in the matrix they secrete and are converted to **osteocytes**, the cells found within Haversian systems.

Bones of the Skeleton

The human skeleton can be divided into 2 parts: the **axial skeleton** and the **appendicular skeleton** (fig. 41.2).

Axial Skeleton The axial skeleton includes the skull, the vertebrae, the ribs, and the sternum. The *skull,* or cranium, is composed of many bones fitted tightly together in adults. In newborns, certain bones are not completely formed and instead are joined by membranous regions called *fontanels,* all of which usually close by the age of 16 months. The bones of the skull contain the *sinuses,* air spaces lined by mucous membrane. Two of these, called the mastoid sinuses, drain into the middle ear. Mastoiditis, a condition that can lead to deafness, is an inflammation of these sinuses. Whereas the skull protects the brain, the several bones of the face join together to support and protect the special sense organs and to form the jawbones.

The *vertebral column* extends from the skull to the pelvis and forms a dorsal backbone, which protects the spinal cord (fig. 41.3). Normally, the vertebral column has 4 curvatures, which provide more resiliency and strength than a straight column could. It is composed of many parts called *vertebrae,* which are held together by bony facets, muscles, and strong ligaments. The vertebrae are named according to their location in the body.

Disks between the vertebrae act as padding (fig. 41.3); they prevent the vertebrae from grinding against one another and also absorb shock caused by movements such as running, jumping, and even walking. The disks allow motion between the vertebrae so that we can bend forward, backward, and from side-to-side. Unfortunately, these disks weaken with age and may slip or even rupture. Pain results when the damaged disk presses against the spinal cord and/or spinal nerves. The body may heal itself or the disk can be removed surgically. If surgery occurs, the 2 adjacent vertebrae can be fused together, but this limits the flexibility of the body.

The vertebral column, directly or indirectly, serves as an anchor for all the other bones of the skeleton. All the *ribs* connect directly to the thoracic vertebrae in the back, and all but 2 pairs connect either directly or indirectly via shafts of cartilage to the *sternum* (breastbone) in the front. The lower 2 pairs of ribs are called "floating ribs" because they do not attach to the sternum.

Appendicular Skeleton The appendicular skeleton consists of the bones within the pectoral and pelvic girdles and the attached appendages. The pectoral (shoulder) girdle and appendages (arms and hands) are specialized for flexibility, while the pelvic girdle (innominate bones) and appendages (legs and feet) are specialized for strength.

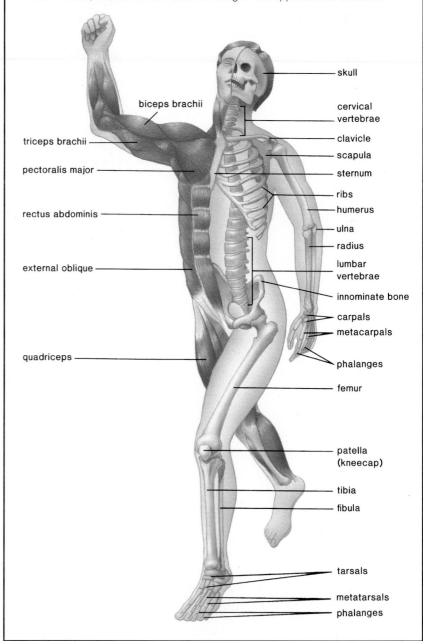

Figure 41.2

Major bones (*right*) and muscles (*left*) of the human body. The axial skeleton, composed of the skull, the vertebral column, the sternum, and the ribs, lies in the midline; the rest of the bones belong to the appendicular skeleton.

The components of the *pectoral girdle* are loosely linked by ligaments rather than firm joints (fig. 41.4). Each *clavicle* (collarbone) connects with the sternum in front and the *scapula* (shoulder blade) behind, but the scapula is freely movable and is held in place by muscles. This allows it to follow the movements of the arm freely. The single long bone in the upper arm, the *humerus,* has a smooth, round head that fits into a socket on the scapula (fig. 41.4). The socket, however, is very shallow and much smaller than the head of the humerus. Although this means that the

Animal Structure and Function

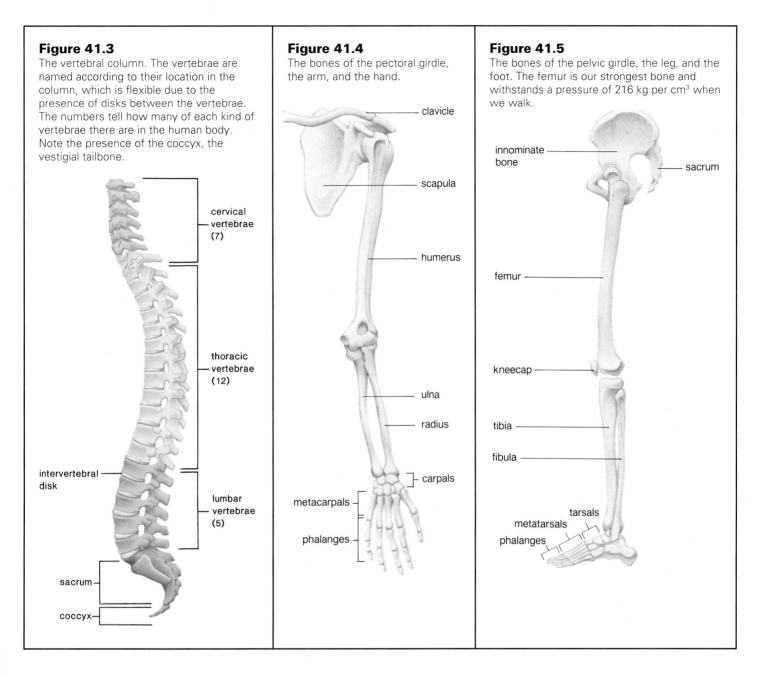

Figure 41.3
The vertebral column. The vertebrae are named according to their location in the column, which is flexible due to the presence of disks between the vertebrae. The numbers tell how many of each kind of vertebrae there are in the human body. Note the presence of the coccyx, the vestigial tailbone.

cervical vertebrae (7)

thoracic vertebrae (12)

intervertebral disk

lumbar vertebrae (5)

sacrum

coccyx

Figure 41.4
The bones of the pectoral girdle, the arm, and the hand.

clavicle

scapula

humerus

ulna

radius

carpals

metacarpals

phalanges

Figure 41.5
The bones of the pelvic girdle, the leg, and the foot. The femur is our strongest bone and withstands a pressure of 216 kg per cm^3 when we walk.

innominate bone

sacrum

femur

kneecap

tibia

fibula

tarsals

metatarsals

phalanges

arm can move in almost any direction, there is little stability. Therefore this is the joint that is most apt to dislocate. The opposite end of the humerus meets the 2 bones of the lower arm, the *ulna* and the *radius,* at the elbow. (The prominent bone in the elbow is the topmost part of the ulna.) When the arm is held so that the palm is turned frontward, the radius and the ulna are about parallel to one another. When the arm is turned so that the palm is next to the body, the radius crosses in front of the ulna, a feature that contributes to the easy twisting motion of the lower arm.

The many bones of the hand increase its flexibility. The wrist has 8 *carpal* bones, which look like small pebbles. From these, 5 *metacarpal* bones fan out to form a framework for the palm. The metacarpal bone that leads to the thumb is placed in such a way that the thumb can reach out and touch the other digits. (*Digits* is a term

that refers to either fingers or toes.) Beyond the metacarpals are the *phalanges,* the bones of the fingers and the thumb. The phalanges of the hand are long, slender, and lightweight.

The *pelvic girdle* consists of 2 heavy, large *innominate* bones, or hipbones (fig. 41.5). The innominate bones are anchored to the *sacrum,* and together these bones form a hollow cavity, the pelvis. The weight of the body is transmitted through the pelvis to the legs and then onto the ground. The largest bone in the body is the *femur,* or thigh bone. Although the femur is a strong bone, it is doubtful that the femurs of a fairy-tale giant could support the increase in weight. If a giant were 10 times taller than an ordinary human being, he would also be about 10 times wider and thicker, making him weigh about 1,000 times as much. This amount of weight would break even giant-size femurs.

Table 41.2
The Major Bones of the Skeleton

Part	Bones
Axial skeleton	Skull, vertebral column, sternum, ribs
Appendicular skeleton	
Pectoral girdle	Clavicle, scapula
Arm	Humerus, ulna, radius
Hand	Carpals, metacarpals, phalanges
Pelvic girdle	Innominate bone
Leg	Femur, tibia, fibula
Foot	Tarsals, metatarsals, phalanges

In the lower leg the larger of the 2 bones, the *tibia*, has a ridge we call the shin (fig. 41.5). Both of the bones of the lower leg have a prominence, which contributes to the ankle—the tibia on the inside of the ankle and the *fibula* on the outside of the ankle. Although there are 7 *tarsal* bones in the ankle, only one receives the weight and passes it on to the heel and the ball of the foot. If you wear high-heeled shoes, your weight is thrown even further toward the front of the foot. The *metatarsal* bones form the arches of the foot—a longitudinal arch from the heel to the toes and a transverse arch across the foot. These provide a stable, springy base for the body. If the tissues that bind the metatarsals together become weakened, flatfeet are apt to result. The bones of the toes are called *phalanges,* just like those of the fingers, but in the foot the phalanges are stout and extremely sturdy.

The axial and appendicular skeletons contain the bones that are listed in table 41.2.

Joints of the Skeleton

Bones are linked at the joints, which are often classified according to the amount of movement they allow. Some bones, such as those that make up the cranium, are sutured together; they are *immovable*. Other joints are *slightly movable,* such as the joints between the vertebrae. The vertebrae are separated by disks, described earlier, which increase their flexibility. Similarly, the 2 hipbones are slightly movable where they are ventrally joined by cartilage. Owing to hormonal changes, this joint becomes more flexible during late pregnancy, which allows the female pelvis to expand during childbirth.

Most joints are *freely movable* or *synovial joints*, in which the 2 bones are separated by a cavity. **Ligaments,** which are composed of fibrous connective tissue, form a capsule, which binds the 2 bones to each other, holding them in place. In a "double-jointed" individual, the ligaments are unusually loose. The joint capsule is lined by synovial membrane, which produces *synovial fluid,* a lubricant for the joint.

Figure 41.6
The knee joint, an example of a freely moveable or synovial joint. Notice that the cavity between the bones is encased by ligaments and lined by synovial membrane. The kneecap protects the joint. Synovial joints are subject to arthritis. In rheumatoid arthritis, the synovial membrane becomes inflamed and thickens. Degenerative changes, which make the joint almost immovable and very painful to use, take place. There is evidence that these effects are brought on by an autoimmune reaction. In old-age arthritis, or osteoarthritis, the cartilage at the ends of the bones disintegrates and the ends of the bones become rough and irregular. This type of arthritis is most likely to affect the joints that have received the greatest use over the years.

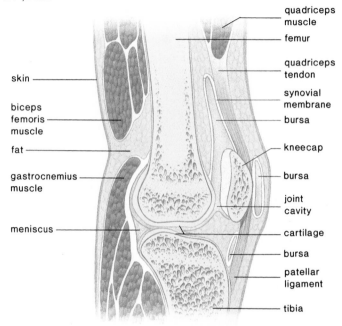

The knee is an example of a synovial joint (fig. 41.6). In the knee, as in other freely movable joints, the bones are capped by cartilage, but the knee also has crescent-shaped pieces of cartilage between the bones, called *menisci* (meniscus, sing.). These add stability, helping to support the weight placed on the knee joint. Unfortunately, athletes often injure the menisci, an injury known as torn cartilage. The knee joint also contains 13 fluid-filled sacs called *bursae* (bursa, sing.), which ease friction between tendons and ligaments and between tendons and bones. Inflammation of bursae is called bursitis. Tennis elbow is a form of bursitis.

There are other types of synovial joints. The knee and elbow joints are *hinge joints* because, like a hinged door, they largely permit movement in one direction only. More movable are the ball-and-socket joints; for example, the ball of the femur fits into a socket on the innominate bone. *Ball-and-socket joints* allow movement in all planes and even a rotational movement.

Joints are classified according to degree of movement. Some joints are immovable, some are slightly movable, and some are freely movable.

Figure 41.7

Figure 41.8

Figure 41.7
Antagonistic muscle pairs. Muscles can exert force only by shortening. Movable joints are supplied with double sets of muscles, which work in opposite directions. As indicated by the arrows, flexor muscles move the lower limb toward the body and extensor muscles move the lower limb away from the body. *a.* Muscles in an insect's leg as an example of antagonistic muscles attached to the inside of an arthropod exoskeleton. *b.* Muscles in the human leg as an example of antagonistic muscles attached to a vertebrate endoskeleton.

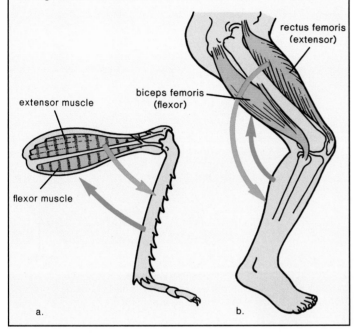

Figure 41.8
Laboratory study of muscle fiber versus whole muscle contraction. *a.* When a muscle fiber is electrically stimulated in the laboratory, it may contract. At first the stimulus may be so weak that no contraction occurs, but as soon as the strength of the stimulus reaches the threshold, the muscle fiber contracts and then relaxes. This action, which is called a muscle fiber twitch, is divided into 3 stages as shown. Increasing the strength of the stimulus does not change the strength of the muscle fiber's contraction; therefore, it is said that a fiber obeys the all-or-none law. If a muscle fiber is given 2 stimuli in quick succession, it does not respond to the second stimulus because it is still recovering from the first stimulus. *b.* A whole muscle behaves quite differently. If the strength of the stimulus is increased, the strength of the response increases. This is because more and more fibers are being stimulated to contract. Also, quickly repeated stimuli to a whole muscle can result in a summation of twitches. This occurs because more fibers can begin to contract while others are relaxing. The muscle as a whole exhibits a greater and greater degree of contraction until tetanus, a sustained contraction, is reached. Eventually the muscle suffers fatigue and begins to relax. Fatigue in isolated muscles is caused by depletion of ATP.

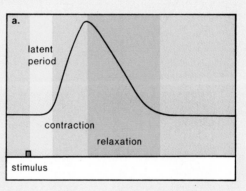

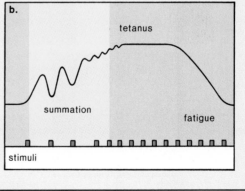

Muscular System

In chapter 33, we discussed the 3 types of muscle in the body of animals. It is skeletal muscle contraction that allows an animal to move, smooth muscle contraction that pushes food along in the digestive tract, and cardiac muscle that allows the heart to pump the blood (fig. 33.6).

Extensive experimental work has been done to understand vertebrate skeletal muscle contraction. It is assumed that the contraction of the other types of muscles is similar to this type of muscle.

Whole Muscle Anatomy and Physiology

Skeletal muscles, which make up over 40% of the body's weight, are attached to the skeleton by **tendons,** made of fibrous connective tissue. When muscles contract, they shorten. Therefore, muscles can only pull; they cannot push. Because of this, skeletal muscles must work in *antagonistic pairs.* For example, one muscle of an antagonistic pair bends the joint and brings the limb toward the body. The other muscle then straightens the joint and extends the limb. Figure 41.7 illustrates this principle and compares the actions of muscles in the limb of an arthropod, in which the muscles are attached to an exoskeleton, with those of a human, in whom the muscles are attached to an endoskeleton.

It is possible to study the contraction of individual whole muscles in the laboratory. Customarily, the calf muscle is removed from a frog and mounted so one end is fixed and the other is movable. The mechanical force of contraction is transduced into an electrical current recorded by an apparatus called a *physiograph* (fig. 41.8). The resulting pattern is called a *myogram.*

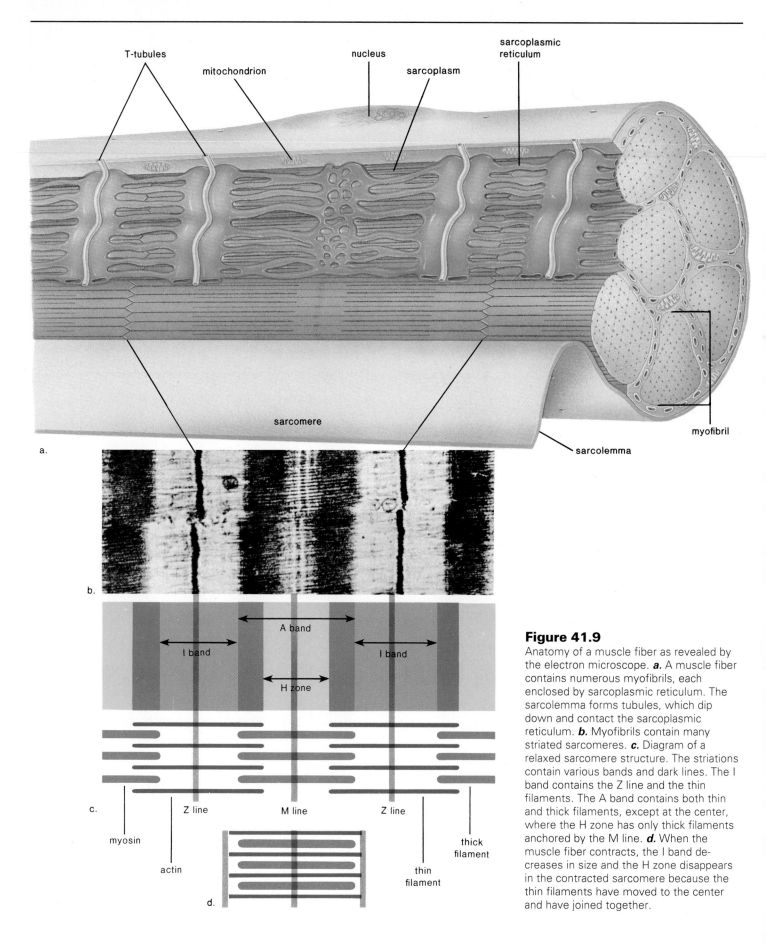

T-tubules

mitochondrion

nucleus

sarcoplasm

sarcoplasmic reticulum

myofibril

sarcolemma

a.

b.

sarcomere

A band

I band

I band

H zone

c.

myosin

actin

Z line

M line

Z line

thick filament

thin filament

d.

Figure 41.9

Anatomy of a muscle fiber as revealed by the electron microscope. *a.* A muscle fiber contains numerous myofibrils, each enclosed by sarcoplasmic reticulum. The sarcolemma forms tubules, which dip down and contact the sarcoplasmic reticulum. *b.* Myofibrils contain many striated sarcomeres. *c.* Diagram of a relaxed sarcomere structure. The striations contain various bands and dark lines. The I band contains the Z line and the thin filaments. The A band contains both thin and thick filaments, except at the center, where the H zone has only thick filaments anchored by the M line. *d.* When the muscle fiber contracts, the I band decreases in size and the H zone disappears in the contracted sarcomere because the thin filaments have moved to the center and have joined together.

Muscle Fiber Anatomy and Physiology

Our understanding of the microscopic anatomy and the physiology of muscle contraction is in part due to the research of 2 English scientists. In the late 1950s, A. F. Huxley studied muscle contraction mainly using the light microscope. H. E. Huxley studied the fine structure of skeletal muscle with the electron microscope.

A whole skeletal muscle is composed of a number of bundles of **muscle fibers.** A muscle fiber (cell) has some unique anatomical characteristics (fig. 41.9). For one thing, it has a T (for transverse) system. The **sarcolemma,** or plasma membrane, forms *T-tubules* that penetrate, or dip down into, the cell so that they come into contact, but do not fuse, with expanded portions of modified endoplasmic reticulum, termed the **sarcoplasmic reticulum.** The expanded portions of the sarcoplasmic reticulum store calcium (Ca⁺⁺) ions, which are essential for muscle contraction. The sarcoplasmic reticulum encases hundreds and sometimes even thousands of myofibrils, the contractile elements of muscle fibers.

Myofibrils

Myofibrils are cylindrical in shape and run the length of the muscle fiber. The light microscope shows that a myofibril has light and dark bands called striations (fig. 33.4*b*). The electron microscope shows that these striations are dependent on the placement of protein filaments within units called **sarcomeres** (fig. 41.9*b*). Notice that a sarcomere extends between 2 dark lines called Z lines. The placement of thick and thin filaments in a sarcomere creates the bands and zones of a sarcomere (fig. 41.9*c*). The I band is light because it contains only thin filaments; the A band is dark because it contains both thin and thick filaments except at the center in the lighter H zone, where only thick filaments are found.

The thick filaments of a sarcomere are made up of a protein called **myosin,** and the thin filaments are made up of a protein called **actin.**

Sarcomere Contraction

A muscle fiber contracts when the sarcomeres within the myofibrils shorten. When sarcomeres shorten, the actin (thin) filaments slide past the myosin (thick) filaments and approach one another (fig. 41.9*d*). This causes the I band to get smaller and the H zone to almost or completely disappear. The movement of actin filaments in relation to myosin filaments is called the *sliding filament theory* of muscle contraction. Notice that after the sarcomere contracts, the filaments are still the same length.

The overall formula for muscle contraction can be represented as follows:

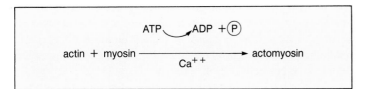

The participants in this reaction have the functions listed in table 41.3. Even though it is the actin filaments that slide past myosin filaments, it is the myosin filaments that do the work. In the presence of calcium (Ca⁺⁺) ions and ATP, portions of a myosin

Table 41.3
Muscle Contraction

Name	Function
Actin filaments	Slide past myosin
Myosin filaments	a. Enzyme that splits ATP b. Pulls actin by means of cross-bridges
Ca⁺⁺	Needed for actin to bind to myosin
ATP	Supplies energy for bonding between actin and myosin = actomyosin

Figure 41.10

Sliding filament theory. *a.* Relaxed sarcomere. *b.* Contracted sarcomere. Note that during contraction, the I band and H zone decrease in size. This indicates that the thin filaments slide past the thick filaments. Even so, the thick filaments do the work by pulling the thin filaments by means of cross-bridges.

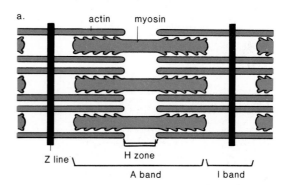

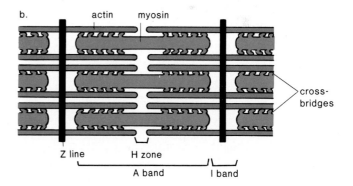

filament called *cross-bridges* bend backward and attach to an actin filament (fig. 41.10). (Each cross-bridge binds to an actin filament at a cross-bridge binding site; actomyosin represents this in the preceding formula.) After attaching, the cross-bridges bend forward and the actin filament is pulled along. Now ATP is broken down by myosin, and detachment occurs. Notice that myosin is not only a structural protein, it is also an ATPase enzyme. The cross-bridges attach and detach some 50–100 times as the thin filaments are pulled to the center of a sarcomere.

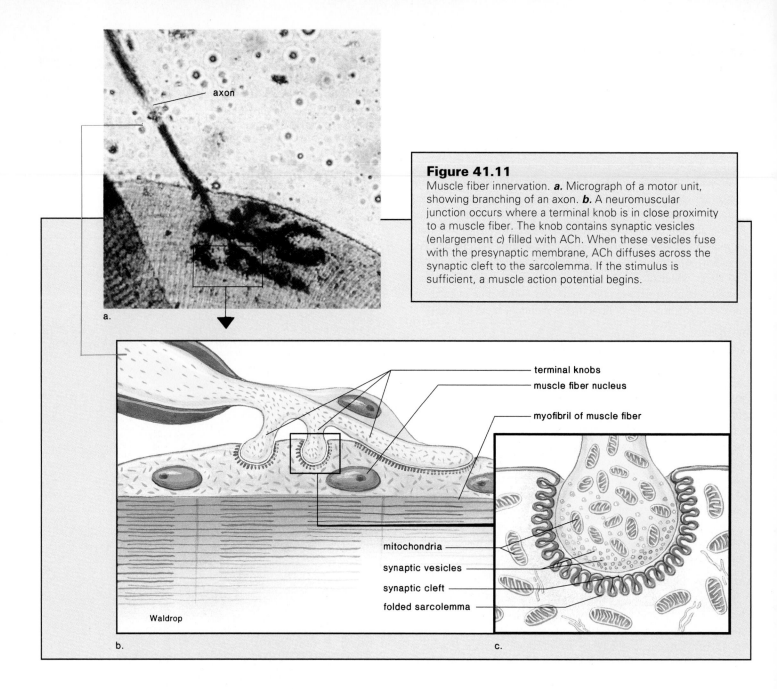

Figure 41.11
Muscle fiber innervation. **a.** Micrograph of a motor unit, showing branching of an axon. **b.** A neuromuscular junction occurs where a terminal knob is in close proximity to a muscle fiber. The knob contains synaptic vesicles (enlargement *c*) filled with ACh. When these vesicles fuse with the presynaptic membrane, ACh diffuses across the synaptic cleft to the sarcolemma. If the stimulus is sufficient, a muscle action potential begins.

axon

a.

terminal knobs
muscle fiber nucleus
myofibril of muscle fiber

mitochondria
synaptic vesicles
synaptic cleft
folded sarcolemma

Waldrop

b.

c.

The sliding filament theory states that actin filaments slide past myosin filaments because myosin has cross-bridges, which pull the actin filaments inward.

It is obvious from our discussion that ATP provides the energy for muscle contraction. To ensure a ready supply of ATP, muscle fibers contain *creatine phosphate* (phosphocreatine), a storage form of high-energy phosphate. Creatine phosphate does not directly participate in muscle contraction. Instead, it is used to regenerate ATP by the following reaction:

$$\text{creatine} \sim \text{P} + \text{ADP} \longrightarrow \text{ATP} + \text{creatine}$$

Oxygen Debt

When all of the creatine phosphate is depleted and no oxygen (O_2) is available for aerobic respiration, a muscle fiber can generate ATP by using fermentation, an anaerobic process. Fermentation, which is apt to occur during strenuous exercise, can supply ATP for only a short time because of lactate buildup. Lactate is a metabolic poison that produces muscle aching and fatigue for a minute or so before it is broken down.

We have all had the experience of having to continue deep breathing for a few minutes following strenuous exercise. We do this because we are in *oxygen debt*—oxygen is needed to complete the metabolism of lactate that has accumulated during exercise. Lactate is transported to the liver, where one-fifth of it is completely broken down to carbon dioxide and water. The

Animal Structure and Function

resulting buildup of ATP by the electron transport system provides the energy to convert the remaining four-fifths of lactate back to glucose.

Muscle contraction requires a ready supply of ATP. Creatine phosphate is used to generate ATP rapidly. If oxygen is in limited supply, fermentation produces ATP and results in oxygen debt.

Innervation

Muscles are innervated; that is, nerve impulses cause muscles to contract. Each motor axon of a nerve branches to several muscle fibers, and collectively, these muscle fibers are called a motor unit. Each branch has several terminal knobs, where there are synaptic vesicles filled with the neurotransmitter acetylcholine (ACh). The region where a terminal knob lies in close proximity to the sarcolemma of a muscle fiber is called a **neuromuscular junction**

(fig. 41.11). A neuromuscular junction has the same components as a synapse; a presynaptic membrane, a synaptic cleft, and a postsynaptic membrane. In this junction, however, the postsynaptic membrane is a portion of the sarcolemma of a muscle fiber.

Nerve impulses cause synaptic vesicles to merge with the presynaptic membrane and release ACh into the synaptic cleft. When ACh reaches the sarcolemma, it is depolarized. The result is a *muscle action potential* that spreads over the sarcolemma and down the T tubules (fig. 41.9a) to where calcium (Ca++) ions are stored in the sarcoplasmic reticulum. Now calcium ions are released, and they diffuse into the sarcoplasm, where they participate in muscle contraction.

It is now necessary to consider the structure of a thin filament in more detail. Figure 41.12 shows the placement of 2 other proteins associated with a thin filament (the double row of twisted globular actin molecules). Threads of tropomyosin wind about a thin filament, and troponin occurs at intervals along the threads. After calcium ions are released, they combine with troponin. After binding occurs, the tropomyosin threads shift their position, and the cross-bridge binding sites are exposed.

The thick filament is a bundle of myosin molecules, each having a globular head. Each head is a cross-bridge that has an ATP-binding site. After ATP attaches to ATP-binding sites, the cross-bridges bend backward and attach to the cross-bridge binding sites on the actin filaments. After attachment occurs, the cross-bridges bend forward, pulling the actin filaments a short distance. Then the myosin heads break down ATP, and detachment of the cross-bridges occurs. The actin filaments move nearer the center of the sarcomere each time the cycle is repeated.

The movement of the actin filaments causes muscle contraction. Contraction ceases when nerve impulses no longer stimulate the muscle fiber. With the cessation of a muscle action potential, calcium ions are pumped back into the sarcoplasmic reticulum by active transport. Relaxation now occurs.

A neuromuscular junction functions like a synapse except that a muscle action potential causes calcium ions to be released from sarcoplasmic reticulum, and thereafter muscle contraction occurs.

Figure 41.12

Detailed structure and function of sarcomere contraction. After calcium (Ca++) ions are released from the sarcoplasmic reticulum, they combine with troponin, a protein that occurs periodically along tropomyosin threads. This causes the tropomyosin threads to shift their position so that cross-bridge binding sites are revealed along the actin filaments. The myosin filament extends its globular heads, forming cross-bridges, which bind to these sites. The breakdown of ATP by myosin causes the cross-bridges to detach and to reattach farther along the actin. In this way, the actin filaments are pulled along past the myosin filaments.

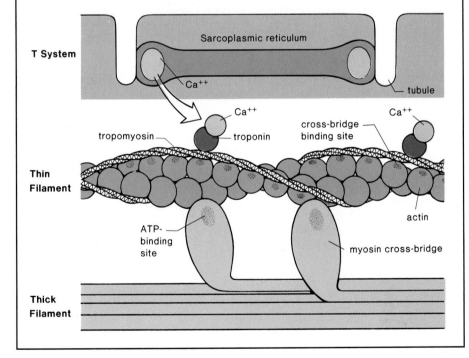

Summary

1. The 3 types of skeletons that are found in the animal kingdom are the hydrostatic skeleton (flatworms and segmented worms), the exoskeleton (arthropods), and the endoskeleton (vertebrates).

2. A rigid skeleton gives support to the body, helps protect internal organs, and assists movement. In humans, the skeleton is also a storage area for calcium and phosphorus salts and the site of blood cell production.

3. Bone is constantly being renewed; osteoclasts break down bone and osteoblasts build new bone. Osteocytes are in the lacunae of Haversian systems.

4. The human skeleton is divided into 2 parts: (1) the axial skeleton, which is made up of the skull, ribs, and vertebrae, and (2) the appendicular skeleton, which is composed of the appendages and their girdles.

5. Joints are classified as immovable, like those of the cranium; slightly movable, like those between the vertebrae; and freely movable or synovial joints, like those in the knee and hip. In synovial joints, ligaments bind the 2 bones together, forming a capsule in which there is synovial fluid.

6. Skeletal muscles can only pull bones; they cannot push. Therefore, they work in antagonistic pairs.

7. Muscle fibers contain a T system: the sarcolemma forms tubules, which contact the sarcoplasmic reticulum. Here calcium ions are stored.

8. The sarcoplasmic reticulum encases myofibrils, long contractile units having longitudinal subunits called sarcomeres. The anatomy of the sarcomere shows the placement of the protein filaments actin and myosin. It is the placement of these that results in the striations of skeletal muscle. Actin (thin) filaments are attached to the Z line, and both actin and myosin (thick) filaments are in the A band. The H zone contains only myosin filaments.

9. The sliding filament theory of muscle contraction says that the myosin filaments have cross-bridges, which attach to and detach from the actin filaments causing them to slide and the sarcomere to shorten. The H zone disappears as the actin filaments approach one another.

10. ATP is the only source of energy for muscle contraction. The globular heads of myosin break down ATP when they detach from the actin filament. Creatine phosphate is a storage form of high-energy phosphate in muscle tissue. When there is no more freely available ATP, fermentation occurs and the organism goes into oxygen debt.

11. Innervation of a muscle fiber begins at a neuromuscular junction. Here synaptic vesicles release ACh into the synaptic cleft. When the sarcolemma receives the ACh, a muscle action potential moves down the T system to the calcium-storage sacs.

12. After calcium ions are released, contraction occurs because calcium combines with troponin, a protein that occurs on tropomyosin threads on the actin filament. This causes the threads to shift their position so that myosin binding sites on actin become available. When calcium ions are actively transported back into the storage sacs, muscle relaxation occurs.

Writing Across the Curriculum

In order to practice writing skills, students should write out the answers to any or all of the study questions and the critical thinking questions. The study questions are sequenced in the same order as the text. Suggested answers to the critical thinking questions are in appendix D.

Study Questions

1. What are the 3 types of skeletons found in the animal kingdom, how do they differ, and what are some animals having each type?
2. Give several functions of the skeletal system in humans.
3. Distinguish between the axial and appendicular skeletons.
4. List the bones that form the pectoral and pelvic girdles.
5. Describe the anatomy of a long bone, and tell how bone is continually being rejuvenated.
6. Name the different types of joints, and give an example of each type.
7. Why do muscles act in antagonistic pairs?
8. Describe the microscopic anatomy of a muscle fiber and the structure of a sarcomere. What is the sliding filament theory?
9. Discuss the availability and the specific role of ATP during muscle contraction. What is oxygen debt and how is it repaid?
10. What causes a muscle action potential? How does the muscle action potential bring about sarcomere and muscle fiber contraction?

Animal Structure and Function

Objective Questions

For questions 1-4, match the following items with the correct locations given in the key.

Key:

   a. upper arm
   b. lower arm
   c. pectoral girdle
   d. pelvic girdle
   e. upper leg
   f. lower leg

1. ulna
2. tibia
3. clavicle
4. femur
5. Spongy bone
   a. contains Haversian systems.
   b. contains red marrow where blood cells are formed.
   c. lends no strength to bones.
   d. All of these.
6. Which of these is mismatched?
   a. slightly movable joint—vertebrae
   b. hinge joint—hip joint
   c. synovial joint—elbow
   d. immovable joint—sutures in cranium
7. In a muscle fiber
   a. the sarcolemma is connective tissue holding the myofibrils together.
   b. the T system contains calcium storage sacs.
   c. both filaments have cross-bridges.
   d. All of these.

8. When muscles contract
   a. sarcomeres increase in size.
   b. myosin slides past actin.
   c. the H zone disappears.
   d. calcium is taken up by calcium storage sacs.
9. Which of these is a source of energy for muscle contraction?
   a. ATP
   b. creatine phosphate
   c. lactic acid
   d. Both a and b.

10. Nervous stimulation of muscles
   a. occurs at a neuromuscular junction.
   b. results in an action potential that travels down the T system.
   c. causes calcium to be released from storage sacs.
   d. All of these.
11. Give the function of each participant in the following reaction.

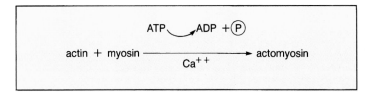

12. Label this diagram of a sarcomere.

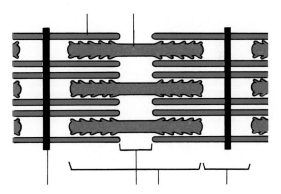

Concepts and Critical Thinking

1. *A bony skeleton determines the shape of an animal.*

How does the skeleton aid adaptation to the environment?

2. *Movement is fundamental to the nature of animals.*

Why is movement in animals more essential than in plants?

3. *Locomotion requires energy.*

How do animals acquire the energy for locomotion, and specifically how is energy used to bring about locomotion?

Selected Key Terms

axial skeleton (ak'se-al skel'ē-ton) 666
appendicular skeleton (ap"en-dik'u-lar skel'ē-ton) 666
muscle fiber (mus'el fi'ber) 671

sarcolemma (sar"ko-lem'ah) 671
sarcoplasmic reticulum (sar"ko-plaz'mik rē-tik'u-lum) 671
myofibril (mi'o-fi'bril) 671
sarcomere (sar"ko-mēr) 671

myosin (mi'o-sin) 671
actin (ak'tin) 671
neuromuscular junction (nu"ro-mus'ku-lar junk'shun) 673

42

Endocrine System

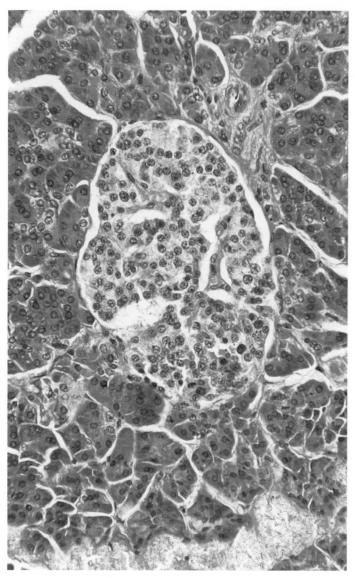

The pancreatic islets are specialized regions of the pancreas, an endocrine gland. The islets secrete the hormone insulin when the blood glucose level is high. Insulin increases the passage of glucose into cells, restoring the normal level. The endocrine glands regulate and coordinate body functions. Magnification, X250.

Your study of this chapter will be complete when you can

1. discuss invertebrate endocrine systems;
2. describe the location of each of the major endocrine glands of the human body;
3. chemically distinguish between the 2 major types of vertebrate hormones, and tell how each type brings about its effect on the cell;
4. contrast the ways in which the hypothalamus controls the posterior and anterior pituitary;
5. list the most important hormones produced by each endocrine gland, and describe their most important effects;
6. give examples of medical conditions associated with the overproduction or underproduction of specific hormones;
7. explain the concept of control by negative feedback, and give an example that involves the hypothalamus, anterior pituitary, and a gland controlled by the anterior pituitary; also give examples that do not involve this 3-tiered system;
8. discuss the relationship between the nervous system and the endocrine system; give examples to show that there are regions of overlap between their functions;
9. discuss the concept of environmental signals in general, and group environmental signals into 3 categories; discuss why the term *hormone* could be broadened to include all these categories.

The **endocrine system**, along with the nervous system, coordinates the various activities of body parts. An animal must actively find food, avoid predators, and find a mate. The internal environment must be maintained within certain limits as food is digested, nutrient molecules and oxygen are distributed to the cells, and wastes eliminated. Both the endocrine and nervous systems utilize chemical messengers to coordinate these activities. In chapter 39, we discussed the neurotransmitters that are released by one neuron and influence the excitability of other neurons. In contrast, the endocrine system utilizes hormones, chemical messengers that are typically released directly into the bloodstream.

The nervous system reacts quickly to external and internal stimuli; you rapidly pull your hand away from a hot stove, for example. The endocrine system is somewhat slower because it takes time for a hormone to travel through the circulatory system to its target organ. You might think from the use of this terminology that the hormone is seeking out a particular organ, but quite the contrary, the organ is awaiting the arrival of the hormone. Cells that can react to a hormone have specific receptors, which combine with the hormone in a lock-and-key manner. Therefore, certain cells respond to one hormone and not to another, depending on their receptors.

Sometimes a hormone is defined as a chemical produced by one set of cells that affects a different set. The problem with this definition of it, however, is that it would allow us to categorize all sorts of chemical messengers as hormones, even neurotransmitters! We will have more to say about this later, but a certain amount of overlap between the nervous and endocrine systems is to be expected. The systems evolved together, and at the same time, no doubt making occasional use of the same chemical messengers and communicating not only with other systems but with each other as well. Later, we will give several examples of such associations between the 2 systems.

Both the nervous and endocrine systems function to coordinate body parts and activities. Perhaps since they evolved at the same time, a certain amount of overlap between the chemical messengers utilized and the types of activities performed by the 2 systems is to be expected.

Invertebrate Endocrine Systems

In the first half of this century, biologists showed that many invertebrates have an endocrine system. In invertebrates, hormones are produced only by neurosecretory cells, specialized neurons capable of synthesizing and secreting hormones. Neurosecretions are discharged directly into the blood. This lends support to the hypothesis that hormones first evolved as neurosecretions and only later did nonnervous endocrine glands appear, especially among the vertebrates.

Neurosecretions have been extensively studied in insects. Insects, being arthropods, have an exoskeleton. Typically, they undergo a series of larval stages marked by molting before

Figure 42.1

Three hormones are involved in insect development. Brain hormone controls the secretion of ecdysone, the molt and maturation hormone. Ecdysone causes the insect to pupate if the level of juvenile hormone is low, and to metamorphose if juvenile hormone is absent.

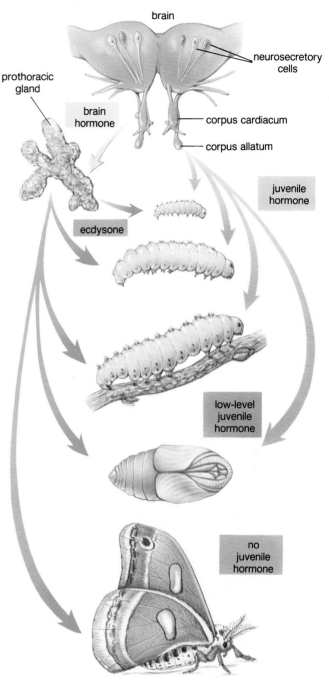

they pupate. V. B. Wigglesworth, who performed experiments on a bloodsucking bug (*Rhodnius*) in the 1930s, showed that the brain is necessary for maturation to take place because it produces a hormone appropriately called *brain hormone* (fig. 42.1). Brain hormone stimulates the prothoracic gland, which lies in the region just behind the head, to secrete *ecdysone*, a steroid hormone. This hormone is also called the molt and maturation hormone because it promotes both molting and maturation. During larval stages, however, response to ecdysone is modified by the action of *juvenile hormone*. If juvenile hormone is present, the insect molts into another larval form; if it is minimally present, the insect pupates; and if it is absent, the insect undergoes metamorphosis into an adult.

Some plants produce compounds that are similar or identical to either ecdysone or juvenile hormone. These compounds apparently protect the plants by disrupting the development of the insects that feed upon them. Work is underway to extract these compounds and to use them as insecticides.

Vertebrate Endocrine Systems

Vertebrate endocrine glands can be contrasted with exocrine glands, which have ducts and secrete their products into these ducts for transport into body cavities; for example, the salivary glands send saliva into the mouth by way of the salivary ducts. Endocrine glands are ductless; they secrete their hormones directly into the bloodstream for distribution throughout the body.

The human endocrine system can be used to exemplify the vertebrate endocrine system (fig. 42.2). Even so, there are marked differences in the vertebrate hormones. For example, prolactin in humans stimulates female breasts to secrete milk, but

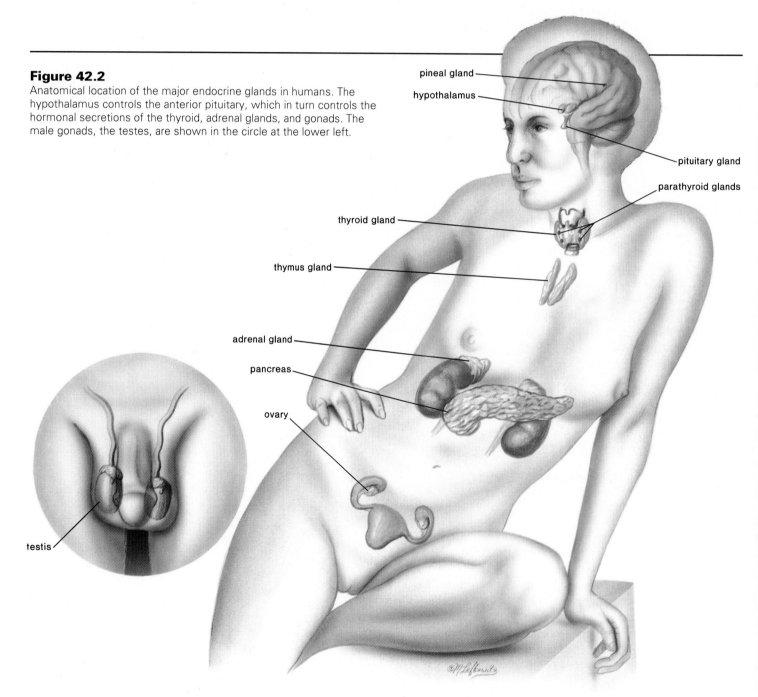

Figure 42.2

Anatomical location of the major endocrine glands in humans. The hypothalamus controls the anterior pituitary, which in turn controls the hormonal secretions of the thyroid, adrenal glands, and gonads. The male gonads, the testes, are shown in the circle at the lower left.

pineal gland

hypothalamus

pituitary gland

parathyroid glands

thyroid gland

thymus gland

adrenal gland

pancreas

ovary

testis

Animal Structure and Function

in pigeons it stimulates the secretion of crop milk, a product of the gut. Some hormones have an entirely different function; for example, thyroxin in humans stimulates metabolism, but in frogs it induces metamorphosis or the transformation of tadpole to adult. Other hormones are species specific, that is, they function only in one species.

Mechanism of Hormonal Action

Table 42.1 indicates that many of the human hormones are proteins or peptides. These molecules are coded for by genes and are synthesized at the ribosomes (see fig. 16.13). Eventually, they are packaged into vesicles at the Golgi apparatus and are secreted at the plasma membrane. There are some other hormones, called catecholamines, that are derived from the amino acid tyrosine; their production requires only a series of metabolic reactions within the cytoplasm. From the viewpoint of hormonal action, we can group the protein, peptide, and catecholamines together and refer to them as *peptide hormones*.

The hormones produced by the adrenal cortex, the ovaries, and the testes are steroids. *Steroid hormones* are derived from cholesterol (see fig. 4.13) by a series of metabolic reactions. These hormones are stored in fat droplets in the cell cytoplasm until their release at the plasma membrane.

Peptide Hormones

The mode of action of peptide hormones was discovered in the 1950s by Earl W. Sutherland, who studied the effects of epinephrine on liver cells. Sutherland received a Nobel Prize for his hypothesis that when this hormone binds to a cell-surface receptor, the resulting complex leads to the activation of an enzyme that produces **cyclic AMP** (fig. 42.3a). Cyclic AMP (cAMP) is a compound made from ATP, but it contains only one phosphate group, which is attached to adenosine at 2 locations.

Since peptide hormones never enter the cell, they are sometimes called the first messengers, while cAMP, which sets the metabolic machinery in motion, is called the second messenger (fig. 42.3). Other messengers have been discovered in cells. Inositol triphosphate (IP_3) is a second messenger that causes the release of calcium (Ca^{++}) ions in muscle cells, for example. Since calcium goes on to activate other proteins, it is called the third messenger.

cAMP operates in a second messenger system that sets an *enzyme cascade* into motion. cAMP activates only one particular enzyme in the cell, but this enzyme in turn activates another, and so forth. The activated enzymes, of course, can be used repeatedly. Therefore, at every step in the enzyme cascade, more and more reactions take place—the binding of a single hormone molecule eventually results in a 1,000-fold response.

Steroid Hormones

Steroid hormones do not bind to cell-surface receptors; they can enter the cell freely (fig. 42.3b). For example, scans of the reproductive organs clearly show that the hormones estrogen and progesterone enter the cells of the reproductive organs. Once inside, steroid hormones bind to receptors in the cytoplasm. The hormone-receptor complex then enters the nucleus, where it binds

---

Figure 42.3

Cellular activity of hormones. *a.* Peptide hormones combine with receptors located in the plasma membrane. This promotes the production of cAMP, which in turn leads to activation of a particular enzyme. *b.* Steroid hormones pass through the plasma membrane to combine with receptors; the complex activates certain genes, leading to protein synthesis.

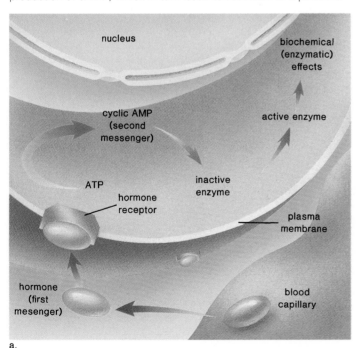

a.

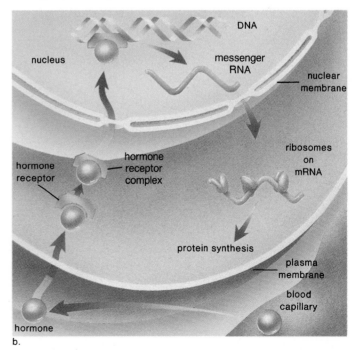

b.

Table 42.1
Human Endocrine Glands and Their Hormones

Endocrine Gland	Hormone Released	Chemical Structure of Hormone	Chief Function of Hormone
Hypothalamus	Hypothalamic-releasing hormones Hypothalamic-release-inhibiting hormones	Peptides	Regulate anterior pituitary hormones
Anterior pituitary	Thyroid-stimulating (TSH, thyrotropic)	Glycoprotein	Stimulates thyroid
	Adrenocorticotropic (ACTH)	Polypeptide	Stimulates adrenal cortex
	Gonadotropic follicle-stimulating (FSH)	Glycoprotein	Controls egg and sperm production
	luteinizing (LH)		Controls sex hormone production
	Prolactin (Pr)	Protein	Stimulates milk production and secretion
	Growth (GH, somatotropic)	Protein	Stimulates cell division, protein synthesis, and bone growth
	Melanocyte-stimulating (MSH)	Polypeptide	Unknown function in humans; regulates skin color in lower vertebrates
Posterior pituitary (storage of hypothalamic hormones)	Antidiuretic (ADH, vasopressin)	Peptide	Stimulates water reabsorption by kidneys
	Oxytocin	Peptide	Stimulates uterine muscle contraction and release of milk by mammary glands
Thyroid	Thyroxin and triiodothyronine	Iodinated amino acid	Increases metabolic rate; helps regulate growth and development
	Calcitonin	Peptide	Lowers blood calcium level
Parathyroids	Parathyroid (PTH)	Polypeptide	Raises blood calcium level
Adrenal cortex	Glucocorticoids (cortisol)	Steroids	Raise blood glucose level; stimulate breakdown of protein
	Mineralocorticoids (aldosterone)	Steroids	Stimulate kidneys to reabsorb sodium and to excrete potassium
	Sex hormones	Steroids	Stimulate development of secondary sex characteristics (particularly in male)
Adrenal medulla	Epinephrine and norepinephrine	Catecholamines	Stimulate fight or flight reactions; raise blood glucose level
Pancreas	Insulin	Polypeptide	Lowers blood glucose level; promotes formation of glycogen, proteins, and fats
	Glucagon	Polypeptide	Raises blood glucose level; promotes breakdown of glycogen, proteins, and fats
Gonads			
Testes	Androgens (testosterone)	Steroids	Stimulate spermatogenesis; develop and maintain secondary male sex characteristics
Ovaries	Estrogen and progesterone	Steroids	Stimulate growth of uterine lining; develop and maintain secondary female sex characteristics
Thymus	Thymosins	Peptides	Stimulate production and maturation of T lymphocytes
Pineal gland	Melatonin	Catecholamine	Involved in circadian and circannual rhythms; possibly involved in maturation of sex organs
Gut*	Gastrin	Peptide	Stimulates secretion of gastric juices
	Secretin	Peptide	Stimulates secretion of pancreatic juices
	Cholecystokinin-pancreozymin (CCK-PZ)	Peptide	Stimulates secretion of pancreatic juices and flow of bile from gallbladder

*The hormones produced by the gut are discussed in chapter 36.

Animal Structure and Function

with chromatin at a location that promotes activation of particular genes. Transcription of DNA and translation of mRNA follow. In this manner, steroid hormones lead to protein synthesis.

Steroids act more slowly than peptides because it takes more time to synthesize new proteins than to activate enzymes that are already present in the cell. But steroids have a more *sustained* effect on the metabolism of the cell than do peptide hormones.

> Hormones are chemical messengers that influence the metabolism of the recipient cell. Peptide hormones activate existing enzymes in the cell, and steroid hormones bring about the synthesis of new proteins.

Hypothalamus and the Pituitary Gland

The hypothalamus is the portion of the brain (fig. 39.10) that regulates the internal environment; for example, it helps to control the heart rate, body temperature, and water balance, as well as the glandular secretions of the **pituitary gland.** The pituitary, a small gland about 1 cm in diameter, lies just below the hypothalamus and is divided into 2 portions called the posterior pituitary and the anterior pituitary.

Posterior Pituitary

The posterior pituitary is connected to the hypothalamus by means of a stalklike structure. There are neurons in the hypothalamus that are called neurosecretory cells because they both respond to neurotransmitter substances and produce the hormones that are stored in and released from the posterior pituitary. The hormones pass from the hypothalamus through axons that terminate in the posterior pituitary (fig. 42.4).

The axon endings in the posterior pituitary store **antidiuretic hormone (ADH),** sometimes called vasopressin, and oxytocin. ADH, as discussed in chapter 38, promotes the reabsorption of water from the collecting duct, a portion of the nephrons within the kidneys. The hypothalamus contains other nerve cells that act as a sensor because they are sensitive to the osmolarity of the blood. When these cells determine that the blood is too concentrated, ADH is released into the bloodstream from the axon endings in the posterior pituitary. As the blood becomes dilute, the hormone no longer is released. This is an example of control by negative feedback because the effect of the hormone (dilute blood) acts to shut down the release of the hormone. Negative feedback mechanisms regulate the activities of most endocrine glands.

Oxytocin is the other hormone that is made in the hypothalamus and stored in the posterior pituitary. Oxytocin causes the uterus to contract and is used to artificially induce labor. It also stimulates the release of milk from the mammary glands when a baby is nursing.

> The posterior pituitary stores 2 hormones, ADH and oxytocin, both of which are produced by and released from neurosecretory cells in the hypothalamus.

Figure 42.4
The hypothalamus produces ADH and oxytocin, 2 hormones that are stored in and secreted by the posterior pituitary.

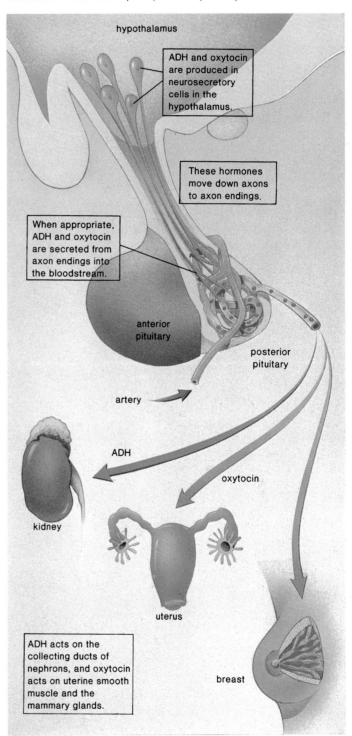

Figure 42.5

The hypothalamus also controls the secretions of the anterior pituitary.

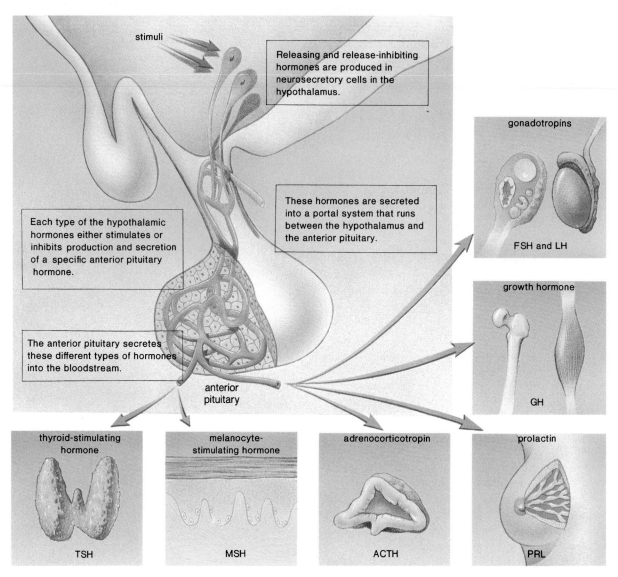

Anterior Pituitary

By the 1930s, it had been determined that the hypothalamus also controls the release of the anterior pituitary hormones. For example, electrical stimulation of the hypothalamus causes the release of anterior pituitary hormones, but *direct* stimulation of the anterior pituitary itself has no effect. Detailed anatomical studies showed that there is a *portal* system consisting of blood vessels connecting a capillary bed in the hypothalamus with one in the anterior pituitary (fig. 42.5). The hypothesis was put forward that a different set of neurosecretory cells in the hypothalamus synthesize a number of different releasing hormones that are sent to the anterior pituitary by way of the vascular portal system. There began a long and competitive struggle between 2 groups of investigators to identify one of these *hypothalamic-releasing hormones,* which

cause the anterior pituitary to release hormones. The group headed by R. Guillemin processed nearly 2 million sheep hypothalami, and the group headed by A. V. Schally processed more than 1 million pig hypothalami until each group announced, in November of 1969, that it had isolated and determined the structure of one of these hormones, a peptide containing only 3 amino acids.

Later, it was found that some of the hypothalamic hormones inhibit the release of anterior pituitary hormones. Therefore, today, it is customary to speak of *hypothalamic-releasing hormones* and *hypothalamic-release-inhibiting hormones.* For example, there is a **gonadotropic-releasing hormone (GNRH)** and a gonadotropic-release-inhibiting hormone (GnRIH). The first hormone stimulates the anterior pituitary to release gonadotropic hormones, and the second inhibits the anterior pituitary from releasing the same hormones.

Figure 42.6

The difference in size between a giant and a dwarf sometimes can be explained by a difference in production of growth hormone by the anterior pituitary. Sandy Allen is one of the world's tallest women due to a higher than usual amount of growth hormone produced by the anterior pituitary.

The anterior pituitary is controlled by the hypothalamic-releasing hormones. These hormones are produced in neurosecretory cells in the hypothalamus and pass to the anterior pituitary by way of a vascular portal system.

The anterior pituitary produces at least 6 different types of hormones, each by a distinct cell type (fig. 42.5). Three of these hormones have a direct effect on the body. **Growth hormone (GH),** or somatotropic hormone, dramatically affects physical appearance since it determines the height of the individual (fig. 42.6). If too little GH is produced during childhood, the individual becomes a pituitary dwarf, and if too much is produced, the individual is a pituitary giant. In both instances, the individual has normal body proportions. On occasion, however, there is overproduction of growth hormone in the adult and a condition called *acromegaly* results. Since only the feet, hands, and face (particularly chin, nose, and eyebrow ridges) can respond, these portions of the body become overly large.

Growth hormone promotes cell division, protein synthesis, and bone growth. It stimulates the transport of amino acids into cells and increases the activity of ribosomes, both of which are essential to protein synthesis. In bones, it promotes growth of the cartilaginous plates and causes osteoblasts to form bone. Evidence suggests that the effects on cartilage and bone may actually be due to hormones called somatomedins, which are released by the liver. Growth hormone causes the liver to release somatomedins.

Prolactin (Pr), is produced in quantity only after childbirth. It causes the mammary glands in the breasts to develop and produce milk. It also plays a role in carbohydrate and fat metabolism.

Melanocyte-stimulating hormone (MSH) causes skin color changes in many fishes, amphibians, and reptiles who have melanophores, special skin cells that produce color variations. The hormone is present in humans, but no one knows what its function might be. It is derived, however, from a molecule that is also the precursor for both ACTH and the anterior pituitary endorphins. These endorphins are structurally and functionally similar to the endorphins produced in brain nerve cells.

Regulation of the Anterior Pituitary

The anterior pituitary is sometimes called the *master gland* because it controls the secretion of some other endocrine glands (fig. 42.5). As indicated in table 42.1, the anterior pituitary secretes the following hormones, which have an effect on other glands:

1. **TSH,** thyroid-stimulating hormone
2. **ACTH,** adrenocorticotropic hormone, which stimulates the adrenal cortex
3. **Gonadotropic hormones** (FSH and LH), which stimulate the gonads—the testes in males and the ovaries in females

TSH causes the thyroid to produce thyroxin; ACTH causes the adrenal cortex to produce cortisol; and gonadotropic hormones cause the gonads to secrete sex hormones. A 3-tiered relationship exists among the hypothalamus, anterior pituitary, and other endocrine glands; the hypothalamus produces releasing hormones, which control the anterior pituitary, and the anterior pituitary produces hormones that control the thyroid, adrenal cortex, and gonads. Figure 42.7 illustrates the negative feedback mechanism that controls the activity of these glands.

The hypothalamus, the anterior pituitary, and the other endocrine glands controlled by the anterior pituitary are all involved in a self-regulating negative feedback loop.

Thyroid Gland

The **thyroid gland** is a large gland located in the neck, where it is attached to the trachea just below the larynx (fig. 42.8a). The 2 hormones produced by the thyroid both contain iodine: **thyroxin,** or T_4, contains 4 atoms of iodine; it is secreted in greater amounts but is less potent than triiodothyronine, or T_3, which has only 3 atoms of iodine. Iodine is actively transported into the thyroid gland, and it may reach a concentration as much as 25 times greater than that of the blood (see fig. 3.4).

Even before the structure of thyroxin was known, it was surmised that the hormone contained iodine because when iodine is lacking in the diet, the thyroid gland enlarges, producing a **goiter** (fig. 42.9). The cause of the enlargement becomes clear if we refer to figure 42.7. When there is a low level of thyroxin in the blood,

Figure 42.7

TRH (thyroid-releasing hormone) stimulates the anterior pituitary, and TSH (thyroid-stimulating hormone) stimulates the thyroid to secrete thyroxin. The level of thyroxin in the body is controlled by negative feedback: *a.* the level of TSH exerts negative feedback control over the hypothalamus; *b.* the level of thyroxin exerts feedback control over the anterior pituitary; and *c.* the level of thyroxin exerts feedback control over the hypothalamus. In this way, thyroxin controls its own secretion. Cortisol and sex hormone levels are controlled in similar ways.

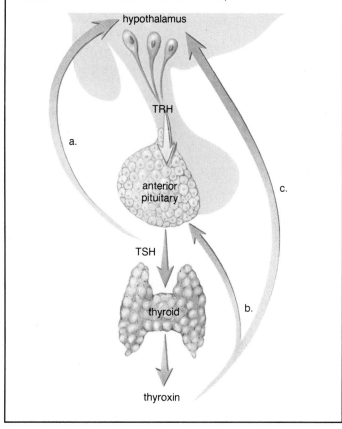

Figure 42.8

a. The thyroid gland is located in the neck in front of the trachea. *b.* The 4 parathyroid glands are embedded in the posterior surface of the thyroid gland, yet the parathyroids and thyroid glands have no anatomical or physiological connection with one another. *c.* Regulation of parathyroid hormone secretion. A low blood level of calcium causes the parathyroids to secrete PTH, which causes the kidneys and gut to retain calcium and osteoclasts to break down bone. The end result is an increased level of calcium in the blood. A high blood level of calcium inhibits secretion of PTH.

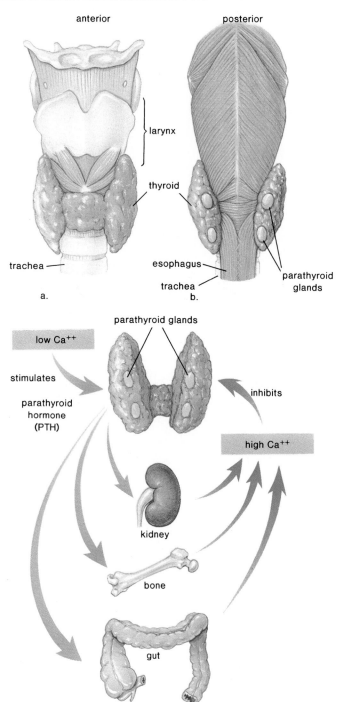

the anterior pituitary is stimulated to produce TSH (thyroid-stimulating hormone). The thyroid responds by increasing in size, but this increase in size is ineffective because active thyroxin cannot be produced without iodine. Later, it was found that goiter can be prevented by consumption of iodized salt.

Thyroxin is necessary in vertebrates for proper growth and development. For example, without thyroxin, frogs do not metamorphose properly and humans do not mature properly. Cretinism occurs in individuals who have suffered from *hypothyroidism* (low thyroid function) since birth. They show reduced skeletal growth, sexual immaturity, and abnormal protein metabolism, which leads to mental retardation.

In general, thyroxin increases the metabolic rate in all cells; the number of respiratory enzymes increases, as does oxygen uptake. Hypothyroidism (too little thyroxin) in adults produces the condition known as myxedema, which is characterized by leth-

Animal Structure and Function

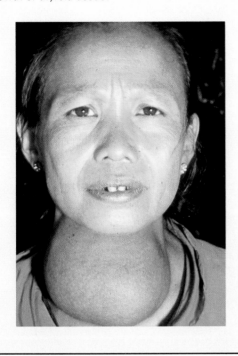

argy, weight gain, loss of hair, slow pulse, and decreased body temperature. In hyperthyroidism (too much thyroxin), the thyroid gland is enlarged and overactive, causing a goiter to form and the eyes to protrude because of edema in the tissues of the eye sockets and swelling of muscles that move the eyes. This type of goiter is called *exophthalmic goiter*. The patient usually becomes hyperactive, nervous, irritable, and suffers from insomnia. Removal or destruction of a portion of the thyroid by means of radioactive iodine sometimes is effective in curing the condition.

In addition to thyroxin, the thyroid gland also produces the hormone *calcitonin*. This hormone lowers the level of calcium in the blood and opposes the action of the parathyroid hormone, which we will discuss in the next section.

> The anterior pituitary produces TSH, a hormone that promotes the production of thyroxin (and triiodothyronine) by the thyroid. Thyroxin helps regulate growth and development in immature animals and speeds up metabolism in all animals.

Parathyroid Glands

The 4 **parathyroid glands** are embedded in the posterior surface of the thyroid gland, as shown in figure 42.8b. They produce a hormone called **parathyroid hormone (PTH).** Under the influence of PTH, the calcium (Ca^{++}) level in the blood increases and the phosphate (PO_4^{---}) level decreases. PTH stimulates the *absorption* of calcium from the gut by activating vitamin D, the *retention* of

calcium (the excretion of phosphate) by the kidneys, and the *demineralization* of the bone by promoting the activity of the osteoclasts, the bone-reabsorbing cells. When the blood calcium level reaches the appropriate level, the parathyroid glands no longer produce PTH (fig. 42.8c).

If PTH is not produced in response to low blood calcium, *tetany* results because calcium plays an important role in both nervous conduction and muscle contraction. In tetany, the body shakes from continuous muscle contraction. The effect is brought about by increased excitability of the nerves, which fire spontaneously and without rest.

> Parathyroid hormone (PTH) maintains a high blood calcium level. Its actions are opposed by calcitonin, which is produced by the thyroid.

Adrenal Glands

Each of the 2 **adrenal glands,** as the name implies (*ad*—near; *renal*—kidney), lies atop a kidney (fig. 42.2). Each consists of an inner portion called the medulla and an outer portion called the cortex. These portions, like the anterior pituitary and the posterior pituitary, have no connection with one another.

Adrenal Medulla

The adrenal medulla secretes *epinephrine* and *norepinephrine* under conditions of stress. They bring about all those responses we associate with the "fight or flight" reaction: the blood glucose level and the metabolic rate increase, as do breathing and the heart rate. The blood vessels in the intestine constrict, and those in the muscles dilate. This increased circulation to the muscles causes them to have more stamina than usual. In times of emergency, the sympathetic nervous system *initiates* these responses but they are *maintained* by secretions from the adrenal medulla.

The adrenal medulla is innervated by one set of sympathetic nerve fibers. Recall that usually there are preganglionic and postganglionic nerve fibers for each organ stimulated. In this instance, what happened to the postganglionic neurons? It appears that the adrenal medulla may have evolved from a modification of the postganglionic neurons. Like the neurosecretory neurons in the hypothalamus, these neurons also secrete hormones into the bloodstream.

> The adrenal medulla releases epinephrine and norepinephrine into the bloodstream. These hormones help us and other animals cope with situations that threaten survival.

Adrenal Cortex

Although the adrenal medulla may be removed with no ill effects, the adrenal cortex is absolutely necessary to life. The 2 major classes of hormones made by the adrenal cortex are the *glucocorticoids* and the *mineralocorticoids*. The cortex also secretes a small amount of male and an even smaller amount of female sex hormones. All of these hormones are steroids.

Of the various glucocorticoids, the hormone responsible for the greatest amount of activity is *cortisol*. Cortisol promotes the hydrolysis of muscle protein to amino acids that enter the blood. This leads to an increased level of glucose when the liver converts these amino acids to glucose. Cortisol also favors metabolism of fatty acids rather than carbohydrates. In opposition to insulin, therefore, cortisol raises the blood glucose level. Cortisol also counteracts the inflammatory response, which leads to the pain and swelling of joints in arthritis and bursitis. The administration of cortisol eases the symptoms of these conditions because it reduces inflammation.

The secretion of cortisol by the adrenal cortex is under the control of the anterior pituitary hormone ACTH. Using the same negative feedback system shown in figure 42.7, the hypothalamus produces a releasing hormone (CRH) that stimulates the anterior pituitary to release ACTH. ACTH in turn stimulates the adrenal cortex to secrete cortisol, which regulates its own synthesis by negative feedback of both CRH and ACTH synthesis.

The secretion of mineralocorticoids, the most significant of which is **aldosterone,** is not under the control of the anterior pituitary. Aldosterone regulates the level of sodium (Na^+) ions and potassium (K^+) ions in the blood, its primary target organ being the kidney, where it promotes renal absorption of sodium and renal excretion of potassium. The level of sodium is particularly important to the maintenance of blood pressure, and the concentration of this ion indirectly regulates the secretion of aldosterone. When the blood sodium level is low, the kidneys secrete renin (fig. 42.10). Renin is an enzyme that converts the plasma protein angiotensinogen to angiotensin I, which becomes angiotensin II in the lungs. Angiotensin II stimulates the adrenal cortex to release aldosterone. This is called the renin-angiotensin-aldosterone system. The effect of this system raises the blood pressure in 2 ways. First, angiotensin constricts the arteries directly, and second, aldosterone causes the kidneys to reabsorb sodium. When blood sodium level is high, water is reabsorbed, and blood volume and pressure are maintained.

Two medical conditions are associated with malfunctioning of the adrenal cortex. In Cushing syndrome, the adrenal cortex is overactive. The muscles of the arms and legs waste away, and the blood glucose level and the blood pressure rise. Hypertension develops. In Addison disease, the adrenal cortex is underactive. There is a reduction in blood pressure and blood glucose levels, and an unexplained, peculiar bronzing of the skin.

There is another hormone involved in regulating blood sodium levels. This hormone, called atrial natriuretic hormone, is produced by the atria of the heart and causes natriuresis, the excretion of sodium. Once sodium is excreted, so is water; therefore, blood volume and blood pressure decrease.

> Cortisol, which raises the blood glucose level, and aldosterone, which raises the blood sodium level, are 2 hormones secreted by the adrenal cortex.

Pancreas

The **pancreas** is a long organ that lies transversely in the abdomen between the kidneys and near the duodenum of the small intestine (fig. 42.11). It is composed of 2 types of tissues—one of these produces and secretes *digestive* juices that go by way of the

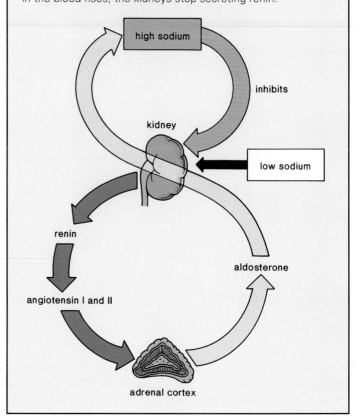

Figure 42.10
Renin-angiotensin-aldosterone system. If the blood level of sodium is low, the kidneys secrete renin. The increased renin acts via the increased production of angiotensin I and II to stimulate aldosterone secretion. Aldosterone promotes reabsorption of sodium by the kidneys; when the sodium level in the blood rises, the kidneys stop secreting renin.

pancreatic duct to the duodenum; the other type, called the *islets of Langerhans* (pancreatic islets), produces and secretes the hormones **insulin** and **glucagon** directly into the blood.

Insulin is secreted when there is a high level of glucose in the blood, which usually occurs just after eating. Insulin has 3 different actions: (1) it stimulates all cells, and in particular fat, liver, and muscle cells to take up and metabolize glucose; (2) it stimulates the liver and muscles to store glucose as glycogen; and (3) it promotes the buildup of fats and proteins and inhibits their use as an energy source. Therefore, these nutrients will be on hand during leaner times. As a result of its activities, insulin lowers the blood glucose level. Glucagon is secreted from the pancreas in between eating, and its effects are opposite to those of insulin; glucagon stimulates the breakdown of stored nutrients and causes the blood glucose level to rise (fig. 42.12).

Diabetes Mellitus

Diabetes mellitus is a fairly common disease caused by a defect in insulin production or utilization. The most common laboratory test for diabetes mellitus (usually called diabetes) is to look for sugar, or glucose, in the urine. Glucose in the urine means

Animal Structure and Function

Figure 42.11

The pancreas is both an exocrine and an endocrine gland. It sends digestive juices to the duodenum by way of the pancreatic duct and the hormones insulin and glucagon into the bloodstream. In 1920, Frederick Banting, a physician who occasionally gave physiology lectures, decided to try to isolate insulin. Investigators before him had been unable to do this because the enzymes in the digestive juices destroyed insulin (a protein) during the isolation procedure. He hit upon the idea of tying off the pancreatic duct, which he knew from previous research would lead to the degeneration only of the cells that produce digestive juices and not of the islets of Langerhans, where insulin is made. J. J. Macleod made a laboratory available to him at the University of Toronto and also assigned a graduate student named Charles Best to assist him. Banting and Best had limited funds and spent that summer working, sleeping, and eating in the lab. By the end of the summer, they had obtained pancreatic extracts that did lower the blood glucose level in diabetic dogs. Macleod then brought in biochemists, who purified the extract. Insulin therapy for the first human patient was begun in January 1922 and large-scale production of purified insulin from pigs and cattle followed. Banting and Macleod received a Nobel Prize for their work in 1923. The amino acid sequence of insulin was determined in 1953. Insulin is presently synthesized using recombinant DNA technology (chapter 18). Banting and Best followed the required steps given in the chart to identify a chemical messenger.

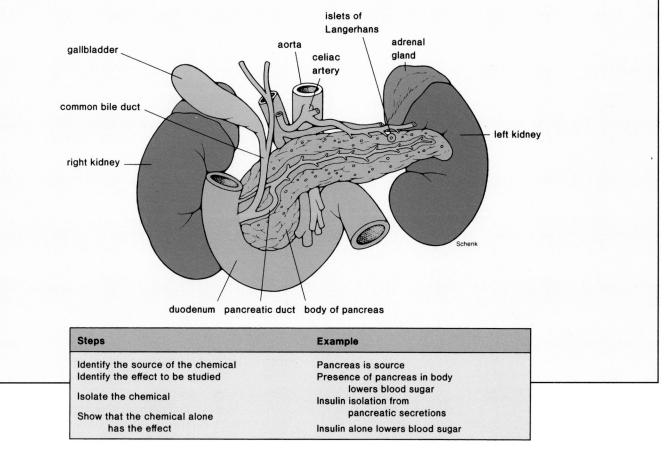

Steps	Example
Identify the source of the chemical	Pancreas is source
Identify the effect to be studied	Presence of pancreas in body lowers blood sugar
Isolate the chemical	Insulin isolation from pancreatic secretions
Show that the chemical alone has the effect	Insulin alone lowers blood sugar

that the blood glucose level is high enough to cause the kidneys to excrete glucose. This indicates that the liver is not storing glucose as glycogen and that cells are using proteins and fats as an energy source due to insufficient production or use of insulin. An individual with untreated diabetes is also extremely thirsty because the solute content of the blood is higher than normal. The breakdown of proteins and fats also leads to rapid weight loss, to the buildup of acids in the blood (acidosis), and to respiratory distress. It is acidosis that can eventually cause coma and death if the diabetic does not receive treatment.

There are 2 forms of diabetes. In Type I diabetes, the pancreas does not produce insulin. This type of diabetes frequently follows a viral infection. Apparently, the infection sets off an autoimmune reaction that destroys the insulin-secreting cells in the islets of Langerhans. Patients with Type I diabetes must either have daily insulin injections or undergo a pancreas transplant. In Type II diabetes, the most common type, the pancreas is producing insulin, but the cells of the body do not respond to it. At first, the cells lack the receptors necessary to detect the presence of insulin, and then later they become incapable of taking up glucose. The best treatment for this type of diabetes is a low-fat and low-sugar diet and regular exercise. If this treatment fails, oral drugs that make the cells more sensitive to the effects of insulin or stimulate the pancreas to make more of it can be prescribed.

Figure 42.12
Insulin and glucagon have contrary effects. When blood glucose level is high, the pancreas secretes insulin. Insulin promotes the storage of glucose as glycogen, the use of glucose as an energy source, and the synthesis of proteins and fats as opposed to their use as energy sources. Therefore, insulin lowers the blood glucose level. When the blood glucose level is low, the pancreas secretes glucagon. Glucagon acts in opposition to insulin in all respects; therefore, glucagon raises the blood glucose level.

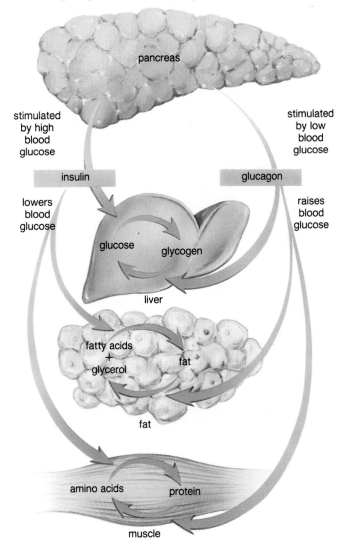

The pancreas secretes insulin, which acts to promote storage of nutrients, thereby lowering the blood glucose level. In diabetes mellitus, either insulin is not produced or it cannot be utilized; the excess glucose in the blood is excreted by the kidneys.

Other Endocrine Glands

Gonads

The gonads are the endocrine glands that produce the hormones that determine sexual characteristics. As we will discuss in detail in the following chapter, the *testes* produce the male sex hormones, the androgens—the most important of which is testosterone—and

Anabolic Steroid Use

 nabolic steroids are essentially the male sex hormone testosterone and its synthetic derivatives. The latter were developed in the 1930s to prevent the atrophy of muscle tissue in patients with debilitating illnesses. In some cases, they are also given to burn victims and surgical patients to speed recovery. Steroids were first used for nonmedical reasons in World War II, when German doctors gave them to soldiers, perhaps to make them more aggressive. Following the war, the Soviet Union began dispensing steroids to athletes, and when the results were observed by a U.S. physician, he recommended that the U.S. weight-lifting team use them. In the 1960s, when their dangerous side effects became apparent, steroid therapy fell into disfavor, and in 1973 the Olympic Committee added anabolic steroids to its list of banned substances (fig. 42.A). Although the FDA (Food and Drug Administration) has also banned most steroids except for limited medical use, they are not *controlled* substances like cocaine and heroin. Presently, they are smuggled into the United States from Mexico and to a lesser degree, Eastern Europe; this illicit trafficking has become a $100-million-a-year business. Congress has passed an antidrug bill that makes the selling of steroids a felony offense with jail terms of 3 years for dispensing to adults and 6 years for dispensing to children.

There is great concern because these drugs, originally taken primarily by professional athletes, are now being more widely used, even by children, to increase muscle size. The latest estimate is that 1–3 million Americans use steroids, a figure that has increased steadily since the early 1970s. No one has really conclusively determined the long-term harmful effects of steroid use because researchers are reluctant to do such studies on humans. The accompanying figure indicates the major side effects that have been observed in users. Often these side effects can be understood in view of the fact that testosterone

the *ovaries* produce estrogen and progesterone, the female sex hormones. The secretion of these hormones is under the control of the anterior pituitary.

The sex hormones bring about the secondary sex characteristics of males and females. Among other traits, males have greater muscle strength than do females. As discussed in this chapter's reading, athletes and others sometimes take so-called *anabolic steroids*, which are synthetic steroids that mimic the action of testosterone, to improve their strength and physique. Unfortunately, this practice is accompanied by harmful side effects.

Thymus Gland

The *thymus* gland extends from below the thyroid gland in the neck into the thoracic cavity. This organ grows during childhood but gradually decreases in size after puberty. Lymphocytes that have passed through the thymus are transformed into T cells. Colonies of T cells in the lymph nodes and other organs are apparently able to produce new T cells when stimulated by various thymus hormones called *thymosins*. There is hope that these hormones can be used in conjunction with lymphokine therapy to restore or stimulate T cell function in patients suffering from AIDS or cancer.

Animal Structure and Function

is the male sex hormone that maintains the secondary sex characteristics of males, such as the development of the testes and penis, a deep voice, and facial hair. Anabolic steroid use can also apparently cause death. Steroids have been implicated in the deaths of young athletes from liver cancer and a type of kidney tumor. Furthermore, because steroids cause the body to retain fluid, users often take diuretics; excessive amounts of these can rob the body of its proper electrolyte balance, resulting in heart attack.

It has come to light that steroids may be addictive and might cause psychological problems. Psychotic side effects, sometimes called "roid mania," have been observed in abusers who experience wild aggression and delusions. For example, it has been reported that one user had a friend videotape him while he deliberately drove a car into a tree at 35 mph. Psychiatrists feel that the magnitude of the mental effects of steroids is only now being realized.

Figure 42.A
The effects of anabolic steroid use.

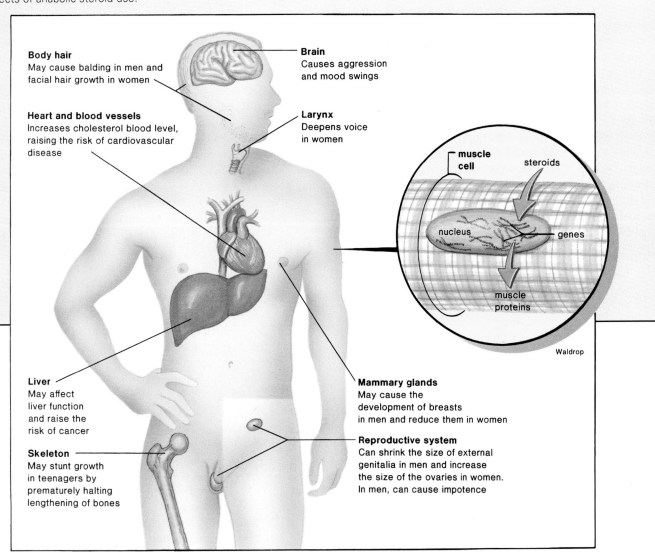

Body hair
May cause balding in men and facial hair growth in women

Brain
Causes aggression and mood swings

Heart and blood vessels
Increases cholesterol blood level, raising the risk of cardiovascular disease

Larynx
Deepens voice in women

Liver
May affect liver function and raise the risk of cancer

Skeleton
May stunt growth in teenagers by prematurely halting lengthening of bones

Mammary glands
May cause the development of breasts in men and reduce them in women

Reproductive system
Can shrink the size of external genitalia in men and increase the size of the ovaries in women. In men, can cause impotence

muscle cell

steroids

nucleus

genes

muscle proteins

Waldrop

Pineal Gland

The *pineal gland* in amphibians and fishes is located near the skin surface, where it functions as a "third eye," transforming light signals into nerve impulses sent to the brain. In birds and mammals, the pineal gland is located in the brain; in birds it still receives light signals directly, but in mammals the eyes communicate with the pineal gland. In any case, at night when light is absent, the pineal gland produces melatonin, a hormone that is involved in both circadian and circannual rhythms. For example, in amphibians it causes a blanching of the skin and in birds it causes roosting, which are nighttime events. Daily cycles such as these are called circadian rhythms.

Endocrine System

The reproductive cycles of most temperate zone animals are related to the seasons. Melatonin causes the reproductive organs to shrink in size. Apparently the reduced production of melatonin during the shorter nights of summer promotes enlargement of reproductive organs. Therefore, mating occurs in the fall and young are born in the spring when food is available to nourish them. Yearly rhythms such as these are called circannual rhythms.

The function of melatonin in humans is not known; however, children with brain tumors that destroy the pineal gland experience early puberty. Therefore, it's possible that melatonin is also involved in human sexual development.

Still Other Glands

Organs traditionally not considered to be endocrine glands have been found to secrete hormones into the bloodstream. For example, as mentioned on page 630, the heart produces atrial natriuretic hormone, a peptide that helps to regulate the water and salt balance in kidneys. Also, in chapter 36, we discussed the hormones produced by the stomach and the small intestine.

A number of different types of organs and cells have been discovered to produce peptide *growth factors* that stimulate cell division and mitosis. They are like hormones in that they act on cell types that have specific receptors to receive them. Some, including lymphokines (p. 579) and blood cell growth factors (p. 563), are released into the blood; others diffuse to nearby neighboring cells, and still others are self-stimulating. The latter are of interest because they play a role in the formation of cancer, as discussed on page 271. Other growth factors are as follows:

Platelet-derived growth factor is released from platelets and also many other cell types. It helps in wound healing and causes an increase in the number of fibroblasts, smooth muscle cells, and certain cells of the nervous system.
Epidermal growth factor and *nerve growth factor* stimulate the cells indicated by their names and many others also. These growth factors are also important in wound healing.
Tumor angiogenesis factor stimulates the formation of capillary networks and is released by tumor cells. One treatment for cancer is to prevent the activity of this growth factor.

Growth factors that simply diffuse to nearby cells and act locally do not fit the standard definition of a hormone. **Prostaglandins (PG)** are another class of chemical messengers that also are produced and act locally. They are derived from fatty acids stored in plasma membranes as phospholipids. When a cell is stimulated by reception of a hormone or even by trauma, a series of synthetic reactions takes place in the plasma membrane, and PG first is released into the cytoplasm and then secreted from the cell. There are many different types of prostaglandins produced by many different tissues. In the uterus, certain prostaglandins cause muscles to contract; therefore, they are implicated in the pain and discomfort of menstruation in some women. (Antiprostaglandin therapy is useful in these cases.) On the other hand, certain prostaglandins are used to treat ulcers because they reduce gastric secretion, to treat hypertension because they lower blood pressure, and to prevent thrombosis because they inhibit platelet aggregation. Because the different prostaglandins can have contrary effects, however, it has been very difficult to standardize their use, and in most instances, prostaglandin therapy still is considered experimental.

Environmental Signals

In this chapter, we concentrated on describing the functions of the human endocrine glands and their hormonal secretions. We already know that hormones are only one type of chemical messenger or environmental signal between cells. In fact, the concept of the environmental signal now has been broadened to include at least the following 3 different categories of messengers (fig. 42.13):

1. *Environmental signals that act at a distance between individuals.* Many organisms release chemical messengers, called *pheromones,* into the air or in externally deposited body fluids. These are intended to be messages for other members of the species. For example, ants lay down a pheromone trail to direct other ants to food, and the female silkworm moth releases bombykol, a sex attractant that is received by male moth antennae even several miles away. This chemical is so potent that it has been estimated that only 40 out of 40,000 receptors on the male antennae need to be activated in order for the male to respond. Mammals, too, release pheromones; the urine of dogs serves as a territorial marker, for example. Studies are being conducted to determine if humans also have pheromones.

2. *Environmental signals that act at a distance between body parts.* This category includes the endocrine secretions, which traditionally have been called hormones. It also includes the secretions of the neurosecretory cells in the hypothalamus—the production and action of ADH and oxytocin illustrate the close relationship between the nervous system and the endocrine system. Neurosecretory cells produce these hormones, which are released when these cells receive nerve impulses. As another example of the overlap between the nervous and endocrine systems, consider that endorphins on occasion travel in the bloodstream, but they act on nerve cells to alter their membrane potential. Also, norepinephrine is both a neurotransmitter and hormone secreted by the adrenal medulla.

3. *Environmental signals that act locally between adjacent cells.* Neurotransmitter substances belong in this category, as do substances like prostaglandins and growth factors, which are sometimes called local hormones. Also, when the skin is cut, histamine is released by mast cells and promotes the inflammatory response.

Redefinition of a Hormone

Traditionally, a hormone was considered to be a secretion of an endocrine gland that was carried in the bloodstream to a target organ. In recent years, some scientists have broadened the definition of a hormone to include *all* types of chemical messengers. This

Animal Structure and Function

change seemed necessary because those chemicals traditionally considered to be hormones now have been found in all sorts of tissues in the human body. For example, it is impossible for insulin produced by the pancreas to enter the brain because of the blood-brain barrier—a tight fusion of endothelial cells of the capillary walls that prevents passage of larger molecules like peptides. Yet, insulin has been found in the brain. It now appears that the brain cells themselves can produce insulin, which is used locally to influence the metabolism of adjacent cells. Also, some chemicals identical to the hormones of the endocrine system have been found in lower organisms, even in bacteria! A moment's thought about the evolutionary process helps to explain this; these regulatory chemicals may have been present in the earliest cells and only became specialized as hormones as evolution proceeded.

Figure 42.13

The 3 categories of environmental signals. Pheromones are chemical messengers that act at a distance between individuals. Endocrine hormones and neurosecretions are typically carried in the bloodstream and act at a distance within the body of a single organism. Some chemical messengers have local effects only; they pass between cells that are adjacent to one another. This, of course, includes neurotransmitters.

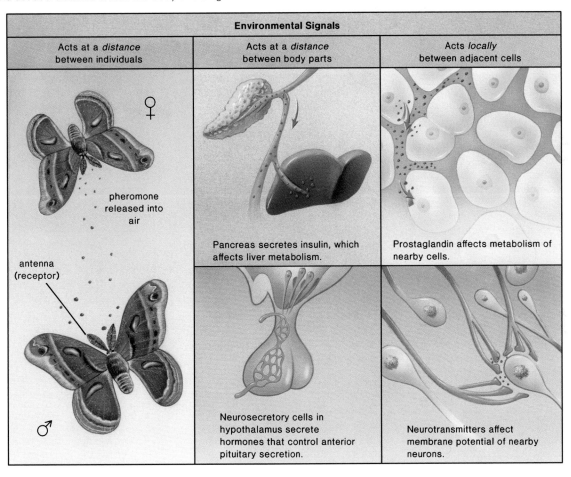

Environmental Signals

Acts at a *distance* between individuals	Acts at a *distance* between body parts	Acts *locally* between adjacent cells
pheromone released into air; antenna (receptor) ♀ ♂	Pancreas secretes insulin, which affects liver metabolism.	Prostaglandin affects metabolism of nearby cells.
	Neurosecretory cells in hypothalamus secrete hormones that control anterior pituitary secretion.	Neurotransmitters affect membrane potential of nearby neurons.

Summary

1. The endocrine glands in humans, which are shown in figure 42.2, produce hormones that are secreted into the bloodstream and circulate about the body until they are received by their target organs.

2. There are 2 major types of compounds used as hormones; most hormones are classified as peptides; a few hormones are steroids.

3. The peptide protein hormones are usually received by a receptor located in the plasma membrane. Most often their reception activates an enzyme that changes ATP to cyclic AMP. Cyclic AMP then activates another enzyme that activates another and so forth. There is an amplification of effect at each step of this enzyme cascade.

4. The steroid hormones are lipid soluble and can pass through plasma membranes. Once inside the cytoplasm, they combine with a receptor

molecule, and the complex moves into the nucleus, where it combines with chromatin. Transcription and translation lead to protein synthesis; in this way, steroid hormones alter the metabolism of the cell.

5. Neurosecretory cells in the hypothalamus produce ADH and oxytocin, which are stored in axon endings in the posterior pituitary until they are released. ADH secretion is controlled by a negative feedback mechanism in which concentrated blood stimulates ADH release and dilute blood inhibits its release. ADH causes the collecting ducts of the kidney to reabsorb water. Oxytocin causes the uterus to contract and milk to be released from the mammary glands.

6. The hypothalamus produces hypothalamic-releasing and hypothalamic-release-inhibiting hormones, which pass to the anterior pituitary by way of a portal system and either stimulate or inhibit the release of a particular anterior pituitary hormone.

7. The anterior pituitary produces at least 6 types of hormones (table 42.1). GH promotes body growth; Pr promotes breast development and production of milk following pregnancy; MSH causes skin-color changes in lower vertebrates; TSH stimulates the thyroid; ACTH stimulates the adrenal cortex to produce corticoids; gonadotropic hormones (FSH and LH) stimulate the gonads. Because the anterior pituitary also stimulates certain other hormonal glands, it is sometimes called the master gland.

8. The thyroid gland produces thyroxin and triiodothyronine, hormones that contain iodine. These hormones play a role in growth and development of immature forms; in mature individuals they increase the metabolic rate. A goiter may develop in individuals who have inadequate iodine in the diet. The thyroid gland also produces calcitonin, which helps lower the blood calcium level.

9. The parathyroid glands raise the level of calcium in the blood by (1) activating vitamin D, which is involved in absorption of calcium from the gut, (2) stimulating retention of calcium by the kidneys, and (3) stimulating calcium removal from the bones. PTH also decreases the blood phosphate level.

10. The adrenal medulla secretes epinephrine and norepinephrine, which bring about responses we associate with the fight or flight reaction.

11. The adrenal cortex produces the glucocorticoids (cortisol) and the mineralocorticoids (aldosterone). Cortisol stimulates hydrolysis of proteins to amino acids that are converted to glucose; in this way, it raises the blood glucose level. It also counteracts the inflammatory reaction. Aldosterone causes the kidneys to reabsorb sodium: when the blood sodium level is low, the kidneys secrete renin, which converts angiotensinogen to angiotensin, a molecule that stimulates the release of aldosterone. The main effect of this renin-angiotensin-aldosterone system is to raise the blood pressure.

12. The pancreas secretes insulin, which lowers the blood glucose level, and glucagon, which has opposite effects to insulin. In Type I diabetes mellitus, insulin-secreting cells in the islets of Langerhans have been destroyed; in Type II diabetes, the cells lack receptors for insulin or do not respond to it. Only Type I diabetes requires daily injections of insulin, while Type II requires a special diet.

13. The gonads produce the sex hormones, which are discussed in chapter 43; the thymus secretes thymosins, which stimulate T lymphocyte production and maturation; the pineal gland produces melatonin, whose function in mammals is uncertain—it may affect development of the reproductive organs.

14. There are 3 categories of chemical messengers: those that act at a distance between individuals (pheromones); those that act at a distance within the individual (traditional endocrine hormones and secretions of neurosecretory cells); and local messengers (such as prostaglandins, growth factors, and neurotransmitters). Since there is great overlap between these categories, perhaps the definition of a hormone should now be expanded to include all of them.

Writing Across the Curriculum

In order to practice writing skills, students should write out the answers to any or all of the study questions and the critical thinking questions. The study questions are sequenced in the same order as the text. Suggested answers to the critical thinking questions are in appendix D.

Study Questions

1. Give a definition of endocrine hormones that includes their most likely source, how they are transported in the body, and how they are received. What does "target" organ mean?
2. Categorize endocrine hormones according to their chemical makeup.
3. Tell how the 2 major types of hormones influence the metabolism of the cell.
4. Give the location in the human body of all the major endocrine glands. Name the hormones secreted by each gland, and describe their chief functions.
5. Explain the relationship of the hypothalamus to the posterior pituitary and to the anterior pituitary.
6. Explain the concept of negative feedback and give an example involving ADH.
7. Give an example of the 3-tiered relationship among the hypothalamus, the anterior pituitary, and other endocrine glands. Explain why the anterior pituitary can be called the master gland.
8. Draw a diagram to explain the contrary actions of insulin and glucagon. Use your diagram to explain the symptoms of Type I diabetes mellitus.
9. Categorize chemical messengers into 3 groups, and give examples of each group.
10. Give examples to show that there is an overlap between the mode of operation of the nervous system and that of the endocrine system. Explain why the traditional definition of a hormone may need to be expanded.

Animal Structure and Function

Objective Questions

Match the hormone in numbers 1–5 to the correct gland in the key.

Key:

a. pancreas
b. anterior pituitary
c. posterior pituitary
d. thyroid
e. adrenal medulla
f. adrenal cortex

1. cortisol
2. GH
3. oxytocin storage
4. insulin
5. epinephrine
6. The anterior pituitary controls the secretion(s) of
    a. both the adrenal medulla and the adrenal cortex.
    b. both cortisol and aldosterone.
    c. thyroxin.
    d. All of these.
7. Peptide hormones
    a. are received by a receptor located in the plasma membrane.
    b. are received by a receptor located in the cytoplasm.
    c. bring about the transcription of DNA.
    d. Both b and c.
8. Aldosterone causes the
    a. kidneys to release renin.
    b. kidneys to reabsorb sodium.
    c. blood volume to increase.
    d. All of these.

9. Diabetes mellitus is associated with
    a. too much insulin in the blood.
    b. too high a level of glucose in the blood.
    c. blood that is too dilute.
    d. All of these.
10. The level of cortisol in the blood controls the secretion of
    a. a releasing hormone from the hypothalamus.
    b. ACTH from the anterior pituitary.
    c. cortisol from the adrenal cortex.
    d. All of these.
11. It is now clear that
    a. there is a discrete distinction between the activities of the nervous and endocrine systems.

b. both the nervous and endocrine systems coordinate the activities of body parts.
    c. the nervous system controls the secretions of all endocrine glands.
    d. All of these.
12. One of the chief differences between pheromones and local hormones is
    a. one is a chemical messenger and the other is not.
    b. the distance over which they act.
    c. one is made by invertebrates and the other is made by vertebrates.
    d. All of these.
13. Label this diagram and explain how a negative feedback system keeps the hormone level constant in the body:

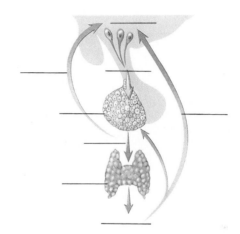

Concepts and Critical Thinking

1. *Hormone levels are maintained by feedback control.*

    Contrast control of neurotransmitter level in the nervous system with control of hormone level in the endocrine system.

2. *The nervous system is fast acting, and the endocrine system is fairly slow moving.*

    Contrast the manner in which the nervous system delivers its message with the way the endocrine system delivers a message.

3. *Hormone levels greatly affect the phenotype.*

    Use the effect of sex hormones to substantiate this concept.

Selected Key Terms

pituitary gland (pi-tu′ĭ-tār″e gland) 681
thyroid gland (thi′roid gland) 683
goiter (goi′ter) 683

parathyroid gland (par″ah-thi′roid gland) 685
adrenal gland (ah-dre′nal gland) 685

pancreas (pan′kre-as) 686
diabetes mellitus (di″ah-be′tēz me-li′tus) 686

43

Reproduction in Animals

Many animals sexually reproduce and invest much energy in the rearing of offspring. Newly hatched birds usually have to be fed before they are able to fly away and seek food for themselves. Complex hormonal and neural regulation are involved in reproductive behavior of parental birds.

Your study of this chapter will be complete when you can

1. contrast asexual reproduction with sexual reproduction, and relate their advantages to environmental conditions;

2. explain why aquatic animals are apt to practice external fertilization, whereas land animals are apt to practice internal fertilization;

3. name and give a function for the organs of the human male reproductive system; trace the path of sperm from the testes to the penis;

4. describe the microscopic anatomy of the testes, including the stages of spermatogenesis; relate the action of the hormones FSH (ICSH) and LH to the functions of the testes;

5. describe hormonal control in the male, mentioning the hypothalamus, anterior pituitary, and testosterone; give several actions of testosterone;

6. name and give a function for the organs of the human female reproductive system;

7. describe the microscopic anatomy of the ovary, including a description of the ovarian cycle;

8. describe the uterine cycle; explain the relationship among the hormones GnRH, FSH, LH, and the female sex hormones; give several functions of estrogen and progesterone;

9. explain the cessation of the ovarian and uterine cycles in the pregnant female;

10. list several means of birth control, and compare their effectiveness in preventing pregnancy;

11. describe the cause and symptoms of the following sexually transmitted diseases: AIDS, genital herpes, genital warts, gonorrhea, chlamydia, and syphilis.

A nimals expend a considerable amount of energy on reproduction. After all, reproduction ensures that the animals' genes are passed on to the next generation. The life cycle of any particular animal comes to an end, but its genes can be perpetuated as long as reproduction has taken place. In evolutionary terms, the most fit organisms are the ones that have produced the most offspring. They are the best adapted to the environment, and it is beneficial to the species that they reproduce more than other members of their cohort.

Patterns of Reproduction

There are 2 fundamental patterns of reproduction: asexual and sexual. We will examine the methodology and particular advantage of each pattern.

Asexual Reproduction

In asexual reproduction, there is only one parent, and the offspring have the same phenotype and genotype as that parent (fig. 43.1). This lack of variation is not a disadvantage as long as the environment stays the same. Organisms that reproduce asexually often have the tremendous advantage of being able to produce a large number of offspring within a limited amount of time.

Only certain methods of asexual reproduction are found among animals. Many flatworms can constrict into 2 halves, each of which can become a new individual. This form of **regeneration** is also seen among sponges and echinoderms. Chopping up a sea star does not kill it; instead, each fragment grows into another animal, and the total number of sea stars increases in the process. Some cnidarians, such as hydras, reproduce by **budding** (see fig. 26.5), during which the new individual arises as an outgrowth (bud) of the parent. Some insects and several other types of arthropods have the ability to reproduce parthenogenetically (fig. 43.1b). **Parthenogenesis** is a modification of sexual reproduction in which an unfertilized egg develops into a complete individual.

Sexual Reproduction

Sexual reproduction involves gametes (sex cells) (fig. 43.1). The gametes may be specialized into eggs or sperm, which are produced by the same or separate individuals. Earthworms (see fig. 27.8) practice cross-fertilization, even though they are *hermaphroditic*—they have both male and female sex organs. Among vertebrates, the sexes tend to be separate, and it is often easy to tell whether an animal is an egg-producing female or a sperm-producing male (fig. 43.2a). When animals reproduce sexually, an offspring inherits half its genes from one parent and the other half from the other parent. Therefore, an offspring often has a different combination of genes than either parent. In this way, variation may be introduced and maintained. Such variation is an advantage to the species if the environment is changing, because an offspring might be better adapted to the new environment than is either parent.

Figure 43.1

Asexual reproduction versus sexual reproduction. ***a.*** In asexual reproduction (*left*), there is only one parent and the offspring tend to be identical to each other and to the parent. Asexual reproduction occurs in various ways: by regeneration, by budding, or by parthenogenesis. In sexual reproduction (*right*), there are most often 2 parents. Each parent contributes one-half of the genes to the offspring either by way of the sperm or the egg. Therefore, the offspring tend to be genetically and phenotypically different from either parent. ***b.*** Stem covered with aphids. In the summer, many generations of aphid females produce up to 100 young, without prior fertilization. In the autumn, however, aphids reproduce sexually and the zygotes overwinter.

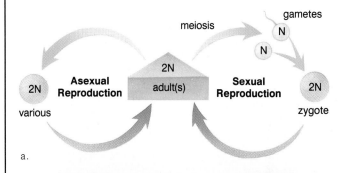

a.

b.

Reproductive Timing

The majority of many sexually reproducing animals has a single breeding period each year. Mating is timed so that the young are born when conditions are favorable for their growth. Many temperate zone mammals mate in the fall and deliver young in the spring, for example. Such reproductive cycles are tied to day-length changes. The length of the day reliably indicates the proper time for reproductive behavior, including migration to distant places. Re-

Figure 43.2

Sexual reproduction among vertebrates. ***a.*** During mating in terrestrial vertebrates, the male usually passes sperm to the female by use of a copulatory organ such as a penis. Note also that it is often possible to tell one sex from the other; in this case because the male lion has a mane. ***b.*** During mating in aquatic vertebrates, both male and female sexes often shed their gametes directly into the water. The watery environment protects the gametes and zygotes from drying out.

a.

b.

searchers have found that melatonin is produced by the pineal gland during the night, and that this hormone shrinks reproductive organs in seasonally breeding animals. The nights lengthen in the winter, and the increased production of melatonin causes reproductive behavior to cease.

External versus Internal Fertilization

Many aquatic animals practice external fertilization. They shed a large number of gametes into the water at the same time and place (fig. 43.2*b*). External fertilization is possible because the water protects the zygote and embryo from drying out. Often the embryo is a swimming larva capable of acquiring its own food.

Internal fertilization is a mechanism practiced by animals that lay a shelled egg or in which the embryo develops for a time within the female parent. Even some aquatic animals practice internal fertilization. In certain sharks, skates, and rays, the pelvic fins are specialized to pass sperm to the female, and in most of these animals, the young develop internally and are born alive. On land, internal fertilization is a necessity because sperm and eggs dry quickly when exposed to air. Often the male has a copulatory organ to transfer sperm to the female. Reptiles and birds lay shelled eggs (see fig. 28.10). In placental mammals, the fetus develops within the uterus, where it receives nourishment by way of the placenta. Humans can be used as an example of reproduction in placental mammals.

> Some animals practice asexual reproduction, especially when environmental conditions are favorable. Most practice sexual reproduction, timed so that the young are born when food is plentiful. Aquatic animals are apt to practice external fertilization, and internal fertilization is common in land animals.

Human Reproduction

The manner in which humans reproduce is similar to that of other mammals and represents an adaptation to the land environment. When the male penis passes sperm to the female, desiccation of not only the male and female gametes but also the subsequent embryo is prevented.

Male Reproductive System

The male reproductive system includes the organs pictured in figure 43.3 and listed in table 43.1. The male gonads are paired testes, which are suspended within the *scrotal sacs* of the scrotum. The testes begin their development inside the abdominal cavity, but they descend into the scrotal sacs as development proceeds. If the testes do not descend, and the male is not treated by administration of hormones or operated on to place the testes in the scrotum, sterility (the inability to produce offspring) results. Sterility in this case results because normal sperm production cannot occur at body temperature; a cooler temperature is required.

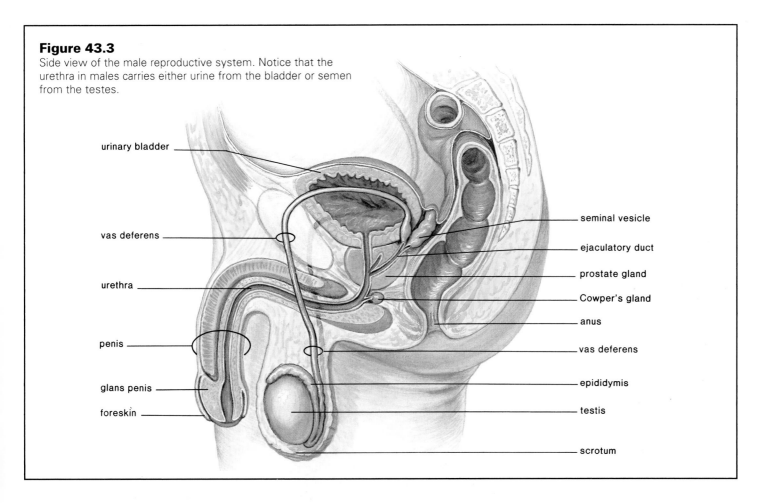

Figure 43.3
Side view of the male reproductive system. Notice that the urethra in males carries either urine from the bladder or semen from the testes.

urinary bladder

vas deferens

urethra

penis

glans penis

foreskin

seminal vesicle

ejaculatory duct

prostate gland

Cowper's gland

anus

vas deferens

epididymis

testis

scrotum

Table 43.1
Male Reproductive System

Organ	Function
Testes	Produce sperm and sex hormones
Epididymides	Maturation and some storage of sperm
Vas deferentia	Conduct and store sperm
Seminal vesicles	Contribute to seminal fluid
Prostate gland	Contributes to seminal fluid
Urethra	Conducts sperm
Bulbourethral glands	Contribute to seminal fluid
Penis	Organ of copulation

Path of Sperm

Sperm produced by the testes mature within the *epididymides* (epididymis, sing.), which are tightly coiled tubules lying just outside the testes. Maturation seems to be required for the sperm to swim to the egg. Once the sperm have matured, they are propelled into the **vas deferentia** (vas deferens, sing.) by muscular contrac-

tions. Sperm are stored in both the epididymides and the vas deferentia. When a male becomes sexually aroused, sperm enter the urethra, part of which is located within the **penis.**

The penis is a cylindrical organ that usually hangs in front of the scrotum. Spongy, erectile tissue containing distensible blood spaces extends through the shaft of the penis (fig. 43.4). During sexual arousal, nervous reflexes cause an increase in arterial blood flow to the penis. This increased blood flow fills the blood space in the erectile tissue, and the penis, which is normally limp (flaccid), stiffens and increases in size. These changes are called *erection*. If the penis fails to become erect, the condition is called *impotency*.

Orgasm in both sexes is characterized by a release of neuromuscular tension, particularly in the genital area. In males, orgasm is obvious because rhythmic muscle contractions compress the urethra and expel semen. This is termed *ejaculation,* after which a male typically experiences a *refractory period* when stimulation does not bring about an erection.

Semen **Seminal fluid** (semen) is a thick, whitish fluid that contains sperm and fluids. As table 43.1 indicates, 3 types of glands contribute fluids to semen. The *seminal vesicles* lie between the bladder and the rectum. Each joins a vas deferens to form an ejaculatory duct that enters the urethra. As sperm pass from the vas deferentia, these vesicles secrete a thick, viscous fluid containing

Figure 43.4

Penis anatomy. ***a.*** Beneath the skin and connective tissue lies the urethra, surrounded by erectile tissue. This tissue expands to form the glans penis, which in uncircumcised males is partially covered by the prepuce (foreskin). ***b.*** Two other columns of erectile tissue in the penis are dorsally located.

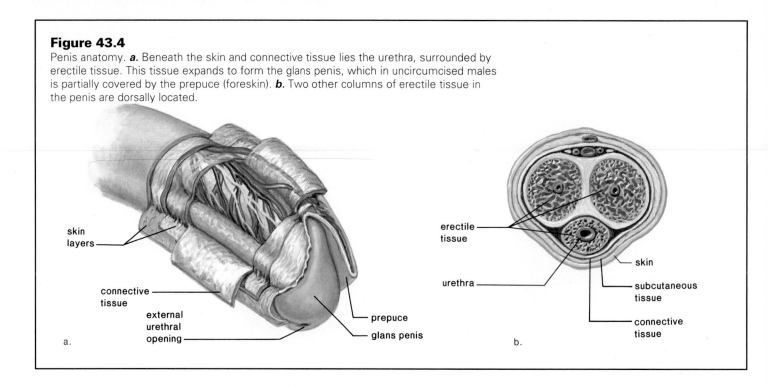

nutrients for possible use by the sperm. Just below the bladder is the *prostate gland,* which secretes a milky alkaline fluid believed to activate or increase the motility of the sperm. In older men, the prostate gland frequently becomes enlarged, thereby constricting the urethra and making urination difficult. Slightly below the prostate gland, on either side of the urethra, is a pair of small glands called *bulbourethral glands,* which have mucous secretions with a lubricating effect. Notice from figure 43.3 that at different times the urethra carries urine from the bladder and semen from the vas deferentia.

> Sperm produced by the testes mature in the epididymides and pass from the vas deferentia to the urethra, where certain glands add seminal fluid prior to ejaculation.

Sperm Production and Hormonal Regulation

A longitudinal section of a testis shows that each testis is composed of compartments called lobules, each of which contains one to 3 tightly coiled *seminiferous tubules* (fig. 43.5*a*). Altogether, these tubules have a combined length of approximately 250 m. A microscopic cross section of a seminiferous tubule shows that it is packed with cells undergoing *spermatogenesis* (fig. 43.5*b*), which involves the process of meiosis (see fig. 11.6). Also present are the Sertoli (or nurse) cells, which support, nourish, and regulate the spermatogenic cells (fig. 43.5*c*).

Mature **sperm,** or spermatozoans, have 3 distinct parts: a head, a middle piece, and a tail (fig. 43.5*d*). The middle piece and tail contain microtubules, in the characteristic 9 + 2 pattern of cilia and flagella. In the middle piece, mitochondria are wrapped around the microtubules and provide the energy for movement. The head

contains a nucleus covered by a cap called the *acrosome,* which stores enzymes needed for fertilization. The human egg is surrounded by several layers of cells and a thick membrane—these acrosome enzymes play a role in allowing a sperm to reach the surface of the egg. The normal human male usually produces several hundred million sperm per day, assuring an adequate number for fertilization to take place. Fewer than 100 ever reach the vicinity of the egg and only one sperm enters an egg.

Hormonal Regulation in Males The hypothalamus has ultimate control of the testes' sexual function because it secretes a releasing hormone (GnRH—gonadotropic-releasing hormone) that stimulates the anterior pituitary to produce the gonadotropic hormones. There are 2 gonadotropic hormones, FSH and LH, in both males and females. In males, FSH promotes spermatogenesis in the seminiferous tubules, which also release the hormone inhibin.

LH in males is sometimes given the name *interstitial cell-stimulating hormone (ICSH)* because it controls the production of testosterone by the interstitial cells, scattered in the spaces between the seminiferous tubules (fig. 43.5*b*). All these hormones are involved in a feedback relationship that maintains the fairly constant production of sperm and testosterone (fig. 43.6).

Male Sex Hormones Testosterone is the main sex hormone in males. It is essential for the normal development and functioning of the organs listed in table 43.1. Testosterone is also necessary for the maturation of sperm.

Testosterone also brings about and maintains the secondary sex characteristics in males that develop at the time of puberty. Testosterone causes growth of a beard, axillary (underarm) hair, and pubic hair. It prompts the larynx and vocal cords to enlarge,

Figure 43.5

Testis and sperm. **a.** Longitudinal section showing lobules containing seminiferous tubules. **b.** Light micrograph of cross section of seminiferous tubules. Magnification, X200. **c.** Diagrammatic representation of spermatogenesis, which occurs in the wall of the tubules. **d.** Mature sperm consist of a head, middle piece, and tail. The nucleus is in the head, capped by an enzyme-containing acrosome.

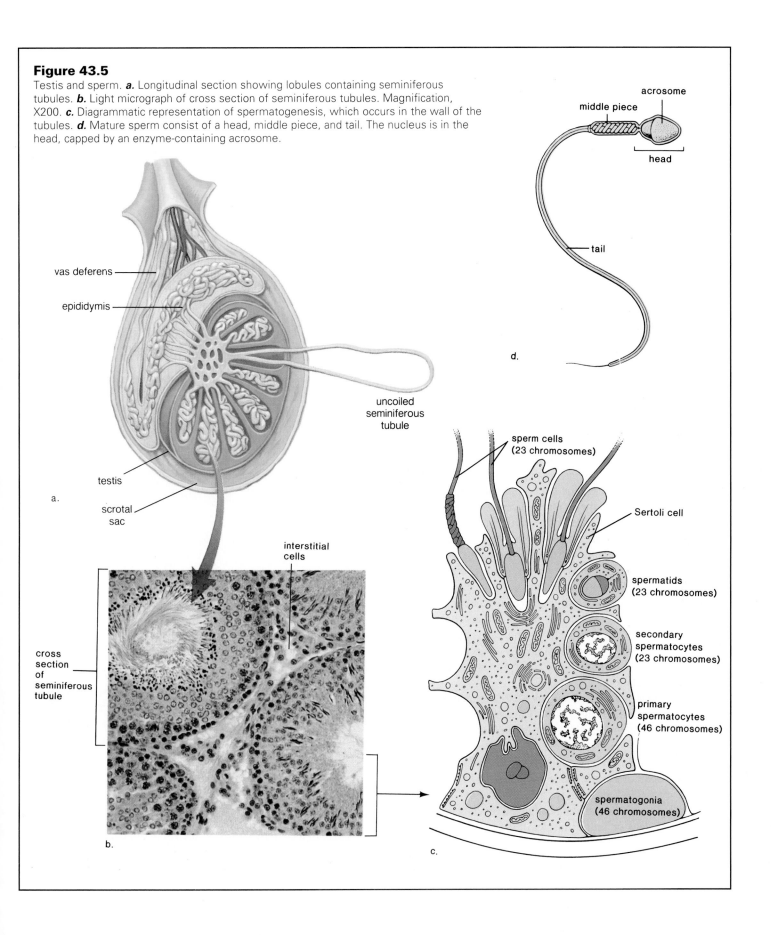

causing the voice to change. It is responsible for the greater muscle strength of males, and this is the reason some athletes take supplemental amounts of anabolic steroids, which are either testosterone or related chemicals. The contraindications of taking anabolic steroids are discussed in this chapter's reading. Testosterone is believed to be largely responsible for the sex drive and may even contribute to the supposed aggressiveness of males.

Testosterone also causes oil and sweat glands in the skin to secrete; therefore, it is largely responsible for acne and body odor. Another side effect of testosterone activity is baldness, as discussed on page 231.

> The hypothalamus, anterior pituitary, and testes are engaged in a feedback system that promotes spermatogenesis and maintains the level of testosterone in the male. Testosterone stimulates development of the male sex organs and secondary sex characteristics.

Female Reproductive System

The female reproductive system includes the ovaries, oviducts, uterus, and vagina (fig. 43.7 and table 43.2). The **ovaries,** which produce an egg each month, lie in shallow depressions, one on each side of the upper pelvic cavity. The **oviducts,** also called uterine or fallopian tubes, extend from the ovaries to the uterus; however, the oviducts are not attached to the ovaries. Instead, they have finger-like projections called *fimbriae* that sweep over the ovaries. When an egg bursts from an ovary during ovulation, it usually is swept into an oviduct by the combined action of the fimbriae and the beating of cilia that line the oviducts. Fertilization, if it occurs, takes place in an oviduct, and the developing embryo is propelled slowly by cilia movement and tubular muscle contraction to the uterus. The **uterus** is a thick-walled muscular organ about the size and shape of an inverted pear. When an embryo embeds itself in the uterine lining, called the **endometrium**, the female is pregnant. A small opening at the cervix leads to the vaginal canal. The **vagina** is a tube that makes a 45-degree angle with the small of the back. The mucosal lining of the vagina lies in folds and can extend. This is especially important when the vagina serves as the birth canal, and it also can facilitate intercourse, when the vagina receives the penis during copulation.

The external genital organs of the female are known collectively as the *vulva* (fig. 43.7b). The *mons pubis* and 2 folds of skin called *labia minora* and *labia majora* are on either side of the urethral and vaginal openings. At the juncture of the labia minora is the *clitoris,* which is analogous to the penis in males. The clitoris has a shaft of erectile tissue and is capped by a pea-shaped glans. The many sense receptors of the clitoris allow it to function as a sexually sensitive organ. Orgasm in the female is a release of neuromuscular tension in the muscles of the genital area, vagina, and uterus.

Egg Production and Hormonal Regulation

The ovaries are responsible for egg production, and they also produce the female sex hormones, **estrogen** and **progesterone,** during the ovarian cycle.

Figure 43.6

The hypothalamus-pituitary-testis control relationship. Testosterone acts on various body tissues and also regulates the amount of hypothalamic GnRH being sent to the pituitary. GnRH affects gonadotropic hormone production by the pituitary. LH regulates the amount of testosterone produced, and FSH controls spermatogenesis. The seminiferous tubules release a substance called inhibin, which is also involved in feedback inhibition.

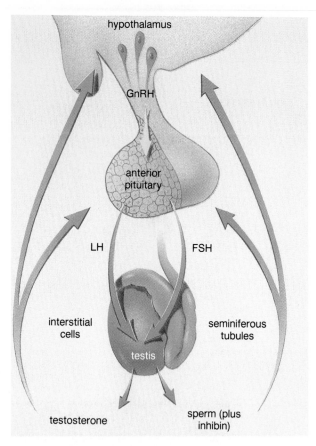

The Ovarian Cycle A longitudinal section through an ovary shows many **follicles,** each containing an oocyte (fig. 43.8). A female is born with as many as 2 million follicles, but the number is reduced to 300,000–400,000 by the time of puberty. Only a small number of follicles (about 400) ever mature, because a female usually produces only one egg per month during her reproductive years.

As the follicle undergoes maturation, it develops from a primary follicle to a secondary follicle to a Graafian follicle. *Oogenesis* is occurring (see fig. 11.6), and a secondary follicle contains a secondary oocyte pushed to one side of a fluid-filled cavity. In a *Graafian follicle,* the fluid-filled cavity increases to the point that the follicle wall balloons out on the surface of the ovary and bursts, releasing the secondary oocyte surrounded by a clear membrane and follicular cells. This is referred to as **ovulation,** and for the sake of convenience, the released oocyte is often called an ovum, or egg. Actually, the second meiotic

Animal Structure and Function

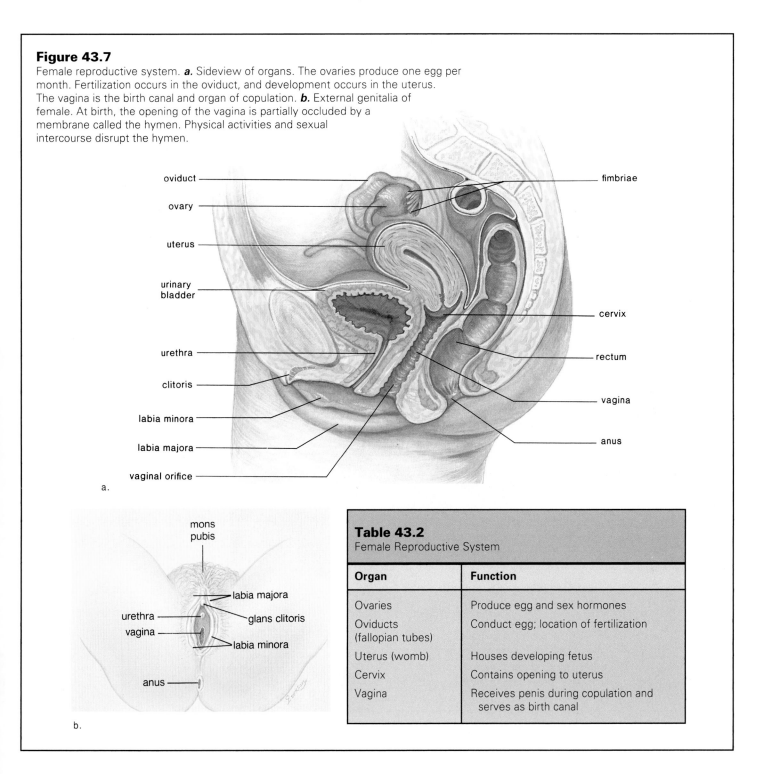

Figure 43.7
Female reproductive system. **a.** Sideview of organs. The ovaries produce one egg per month. Fertilization occurs in the oviduct, and development occurs in the uterus. The vagina is the birth canal and organ of copulation. **b.** External genitalia of female. At birth, the opening of the vagina is partially occluded by a membrane called the hymen. Physical activities and sexual intercourse disrupt the hymen.

Table 43.2
Female Reproductive System

Organ	Function
Ovaries	Produce egg and sex hormones
Oviducts (fallopian tubes)	Conduct egg; location of fertilization
Uterus (womb)	Houses developing fetus
Cervix	Contains opening to uterus
Vagina	Receives penis during copulation and serves as birth canal

division does not take place unless fertilization occurs. In the meantime, the follicle is developing into the **corpus luteum.** If pregnancy does not occur, the corpus luteum begins to degenerate after about 10 days.

The ovarian cycle is under the control of the gonadotropic hormones, **follicle-stimulating hormone (FSH)** and **luteinizing hormone (LH)** (fig. 43.9 and table 43.3). The gonadotropic hormones are not present in constant amounts and instead are secreted at different rates during the cycle. For simplicity's sake, it is convenient to emphasize that during the first half, or *follicular phase,* of the cycle, FSH promotes the development of a follicle, which secretes estrogen. As the estrogen level in the blood rises, it exerts feedback control over the anterior pituitary secretion of FSH so that the follicular phase comes to an end.

Figure 43.8

Anatomy of ovary and follicle. **a.** As a follicle matures, the oocyte enlarges and is surrounded by a mantle of follicular cells and fluid. Eventually, ovulation occurs, the mature follicle ruptures, and the secondary oocyte is released. A single follicle actually goes through all stages in one place within the ovary. **b.** Scanning electron micrograph of a secondary follicle. Magnification, X80.

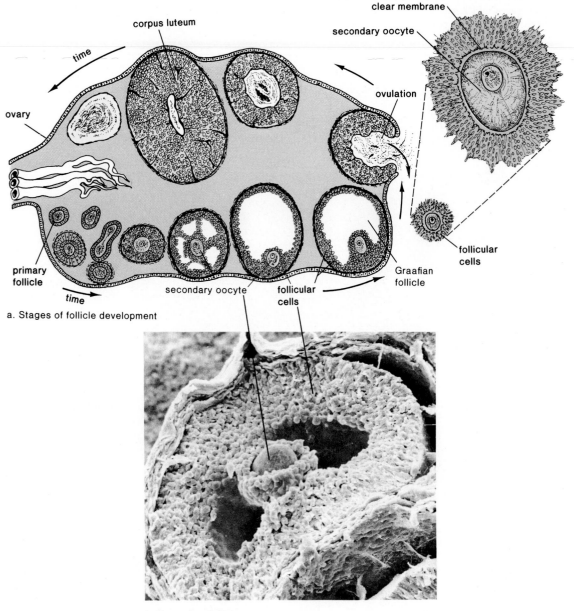

a. Stages of follicle development

b. Secondary follicle

Presumably, the high level of estrogen in the blood also causes the hypothalamus suddenly to secrete a large amount of GnRH. This leads to a surge of LH production by the anterior pituitary and ovulation at about the fourteenth day of a 28-day cycle (fig. 43.10).

During the second half, or *luteal phase,* of the ovarian cycle, it is convenient to emphasize that LH promotes the development of the corpus luteum, which secretes progesterone. As the blood level of progesterone rises, it exerts feedback control over anterior pituitary secretion of LH so that the corpus luteum begins to degenerate. As the luteal phase comes to an end, menstruation occurs.

A negative feedback system involving the hypothalamus and anterior pituitary causes one ovarian follicle per month to produce a secondary oocyte. Following ovulation, the follicle, which secretes estrogen before ovulation, develops into the corpus luteum, which produces progesterone after ovulation.

Animal Structure and Function

Table 43.3
Ovarian and Uterine Cycles (Simplified)

Ovarian Cycle	Events	Uterine Cycle	Events
Follicular phase—Days 1-13 **Ovulation—Day 14***	FSH Follicle maturation Estrogen	Menstruation—Days 1-5 Proliferative phase—Days 6-13	Endometrium breaks down Endometrium rebuilds
Luteal phase—Days 15-28	LH Corpus luteum Progesterone	Secretory phase—Days 15-28	Endometrium thickens and glands are secretory

*Assuming a 28-day cycle

Figure 43.9
Hypothalamic-pituitary-gonad system (simplified) as it functions in
the female. GnRH is a hypothalamic-releasing hormone that
stimulates the anterior pituitary to secrete FSH and LH. These
gonadotropic hormones act on the ovaries. FSH promotes the
development of the follicle that later, under the influence of LH,
becomes the corpus luteum. Negative feedback controls the level of
all hormones involved.

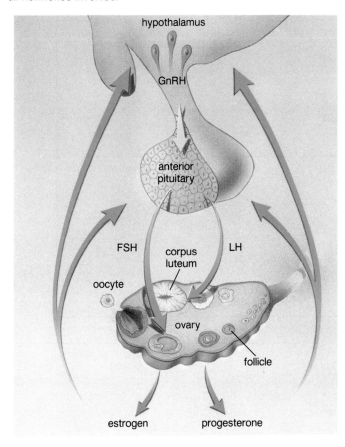

The Uterine Cycle The female sex hormones, estrogen and
progesterone, have numerous functions. The effect these hormones have on the endometrium of the uterus causes the uterus to
undergo a cyclical series of events known as the **uterine cycle**
(table 43.3). Cycles that last 28 days are divided as follows.

During *days 1–5*, there is a low level of female sex
hormones in the body, causing the uterine lining to disintegrate and
its blood vessels to rupture. A flow of blood, known as the *menses*,
passes out of the vagina during a period of **menstruation,** also
known as the menstrual period.

During *days 6–13*, increased production of estrogen by an
ovarian follicle causes the endometrium to thicken and to become
vascular and glandular. This is called the proliferative phase of the
uterine cycle.

Ovulation usually occurs on the fourteenth day of the 28-
day cycle.

During *days 15–28*, increased production of progesterone
by the corpus luteum causes the endometrium to double in thickness and the uterine glands to mature, producing a thick mucoid
secretion. This is called the secretory phase of the uterine cycle.
The endometrium now is prepared to receive the developing
embryo, but if pregnancy does not occur, the corpus luteum
degenerates and the low level of sex hormones in the female body
causes the uterine lining to break down. This is evident, due to the
menstrual discharge that begins at this time. Even while menstruation is occurring, the anterior pituitary begins to increase its
production of FSH and a new follicle begins to mature. Table 43.3
indicates how the ovarian cycle controls the uterine cycle.

At about age 45–50, the ovaries gradually cease to respond to the anterior pituitary hormones. Eventually, no more
follicles are produced. Following this occurrence, called menopause, menstruation ceases entirely.

The female sex hormones, estrogen and progesterone, regulate
the uterine cycle, in which the endometrium first builds up,
becomes secretory, and then is shed (menstruation).

Reproduction in Animals

Figure 43.10

Plasma hormonal levels associated with the ovarian and uterine cycles. During the follicular phase, FSH produced by the anterior pituitary promotes the maturation of a follicle in the ovary. The structure produces increasing levels of estrogen, which causes the endometrial lining of the uterus to thicken. After ovulation and during the luteal phase, LH promotes the development of the corpus luteum. This structure produces increasing levels of progesterone, which causes the endometrial lining to become secretory. Menstruation begins when progesterone production declines to a low level.

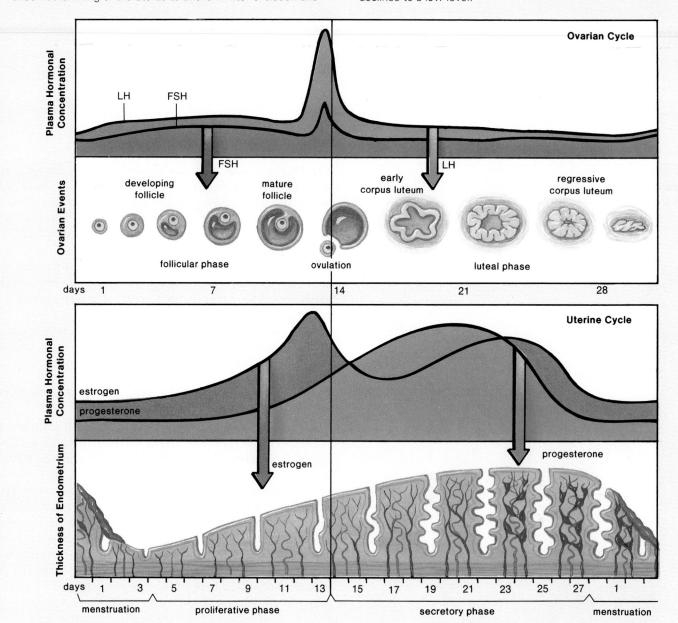

Pregnancy

If fertilization does occur, an embryo begins development even as it travels down the oviduct to the uterus. The endometrium is now prepared to receive the developing embryo, which becomes embedded in the lining several days following fertilization. This process, called *implantation,* causes the female to become pregnant. During implantation, an outer layer of cells surrounding the embryo produces a gonadotropic hormone (*HCG,* or human chorionic gonadotropin) that prevents degeneration of the corpus luteum and instead causes it to secrete even larger quantities of progesterone. The corpus luteum may be maintained for as long as 6 months, even after the placenta is fully developed.

The *placenta* originates from both maternal and fetal tissues. It is the region of exchange of molecules between fetal and maternal blood, although there is rarely any mixing of the 2 types of blood. After its formation, the placenta continues production of

Animal Structure and Function

HCG and begins production of progesterone and estrogen. Progesterone and estrogen have 2 effects. They shut down the anterior pituitary so no new follicles mature, and they maintain the lining of the uterus so the corpus luteum is not needed. There is no menstruation during pregnancy.

HCG is produced in such large quantities that pregnant women excrete considerable amounts of it in their urine. The chemical tests for pregnancy are based on the detection of this hormone in the urine. HCG is so readily available that it is routinely used in many teaching and research laboratories. For example, it is even used to induce ovulation in female frogs to obtain eggs for embryological studies.

Illness of the mother or a failure of the placenta to continue adequate HCG production may occasionally cause corpus luteum activity to decrease. Due to a resultant drop in progesterone levels, menstruation-like breakdown of the endometrium and loss of the implanted embryo is initiated. Although the blood flow is somewhat heavier and longer than normal as miscarriage occurs, the terminated pregnancy may be mistaken for a somewhat delayed menstrual period.

Female Sex Hormones

The female sex hormones, estrogen and progesterone, have many other affects in the body, in addition to their affect on the uterus. Estrogen is largely responsible for the secondary sex characteristics in females, including body hair and fat distribution. In general, females have a more rounded appearance than males because of a greater accumulation of fat beneath the skin. Also, the pelvic girdle enlarges in females, and the pelvic cavity has a larger relative size compared to males. This means that females have wider hips. Both estrogen and progesterone are also required for breast development.

Breasts A female breast contains 15–25 lobules, each with its own mammary duct (fig. 43.11). This duct begins at the nipple and divides into numerous other ducts, which end in blind sacs called *alveoli*. In a nonlactating breast, the ducts far outnumber the alveoli because alveoli are made up of cells that can produce milk.

Milk is not produced during pregnancy. Prolactin is needed for lactation (milk production) to begin, and production of this hormone is suppressed by the feedback inhibition estrogen and progesterone have on the anterior pituitary during pregnancy. It takes a couple of days after delivery of a baby for milk production to begin. In the meantime, the breasts produce a watery, yellowish white fluid called *colostrum,* which has a similar composition to milk but contains more protein and less fat.

Reproductive Concerns

Two primary reproductive concerns are control of pregnancy and avoidance of sexually transmitted diseases.

Control of Pregnancy

Sterility causes a person to have no children, and *infertility* causes a person to have fewer children than he or she desires despite frequent intercourse. The 2 major causes of these conditions in females are blocked oviducts, possibly due to a sexually transmit-

Figure 43.11
Anatomy of breast. The female breast contains lobules consisting of ducts and alveoli. The alveoli are lined by milk-producing cells in the lactating (milk-producing) breast.

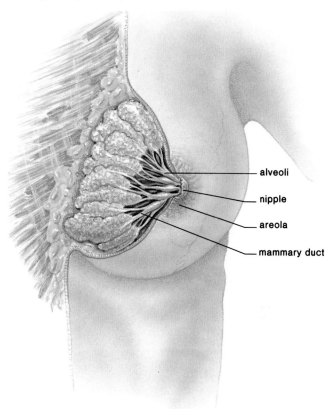

- alveoli
- nipple
- areola
- mammary duct

ted disease, and failure to ovulate due to low body weight. Endometriosis, the spread of uterine tissue beyond the uterus, is also a cause. If no obstruction is apparent and body weight is normal, it is possible to give females HCG extracted from the urine of postmenopausal women. This treatment causes multiple ovulations and sometimes multiple pregnancies.

The most frequent causes of sterility and infertility in males are low sperm count and/or a large proportion of abnormal sperm. Disease, radiation, chemical mutagens, too much heat near the testes, and the use of psychoactive drugs can contribute to this condition.

When reproduction does not occur in the usual manner, couples often seek alternative reproductive methods, which include artificial insemination (sperm are placed in the vagina by a physician), in vitro fertilization (fertilization takes place in laboratory glassware and the zygote is inserted into the uterus of the woman), and surrogate motherhood (a woman has another woman's child).

Sometimes couples wish to prevent a possible pregnancy. Several means of *birth control* are readily available in the United States, and the most common are listed in table 43.4. Effectiveness of the method refers to the number of women per year who will not get pregnant even though they are regularly engaging in sexual

Table 43.4
Common Birth-Control Methods

Name	Procedure	Methodology	Effectiveness*	Action Needed	Risk
Vasectomy	Vas deferentia are cut and tied	No sperm in semen	Almost 100%	Sexual freedom	Irreversible sterility
Tubal ligation	Oviducts are cut and tied	No eggs in oviduct	Almost 100%	Sexual freedom	Irreversible sterility
Pill	Medication is taken daily	Shuts down anterior pituitary	Almost 100%	Sexual freedom	Thromboembolism
Norplant	Progesterone implant is placed under skin by physician	Shuts down anterior pituitary	99.7%	Sexual freedom	_____
IUD	Inserted into uterus by physician	Prevents implantation	More than 90%	Sexual freedom	Infection
Sponge and diaphragm	Inserted into vagina at cervix	Blocks entrance of sperm into uterus	With jelly about 90%	Must be inserted each time before intercourse	_____
Condom	Sheath is placed over erect penis	Traps sperm	About 85%	Must be placed on penis at time of intercourse	_____
Coitus interruptus (withdrawal)	Male withdraws penis before ejaculation	Prevents sperm from entering vagina	About 80%	Intercourse must be interrupted before ejaculation	_____
Jellies, creams, foams	Inserted into vagina	Spermicidal chemicals kill a large number of sperm	About 75%	Must be inserted before intercourse	_____
Rhythm method	Determine day of ovulation by record keeping; testing by various methods	Avoid day of ovulation	About 70%	Limits sexual activity	_____

*Effectiveness is the average percentage of women who did not become pregnant in a population of 100 sexually active women using the technique for one year.

intercourse. For example, with the least effective method given in table 43.4, 70 out of 100, or 70% of sexually active women will not get pregnant, while 30 women will get pregnant, within a year. If no birth-control method is used, it is expected that 80 out of 100 women will be pregnant within a year.

Sexually Transmitted Diseases

Sexually transmitted diseases (STDs) are caused by organisms ranging from viruses to arthropods; however, we will discuss only certain STDs caused by viruses and bacteria. Unfortunately, for unknown reasons, humans cannot develop good immunity to any of the STDs. Therefore, prompt and proper medical treatment should be received when exposed to an STD. A condom serves as protection against spreading STDs. The concomitant use of a spermicide containing nonoxynol-9 gives added protection.

STDs of Viral Origin
AIDS (acquired immunodeficiency syndrome), genital herpes, and genital warts are all viral infections. AIDS is discussed later in this chapter.

Genital herpes is caused by herpes simplex virus, of which type 1 usually causes cold sores and fever blisters, while type 2 more often causes genital herpes. Many times an infected person has no symptoms, but if symptoms are present, there are painful ulcers on the genitals that heal and then reappear. The ulcers may be accompanied by fever, pain upon urination, and swollen lymph nodes. At this time, the individual has an increased risk of acquiring an AIDS infection. Exposure in the birth canal can cause an infection in the newborn, which leads to neurological disorders and even death. Birth by cesarean section prevents this possibility.

Genital warts are caused by the human papilloma viruses (HPVs). Many times, carriers do not have any sign of warts or only flat lesions may be present. If visible warts are removed, they may recur. HPVs are now associated with cancer of the cervix, as well as tumors of the vulva, the vagina, the anus, and the penis. Some researchers believe that the viruses are involved in 90%–95% of all cases of cancer of the cervix.

STDs of Bacterial Origin
Gonorrhea, chlamydia, and syphilis are STDs caused by bacteria. Therefore, they are curable by antibiotic therapy.

Gonorrhea is caused by the bacterium *Neisseria gonorrheae*. Diagnosis in the male is not difficult, as long as he displays the typical symptoms of pain upon urination and a thick, greenish yellow urethral discharge. In males and females, a latent infection leads to pelvic inflammatory disease (PID), in which the vas deferentia or oviducts are affected. As the inflamed tubes heal, they may become partially or completely blocked by scar tissue, resulting in sterility or infertility. If a baby is exposed during the process of birth, an eye infection leading to blindness can result. All newborns are given eye drops to prevent this possibility.

Chlamydia is name for the tiny bacterium that causes it (*Chlamydia trachomatis*). Chlamydia is the most common cause of nongonococcal urethritis (NGU), which is often difficult to distinguish from gonococcal urethritis. Since an infection can also cause PID, physicians routinely prescribe medicines for both gonorrhea and chlamydia at the same time. Chlamydia also causes cervical ulcerations, which increase the risk of acquiring AIDS. If a baby comes in contact with chlamydia during birth, inflammation of the eyes or pneumonia can result.

Syphilis, which is caused by the bacterium *Treponema pallidum,* has 3 stages, which are typically separated by latent periods. In the primary stage, a hard chancre (ulcerated sore with hard edges) appears. In the secondary stage, a rash appears all over the body—even on the palms of the hands and on the soles of the feet. During the tertiary stage, syphilis may affect the cardiovascular and/or nervous system. An infected person may become mentally retarded, blind, walk with a shuffle, or show signs of insanity. *Gummas,* which are large destructive ulcers, may develop on the skin or within the internal organs. Syphilitic bacteria can cross the placenta, causing a baby to be stillborn or have many and various anatomical malformations. Unlike the other STDs discussed, there is a blood test to diagnose syphilis.

> Sexually transmitted diseases include AIDS; herpes, which can recur; genital warts, which lead to cancer of the cervix; gonorrhea and chlamydia, which cause PID; and syphilis, which has cardiovascular and neurological complications if untreated.

AIDS

AIDS is caused by a group of related retroviruses known as HIV (human immunodeficiency viruses). The World Health Organization now estimates that in 9 years, over 40 million persons worldwide will be infected with HIV. Presently in the United States, 90% of those infected are males and 10% are female. In Africa, where heterosexual transmission is more prevalent, an equal number of men and women are infected.

Transmission of AIDS

After HIV enter the blood, they infect helper T lymphocytes (p. 579) of a type known as T4 cells. The T4 cells die and their number declines as the disease progresses. About 28% of AIDS cases are intravenous drug abusers who have contracted the disease after using a contaminated needle. About 3% of AIDS cases are persons who have received infected blood during a blood transfusion.

Figure 43.12
Balance of power between HIV and the immune system during the course of an HIV infection. During the time that a person is an asymptomatic carrier, the concentration of HIV in the body is high, but then the immune system becomes active. While an infected person has ARC, the immune system begins to lose the battle and eventually the battle is lost when a person develops full-blown AIDS. Source: Data from R. R. Redfield and D. S. Burke, "HIV Infection: The Clinical Picture" in *Scientific American*, October 1988.

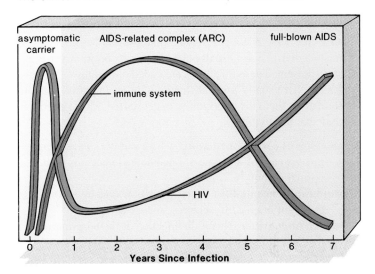

AIDS, however, is primarily a sexually transmitted disease. Semen can contain the virus, but more likely an infected lymphocyte does. About 64% of the total cases in the United States are homosexual men who practice anal intercourse. (About 6% of these are also IV drug abusers.) Unlike the vagina, the epithelial lining of the rectum is a thin, single-celled layer that is easily torn during intercourse. Nevertheless, heterosexual transmission does occur and may become more prevalent as more females become infected. One unhappy side effect to female infection is the fact that viruses and infected lymphocytes can pass to a fetus via the placenta or to an infant via the mother's milk. Presently, infected infants account for about 1% of all AIDS cases.

Symptoms of AIDS

An HIV infection can be divided into 3 stages (fig. 43.12).

Asymptomatic Carrier

Only 1%–2% of those newly infected have mononucleosis-like symptoms that may include fever, chills, aches, swollen lymph glands, and an itchy rash. These symptoms disappear, and there are no other symptoms for 9 months or longer. Although the individual exhibits no symptoms during this stage, he or she is highly infectious. The standard HIV blood test for the presence of antibody becomes positive during this stage.

AIDS Related Complex (ARC)

The most common symptom of ARC is swollen lymph glands in the neck, armpits, or groin that persist for 3 months or more. There is severe fatigue unrelated to exercise or drug use; unexplained persistent or recurrent fevers, often with night sweats; persistent cough not associated with smoking, a cold, or the flu; and persistent diarrhea. Also possible are signs of nervous system impairment, including loss of memory, inability to think clearly, loss of judgment, and/or depression.

When the individual develops non-life-threatening and recurrent infections such as thrush or herpes simplex, it is a signal that full-blown AIDS will occur shortly.

Full-Blown AIDS

In this final stage, there is severe weight loss and weakness due to persistent diarrhea and usually one of several opportunistic infections is present. These infections are called opportunistic because the body can usually prevent them—only an impaired immune system gives them the opportunity to get started. These infections include the following:

***Pneumocystis carinii* pneumonia.** There is not a single documented case of this type of pneumonia in persons with normal immunity.

Toxoplasmic encephalitis. In AIDS patients, this infection leads to loss of brain cells, seizures, and weakness.

***Myobacterium avium*.** This is an infection of the bone marrow that leads to a decrease in red blood cells, white blood cells, and platelets.

Kaposi's sarcoma. A cancer of the blood vessels that causes reddish purple, coin-size spots and lesions on the skin.

Treatment

The drug zidovudine (also called azidothymidine, or AZT) and dideoxyinosine (DDI) prevent HIV reproduction in cells. Proteases are enzymes HIV needs to bud from the host cell; researchers are hopeful that a protease inhibitor drug will soon be available.

A number of different types of vaccines are in, or are expected to be in, human trials. Several of these are subunit vaccines that utilize genetically engineered proteins that resemble those found in HIV. For example, HIV-1, the cause of most AIDS cases in the United States, has an outer envelope molecule called GP120 (fig. 43.13). When GP120 combines with a CD4 molecule that projects from a helper T lymphocyte, the virus enters the cell. There are subunit vaccines that make use of GP120. An entirely different approach is being taken by Jonas Salk, who developed the polio vaccine. His vaccine utilizes whole HIV-1 killed by treatment with chemicals and radiation. So far, this vaccine has been found to be effective against experimental HIV-1 infection in chimpanzees, and clinical trials will occur soon.

Figure 43.13

An HIV-1 has an envelope molecule called GP120 that allows it to attach to CD4 molecules that project from a T4 cell. Infection of the T4 cell follows, and HIV eventually buds from the infected T4 cell. If the immune system can be trained by the use of a vaccine to attack and destroy all cells that bear GP120, a person would not be able to be infected with HIV-1.

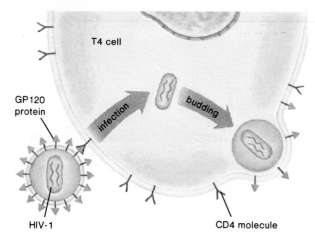

AIDS Prevention

Shaking hands, hugging, social kissing, coughing or sneezing, and swimming in the same pool do not transmit the AIDS virus. You cannot get AIDS from inanimate objects such as toilets, doorknobs, telephones, office machines, or household furniture.

The following behaviors will help prevent the spread of AIDS:

1. Do not use alcohol or drugs in a way that prevents you from being in control of your behavior. Especially, do not inject drugs into veins, but if you are an intravenous drug user and cannot stop your behavior, always use a sterile needle for injection or one cleansed by bleach.

2. Refrain from multiple sex partners, especially with homosexual or bisexual men or intravenous drug users of either sex. Either abstain from sexual intercourse or develop a long-term monogamous (always the same partner) sexual relationship with a partner who is free of HIV and is not an intravenous drug user.

3. If you are uncertain about your partner, always use a latex condom. Follow the directions, and also use a spermicide containing nonoxynol-9, which kills viruses and virus-infected lymphocytes. The risk of contracting AIDS is greater in persons who already have a sexually transmitted disease.

Summary

1. Asexual reproduction may quickly produce a large number of offspring exactly like the single parent and is advantageous when environmental conditions are relatively unchanging.

2. Sexual reproduction involves the use of gametes and produces offspring that are slightly different from the parents. This may be advantageous if the environment is changing.

3. Drying out of gametes and zygotes is not a threat for aquatic animals, and they tend to practice external fertilization. Drying out is a problem for land

Animal Structure and Function

animals, and they tend to practice internal fertilization.

4. In human males, sperm are produced in the testes, mature in the epididymides, and may be stored in the vas deferens before entering the urethra, along with seminal fluid (produced by the seminal vesicles, prostate gland, and Cowper's gland), prior to ejaculation during male orgasm, when the penis becomes erect.

5. Spermatogenesis occurs in the seminiferous tubules of the testes, which also produce testosterone in interstitial cells. Testosterone maintains the secondary sex characteristics of males, such as low voice, facial hair, and increased muscle strength.

6. FSH (also called ICSH) from the anterior pituitary stimulates spermatogenesis, and LH stimulates testosterone production. A hypothalamic releasing hormone, GnRH, controls anterior pituitary production and FSH and LH release. The level of testosterone in the blood controls the secretion of GnRH and the anterior pituitary hormones by a negative feedback system.

7. In females, an egg produced by an ovary enters an oviduct, which leads to the uterus. The uterus opens into the vagina. The genital area of women includes the vaginal opening, clitoris, and labia minora and labia majora.

8. In either ovary, one follicle a month matures, produces a secondary oocyte, and becomes a corpus luteum. This is called the ovarian cycle. The follicle and corpus luteum produce estrogen and progesterone, the female sex hormones.

9. A uterine cycle occurs concurrently with the ovarian cycle. In the first half of these cycles (days 1–13 before ovulation), the anterior pituitary produces FSH and the follicle produces estrogen. Estrogen causes the uterine lining to increase in thickness. In the second half of these cycles (days 15–28 after ovulation), the anterior pituitary produces LH and the follicle produces progesterone. Progesterone causes the uterine lining to become secretory. Feedback control of the hypothalamus and anterior pituitary causes the level of estrogen and progesterone to fluctuate. When they are at a low level, menstruation begins.

10. If fertilization occurs, the corpus luteum is maintained because of HCG production. Progesterone production does not cease, and the zygote implants itself in the thick uterine lining.

11. Estrogen and progesterone maintain the secondary sex characteristics of females, including less body hair than males, a wider pelvic girdle, more rounded appearance, and development of breasts.

12. Infertile couples are increasingly resorting to alternative methods of reproduction. Numerous birth-control methods and devices are available for those who wish to prevent pregnancy.

13. Sexually transmitted diseases include AIDS; herpes, which can recur; genital warts, which lead to cancer of the cervix; gonorrhea and chlamydia, which cause PID; and syphilis, which has cardiovascular and neurological complications if untreated.

14. AIDS is caused by HIV, which infect helper T4 cells. As the number of T4 cells declines, the symptoms of AIDS related complex and then full-blown AIDS appear. Drugs are available that prevent HIV reproduction in cells, and it is hoped that a vaccine will one day be available. In the meantime, all persons should take the proper steps to prevent the spread of AIDS.

Writing Across the Curriculum

In order to practice writing skills, students should write out the answers to any or all of the study questions and the critical thinking questions. The study questions are sequenced in the same order as the text. Suggested answers to the critical thinking questions are in appendix D.

Study Questions

1. Give examples of asexual and sexual reproduction among animals. Relate these practices to environmental conditions.
2. Discuss the human reproductive system as an adaptation to life on land.
3. Discuss the anatomy and physiology of the testes. Describe the structure of sperm.
4. Give the path of sperm. What glands contribute fluids to semen?
5. Name the endocrine glands involved in maintaining the sex characteristics of males and the hormones produced by each.
6. Discuss the anatomy and physiology of the ovaries. Describe ovulation.
7. Give the path of the egg. Where do fertilization and implantation occur? Name 2 functions of the vagina.
8. Discuss hormonal regulation in the female by giving the events of the uterine cycle and relating these to the ovarian cycle. In what way is menstruation prevented if pregnancy occurs?
9. Describe at least 3 common sexually transmitted diseases.
10. What means of birth control help prevent the spread of AIDS? What other measures can be taken to protect oneself from AIDS?

Objective Questions

1. Which of these is a requirement for sexual reproduction?
   a. male and female parents
   b. production of gametes
   c. optimal environmental conditions
   d. aquatic habitat
2. Internal fertilization
   a. prevents the drying out of gametes and zygotes.
   b. must take place on land.
   c. is practiced by humans.
   d. Both a and c.
3. Which of these is mismatched?
   a. interstitial cells—testosterone
   b. seminiferous tubules—sperm production
   c. vas deferens—seminal fluid production
   d. penis—erection

4. FSH
   a. occurs in females but not males.
   b. stimulates the seminiferous tubules to produce sperm.
   c. secretion is controlled by GnRH.
   d. Both b and c.
5. Which of these combinations is most likely to be present before ovulation occurs?
   a. FSH, corpus luteum, estrogen, secretory uterine lining
   b. LH, follicle, progesterone, thick uterine lining
   c. FSH, follicle, estrogen, uterine lining becoming thick
   d. LH, corpus luteum, progesterone, secretory uterine lining
6. In tracing the path of sperm, you would mention vas deferens before
   a. testes.
   b. epididymis.
   c. urethra.
   d. uterus.
7. An oocyte is fertilized in the
   a. vagina.
   b. uterus.
   c. oviduct.
   d. ovary.
8. During pregnancy,
   a. the ovarian cycle and uterine cycle occur more quickly than before.
   b. GnRH is produced at a higher level than before.
   c. the ovarian cycle and uterine cycle do not occur.
   d. the female secondary sex characteristics are not maintained.

9. Which of the following means of birth control is most effective in preventing AIDS?
   a. condom
   b. pill
   c. diaphragm
   d. spermicidal jelly
10. Which of these sexually transmitted diseases is mismatched with its cause?
    a. AIDS—bacterial infection of red blood cells
    b. gonorrhea—bacterial infection of genital tract
    c. chlamydia—bacterial infection of genital tract
    d. syphilis—systemic bacterial infection
11. Label this diagram of the male reproductive system and trace the path of sperm:

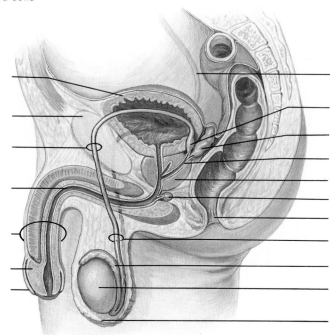

Concepts and Critical Thinking

1. *Successful reproduction on land requires certain adaptations.*

   Contrast the manner in which reptiles are adapted to reproduce on land with the manner in which humans are adapted to reproduce on land.

2. *Reproduction is under hormonal rather than nervous control.*

   Why would you have predicted hormonal rather than nervous control of reproduction?

Selected Key Terms

regeneration (re-jen″er-a′shun) 695
budding (bud′ing) 695
parthenogenesis (par″thē-no-jen′-ē-sis) 695
penis (pe′nis) 697

testosterone (tes-tos′tĕ-rōn) 698
ovary (o′vah-re) 700
oviduct (o′vĭ-dukt) 700
uterus (u′ter-us) 700
vagina (vah-ji′nah) 700

follicle (fol′ĭ-k′l) 700
ovulation (o″vu-la′shun) 700
corpus luteum (kor′pus lu′te-um) 701
uterine cycle (u′ter-īn si′k′l) 703
menstruation (men″stroo-a′shun) 703

Animal Structure and Function

44

Development in Animals

An iguana emerging from its shell, a structure that allows reptiles to develop on land. Within the shelled egg, the extraembryonic membranes perform many necessary functions, and they provide a watery environment. In effect, all vertebrates are surrounded by water as they develop.

Your study of this chapter will be complete when you can

1. name and define 3 processes that occur whenever there is a developmental change;
2. compare the early developmental stages of the lancelet, frog, and chick;
3. explain the germ layer theory of development, and give examples;
4. draw and label a cross section of a typical vertebrate embryo at the neurula stage of development;
5. give evidence that differentiation probably begins soon after formation of the zygote;
6. give evidence that morphogenesis can be accounted for by interactions between tissues;
7. state the stages of embryonic development in humans, and compare these to the development of the chick;
8. list the extraembryonic membranes, and give their function in chicks and in humans;
9. briefly outline the developmental changes in humans from the fetus to adult;
10. define aging, and discuss 3 theories of the aging process.

The study of development concerns the events and processes that occur as a single cell becomes a complex organism. These same processes are also seen as the newly born or hatched organism matures, as lost parts regenerate, as a wound heals, and even during aging. Therefore, it is customary to stress that the study of development encompasses not only embryology (development of the embryo) but these other events as well.

Development requires growth, differentiation, and morphogenesis. When an organism increases in size, we say that it has grown. During *growth,* cells divide, get larger, and divide once again. **Differentiation** occurs when cells become specialized in structure and function. A muscle cell looks and acts quite differently than a nerve cell, for example. **Morphogenesis** goes one step beyond growth and differentiation. It occurs when body parts become shaped and patterned into a certain form. There is a great deal of difference between your arm and leg for example, even though they contain the same types of tissues.

We will discuss these processes as they apply to development of the embryo, but keep in mind that they also occur whenever an organism goes through any developmental change.

Growth, differentiation, and morphogenesis are 3 processes that are seen whenever a developmental change occurs.

Early Developmental Stages

Embryological development begins when the sperm fertilizes the egg. Each gamete has a haploid number of chromosomes, therefore the resulting zygote has the diploid number. Gene expression is required for development to proceed normally, and we will stress this again in later sections.

All chordate embryos go through the same early developmental stages of cleavage, blastulation, gastrulation, and neurulation (fig. 44.1). The presence of *yolk,* which is dense, nutrient material, however, affects the manner in which embryonic cells complete the first 3 stages and hence the appearance of the embryo at the end of each stage. Varying amounts of yolk will result in embryos with different appearances. Table 44.1 indicates the amount of yolk in the 4 embryos discussed in this chapter and relates the amount of yolk to the environment in which the animal develops. The 2 animals (lancelet and frog) that develop in water

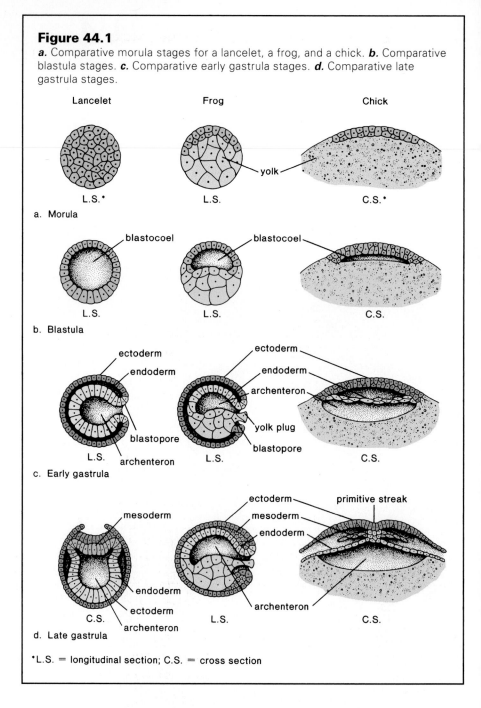

Figure 44.1
a. Comparative morula stages for a lancelet, a frog, and a chick. *b.* Comparative blastula stages. *c.* Comparative early gastrula stages. *d.* Comparative late gastrula stages.

Lancelet Frog Chick

a. Morula

b. Blastula

c. Early gastrula

d. Late gastrula

*L.S. = longitudinal section; C.S. = cross section

Table 44.1
Amount of Yolk in Eggs versus Location of Development

Animal	Yolk	Location of Development
Lancelet	Little	External in water
Frog	Some	External in water
Chick	Much	Within hard shell
Human	Little	Inside mother

Animal Structure and Function

have less yolk than the chick because development in these 2 animals proceeds quickly to a swimming larval stage that can feed itself. But the chick is representative of animals that have solved the problem of reproduction on land, in part, by providing a great deal of yolk within a hard shell. Development continues in the shell until there is an offspring capable of land existence.

Early stages of human development resemble those of the chick embryo, yet this resemblance cannot be related to the amount of yolk because the human egg contains little yolk. But the evolutionary history of these 2 animals can provide an answer for this similarity. Both birds (e.g., chicks) and mammals (e.g., humans) are related to reptiles, and this explains why all 3 groups develop similarly, despite a difference in the amount of yolk in the eggs.

> The amount of yolk affects the manner in which lancelets, frogs, and chicks complete the first 3 stages of development.

Cleavage and Formation of Blastula

Cell division without growth occurs during **cleavage** (fig. 44.2). DNA replication and mitosis occur repeatedly, and the cells get smaller with each division. Chordates, being deuterostomes, have a pattern of cleavage that is radial and indeterminate (see fig. 27.2). The term *radial* means that any plane passing through the major axis will divide the embryo into 2 symmetrical halves. The term *indeterminate* means that the cleavage cells have not differentiated, and therefore their developmental fate is not yet set.

In a lancelet, the cell divisions are equal, and the cells are of uniform size, whereas in a frog, the upper cells are smaller than the lower cells. This difference in size occurs because the upper cells at the animal pole contain little yolk, whereas the lower cells at the vegetal pole contain a large amount of yolk. Cells containing yolk cleave more slowly than those without yolk. The presence of extensive yolk in the chick egg causes cleavage to be incomplete, and only those cells lying on top of the yolk cleave. This means that although cleavage in a lancelet and a frog results in a ball of cells called the morula, no such ball is seen in a chick (fig. 44.1a). Instead, during the morula stage the cells spread out on a portion of the yolk.

The morula is a solid mass of cells, but then a cavity called the **blastocoel** develops (fig. 44.2). This "hollow-ball stage" of development, called the **blastula,** is best exhibited by the lancelet. In the frog, the blastocoel is formed at the animal pole only. The heavily laden yolk cells of the vegetal pole do not participate in this step. In a chick, the blastocoel is created when the cells lift up from the yolk and leave a space between the cells and the yolk (fig. 44.1b).

> Cleavage results in a ball of cells, which becomes the blastula when an internal cavity develops. In lancelets and frogs, the blastula is a hollow ball. In the chick, the cavity is found beneath cells that lie flat atop the great mass of yolk.

Gastrulation and Formation of Germ Layers

The **gastrula** stage is evident in a lancelet when certain cells begin to push, or invaginate, into the blastocoel, creating a double layer of cells (fig. 44.2). The outer layer is called the **ectoderm,** and the inner layer is called the **endoderm.** The space created by invagi-

Figure 44.2

Early development in a lancelet. A lancelet has little yolk as an embryo, and it can be used to exemplify the early stages of development in such animals. Cleavage produces a number of cells that form a cavity. Invagination during gastrulation produces the germ layers, ectoderm, and endoderm. Mesoderm arises from pouches that pinch off from the endoderm.

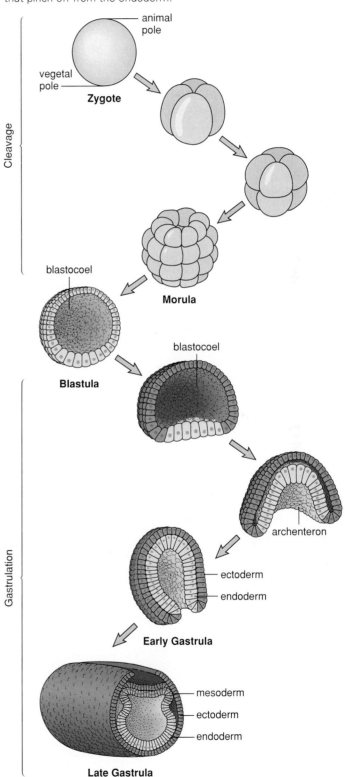

nation becomes the gut and is called either the primitive gut or the archenteron. The pore, or hole, created by invagination is called the blastopore, and in a lancelet, as well as in the other animals discussed here, it eventually becomes the anus (see fig. 27.2).

In the frog, the cells containing yolk do not participate in gastrulation and therefore do not invaginate. Instead, a slitlike blastopore is formed when the animal pole cells begin to invaginate from above. Following this, other animal pole cells move down over the yolk, and the blastopore becomes rounded when these cells also invaginate from below. At this stage, there are some yolk cells temporarily left in the region of the pore; these are called the yolk plug. In the chick, there is so much yolk that endoderm formation does not occur by invagination. Instead, an upper layer of cells differentiates into ectoderm, and a lower layer differentiates into endoderm (fig. 44.1*c*).

Gastrulation is not complete until 3 layers of cells have formed (fig. 44.2). The third, or middle, layer of cells is called the

mesoderm. In a lancelet, this layer begins as outpocketings from the primitive gut (see fig. 27.2). These outpocketings grow in size until they meet and fuse. In effect, then, 2 layers of mesoderm are formed, and the space between them is the coelom.

In the frog, cells from the dorsal lip of the blastopore migrate between the ectoderm and endoderm, forming the mesoderm. Later, a splitting of the mesoderm creates the coelom. In the chick, the mesoderm layer arises by an invagination of cells along the edges of a longitudinal furrow in the midline of the embryo. Because of its appearance this furrow is called the *primitive streak* (fig. 44.1*d*). Later, the newly formed mesoderm will split to give a coelomic cavity.

Ectoderm, mesoderm, and endoderm are called the primary **germ layers** of the embryo, and no matter how gastrulation takes place, the end result is the same: 3 germ layers are formed. It is possible to relate the development of future organs to these germ layers, as is done in table 44.2. Karl E. Von Baer, the nineteenth-century embryologist, first related later development to the early formation of germ layers. This is called the *germ layer theory*.

The 3 embryonic germ layers arise during gastrulation, when cells invaginate into the blastocoel. The development of organs can be related to the 3 germ layers: ectoderm, mesoderm, and endoderm.

Neurula

In chordate animals, newly formed mesoderm cells that lie along the main longitudinal axis of the animal coalesce to form a dorsal supporting rod called the **notochord.** The notochord persists in lancelets (see fig. 27.15), but in frogs, chicks, and humans, it is later replaced by the vertebral column.

The nervous system develops from ectoderm located just above the notochord. At first, a thickening of cells called the *neural plate* is seen along the dorsal surface of the embryo. Then, *neural folds* develop on either side of a neural groove, which becomes the *neural tube* when these folds fuse. Figure 44.3 shows cross sections of frog development to illustrate the formation of the neural tube. At this point, the embryo is called a *neurula*. Later, the anterior end of the neural tube develops into the brain.

Table 44.2
Organs Developed from the 3 Primary Germ Layers

Ectoderm	Mesoderm	Endoderm
Skin epidermis, including hair, nails, and sweat glands	All muscles	Lining of digestive tract, trachea, bronchi, lungs, gallbladder, and urethra
Nervous system, including brain, spinal cord, ganglia, nerves, and sense receptors	Dermis of skin	
	All connective tissue, including bone, cartilage, and blood	Liver
	Blood vessels	Pancreas
Lens and cornea of eye	Kidneys	Thyroid, parathyroid, and thymus glands
Lining of nose, mouth, and anus	Reproductive organs	Urinary bladder
Tooth enamel		

Figure 44.3
Development of neural tube and coelom in a frog embryo.
a. Ectoderm cells that lie above the future notochord (called presumptive notochord) thicken to form a neural plate. *b.* The neural groove and folds are noticeable as the neural tube begins to form. *c.* A splitting of the mesoderm produces a coelom, which is completely lined by mesoderm. *d.* A neural tube and coelom have now developed.

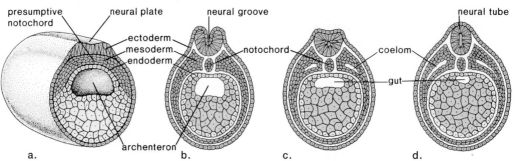

Animal Structure and Function

Midline mesoderm cells that did not contribute to the formation of the notochord now become 2 longitudinal masses of tissue. These 2 masses become blocked off into the somites, which give rise to segmental muscles in all chordates. In vertebrates, the somites also produce the vertebral bones.

With the formation of the nervous system, it is possible to show a generalized diagram (fig. 44.4) of a chordate embryo to illustrate the location of parts. Consideration of this figure with table 44.2 will help you relate the formation of chordate structures and organs to the 3 embryonic layers of cells: the ectoderm, the mesoderm, and the endoderm.

> During neurulation, the neural tube develops just above the notochord. At the neurula stage of development, a cross section of all chordate embryos is similar in appearance.

Differentiation and Morphogenesis

Differentiation accounts for specialization of tissues that have specific functions, and morphogenesis accounts for the formation of organs that have an overall pattern of shape and form. The process of differentiation must start long before different types of cells are recognizable, most likely even with the very first cleavage of the egg.

The reading in chapter 17 (p. 267) describes an experiment in which a tadpole cell nucleus is placed in an enucleated egg. Development proceeds normally, showing that tadpole nuclei are totipotent—they contain all the genetic information required to bring about complete development of the organism. Therefore, differentiation cannot be due to a parceling out of genes into the various embryonic cells. Instead, it must be due to the expression of particular genes, controlled at first by *ooplasmic segregation,* which is the distribution of maternal cytoplasmic contents to the cells of the morula:

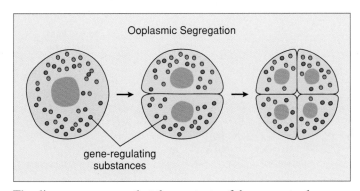

Ooplasmic Segregation

gene-regulating substances

The diagram proposes that the contents of the egg cytoplasm are not distributed uniformly. Following cleavage, therefore, embryonic cells will differ according to the cytoplasmic contents they receive.

The fact that the cytoplasm of an egg is not uniform can be substantiated by considering the egg of a frog. As we mentioned earlier, a frog's egg is polar: there is an animal pole and a vegetal pole (fig. 44.5a). The vegetal pole, which is distinguishable by the presence of yolk, becomes the germ layer called

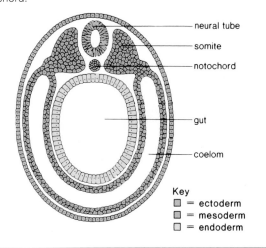

Figure 44.4
Typical cross section of a chordate embryo at the neurula stage. Each of the germ layers, indicated by color (see key), can be associated with the later development of particular parts (see table 44.2). The somites give rise to the muscles of each segment and to the vertebrae, which replace the notochord.

neural tube
somite
notochord
gut
coelom

Key
☐ = ectoderm
☐ = mesoderm
☐ = endoderm

endoderm. Researchers working with the frog *Xenopus* have been able to identify a particular mRNA (called Vg 1) that is localized at the vegetal pole during the process of oogenesis. It is translated into a particular peptide growth factor that perhaps induces the formation of endoderm. **Induction** is the ability of a chemical or a tissue to influence the development of another tissue. The inducing chemical is called a *signal.*

After a frog's egg is fertilized, contents of the egg shift position, and a *gray crescent* appears on the egg opposite the point where the sperm entered (fig. 44.5a). The gray crescent marks the dorsal side of the embryo. It most likely represents a combination of animal and vegetal pole contents, and it is speculated that the gray crescent might contain growth factors from each pole. The dorsal side of the embryo develops a notochord and nervous system—perhaps due to the action of these growth factors.

This line of reasoning is substantiated by an experiment performed by Hans Spemann, who received a Nobel Prize in 1935 for his extensive work in embryology. Spemann showed that if the gray crescent is divided equally by the first cleavage, each experimentally separated daughter cell develops into a complete embryo. If he caused the egg to divide so that only one daughter cell receives the gray crescent, however, only that cell becomes a complete embryo (fig. 44.5b). The other cell gives rise to a tissue mass that lacks any sign of a notochord and nervous system.

Spemann later showed that the gray crescent becomes the dorsal lip of the blastopore, where gastrulation begins. Since this region is necessary for complete development, he called it the primary organizer. The cells closest to Spemann's primary organizer become endoderm, those farther away become mesoderm,

Figure 44.5

Importance of the gray crescent in frog development **a.** The position of the gray crescent can be correlated with the anterior/posterior and dorsal/ventral axes of the body. **b.** The first cleavage normally divides the gray crescent in half and each daughter cell is capable of developing into a complete tadpole. But if only one daughter cell receives the gray crescent, then only that cell can become a complete embryo.

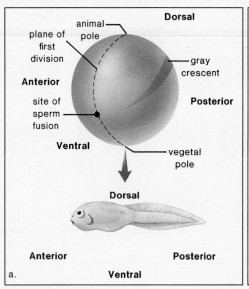

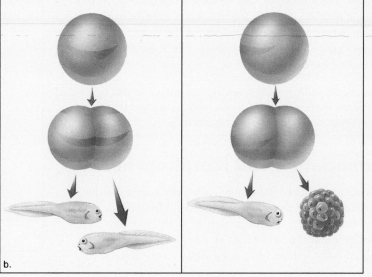

and those farthest away become ectoderm. This suggests that there may be a molecular concentration gradient that acts as a signal to induce germ layer differentiation. A recent experiment performed by Jim Smith of the National Institute of Medical Research in London indicates that the peptide growth factor called activin may play a role in the signaling process. At low concentrations of activin, animal pole cells become epidermis, an ectoderm derived tissue, and at higher concentrations they become muscle and notochord, a mesoderm derived tissue.

Later Spemann and his colleague Hilde Mangold discovered other instances of induction. For example, presumptive (potential) notochord tissue induces the formation of the nervous system (fig. 44.6). If presumptive nervous system tissue, located just above the presumptive notochord, is cut out and transplanted to the belly region of the embryo, it does not form a neural tube. On the other hand, if presumptive notochord tissue is cut out and transplanted beneath what would be belly ectoderm, this ectoderm differentiates into neural tissue. Still other examples of induction are now known. In 1905, Warren Lewis studied the formation of the eye in frog embryos (fig. 44.7). He found that an optic vesicle, which is a lateral outgrowth of developing brain tissue, induces overlying ectoderm to thicken and become a lens. The developing lens in turn induces an optic vesicle to form an optic cup, where the retina develops.

Figure 44.6

Experiment performed by Hans Spemann in 1924. **a.** The presumptive nervous system (blue) does not develop into the neural plate if moved from the normal location. **b.** The presumptive notochord (red) can cause the belly ectoderm to develop into the neural plate (blue). This shows that the notochord induces ectoderm cells to become a neural plate most likely by sending out chemical signals.

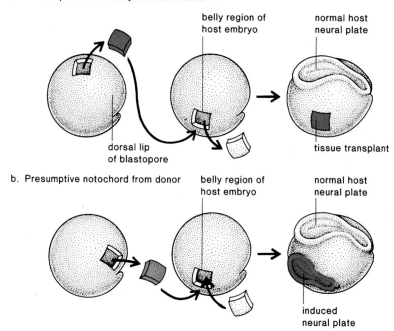

a. Presumptive nervous system from donor

b. Presumptive notochord from donor

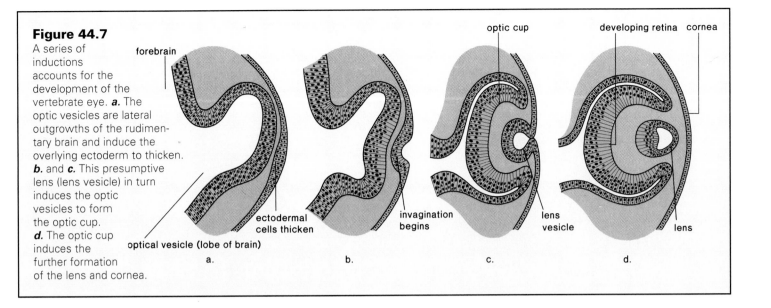

Figure 44.7
A series of inductions accounts for the development of the vertebrate eye. **a.** The optic vesicles are lateral outgrowths of the rudimentary brain and induce the overlying ectoderm to thicken. **b.** and **c.** This presumptive lens (lens vesicle) in turn induces the optic vesicles to form the optic cup. **d.** The optic cup induces the further formation of the lens and cornea.

All these various experiments allow us to envision that as development proceeds, a series of inductive events occurs. A signal activates certain genes, which in turn encode other signals, which activate new genes, which encode new signals, and so forth:

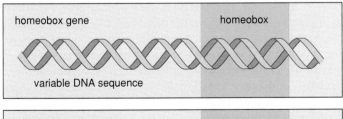

The end result is the orderly process of development from the fertilized egg to the adult.

Homeobox Genes and Pattern Formation

A *homeobox gene,* also called a homeotic gene, is a master gene that controls the activity of many other genes necessary for embryonic development. Each organism has several of these genes. One portion of a homeobox gene (called the homeobox) contains a constant sequence of 180 nucleotides. The sequence of nucleotides in the variable region of a homeobox gene varies from gene to gene. A homeobox gene codes for a homeodomain protein. Each of these has a homeodomain, a sequence of 60 amino acids that is found in all other homeodomain proteins.

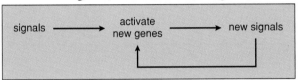

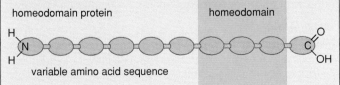

Since homeobox genes and their products contain variable regions, each is different from the other. Their similarity is limited to just the homeobox and homeodomain, respectively.

Homeobox DNA can be radioactively labeled and used as a DNA probe (p. 283) to locate other homeobox genes. This methodology has allowed researchers to determine that almost all eukaryotic organisms contain homeobox genes. The conclusion is that all homeoboxes are derived from an original DNA sequence that has been largely conserved (maintained from generation to generation) because of its importance in the regulation of animal development.

Homeoboxes were discovered by researchers studying homeotic mutations in *Drosophila.* A homeotic mutation causes body parts to be misplaced—a homeotic mutant fly can have 2 pairs of wings or extra legs where antennae should be. Homeobox genes are clearly involved in pattern formation, that is, the shaping of an embryo so that the adult has a normal appearance. Homeobox genes are arranged in a definite order on a chromosome, and the gene closest to the DNA 5′ end is mainly involved in determining the anterior portion of the embryo, while the gene closest to the 3′ end is mainly involved in determining the most posterior portion of the embryo. In *Xenopus,* if the expression levels of the homeobox genes are altered, headless and tailless embryos are produced.

In keeping with the diagram explaining sequential development events above, researchers envision that a homeodomain protein produced by one homeobox gene binds to and turns on the next homeobox gene, and this orderly process determines the overall pattern of the embryo. It appears that homebox genes also establish homeodomain protein gradients that affect the pattern development of specific parts, such as the limbs. Many laboratories are engaged in discovering much more about homeobox genes and homeodomain proteins. We need to know, for example, how homeobox genes are turned on, if cells have receptors for homeodomain proteins, and how their gradient is maintained

Figure 44.8

Evolution of the extraembryonic membranes in the reptile made development on land possible. If an embryo develops in water, the water supplies oxygen, takes away waste, prevents desiccation, and provides a protective cushion. On land some of these functions are performed by the extraembryonic membranes, which must be protected from drying out by an outer shell or by internal development. *a.* Chick within its hard shell. The chorion lies just beneath the shell and carries on gas exchange. The allantois collects nitrogenous wastes. The yolk sac provides nourishment, and the amnion provides a watery environment. *b.* In humans, only the chorion and amnion have comparable functions. The chorion forms the fetal half of the placenta, where exchange with the mother's blood occurs, and the amnion provides a watery environment. The allantoic blood vessels become the umbilical blood vessels, and the yolk sac is the first site of blood cell formation.

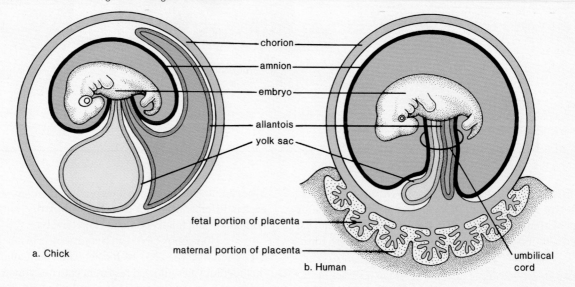

a. Chick

b. Human

chorion
amnion
embryo
allantois
yolk sac
fetal portion of placenta
maternal portion of placenta
umbilical cord

in tissues. The roles of the cytoskeleton and extracellular matrix are also being explored.

> Differentiation and morphogenesis are dependent upon signals (in many cases peptide growth hormones) from neighboring cells. The signals activate homeobox genes, which produce homeodomain proteins, which are themselves signals.

Human Development

Human development is often divided into embryonic development (first 2 mon) and fetal development (3–9 mon). The embryonic period consists of early formation of the major organs, and fetal development is a refinement of these structures.

Of the 3 animals we have previously discussed, human development most resembles that of the chick, no doubt because reptiles are a common ancestor of both birds and mammals. In the chick, the germ layers extend out over the yolk and develop into the *extraembryonic membranes,* so called because they lie outside the embryo. Their placement and function are described in figure 44.8. In humans, the extraembryonic membranes develop earlier than in the chick. In contrast to the chick, the human egg has little yolk, and the nutritional needs of the embryo are supplied by the chorion, which becomes part of the placenta.

Embryonic Development

Cleavage

Following fertilization in the oviduct, the human zygote begins to cleave as it travels down the oviduct to the uterus (fig. 44.9). The first cleavage is complete about a day after fertilization, and subsequent divisions occur at intervals of 8–10 hours. Passage of the embryo through the oviduct takes from 3–3½ days. While the embryo lies free in the uterine cavity, the morula becomes the *blastocyst* (the human equivalent of the blastula), consisting of the *inner cell mass* and a single layer of surrounding cells known as the *trophoblast*. Later, the trophoblast, reinforced by a layer of mesoderm, gives rise to the **chorion,** an extraembryonic membrane. The inner cell mass eventually becomes the fetus.

At the end of the first week, the embryo begins the process of implanting itself in the wall of the uterus. During implantation, the trophoblast secretes enzymes to digest away some of the tissue and blood vessels of the uterine wall. The trophoblast begins to secrete HCG (human chorionic gonadotropin), the hormone that is the basis for the pregnancy test and serves to maintain the corpus luteum past the time it normally disintegrates. Because of this, the endometrium is maintained and menstruation does not occur.

Gastrulation

Gastrulation occurs during the second week. The inner cell mass becomes the *embryonic disk,* consisting of a layer of endodermal

Animal Structure and Function

Figure 44.9
Human development before
implantation. Structures and events
proceed counterclockwise. At ovulation,
the secondary oocyte leaves the ovary.
Fertilization occurs in the oviduct. As
the zygote moves along the
oviduct, it undergoes
cleavage to produce
a morula. The
blastocyst forms
and implants
itself in the
uterine lining.

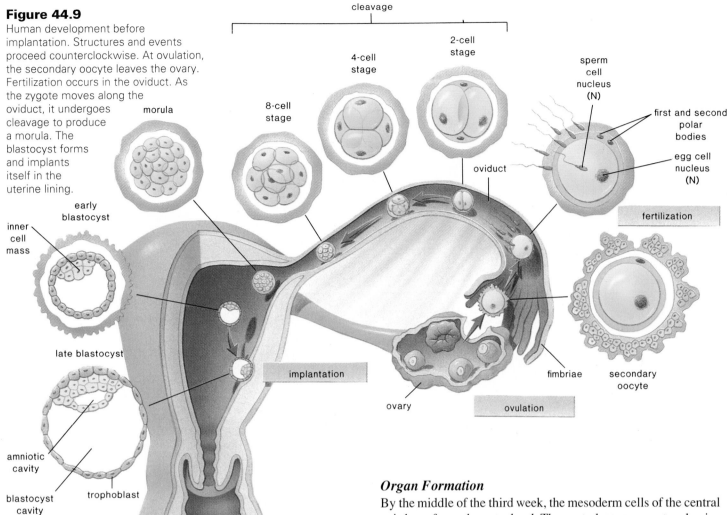

cleavage

2-cell
stage

4-cell
stage

sperm
cell
nucleus
(N)

8-cell
stage

morula

first and second
polar
bodies

oviduct

egg cell
nucleus
(N)

early
blastocyst

inner
cell
mass

fertilization

late blastocyst

implantation

fimbriae

secondary
oocyte

amniotic
cavity

ovary

ovulation

blastocyst
cavity

trophoblast

cells beneath a layer of ectodermal cells (fig. 44.10). The
amnion, an extraembryonic membrane that contains a cavity
filled with amniotic fluid, is now visible above the embryo. The
amniotic fluid cushions the embryo and keeps it from drying out.
At this time we also see the **yolk sac,** which has no nutritive
function but is the first site of blood cell formation.

As gastrulation continues, mesoderm forms along the
length of a primitive streak by invagination of cells between the
ectoderm and endoderm. When the edges of the embryonic disk
fold under, the embryo is converted into a tubular, 3-layered
body. The ectoderm is on the outside, the endoderm is on the
inside, and the mesoderm is between these 2. It is possible to
relate the development of future organs of these germ layers
(table 44.2).

Organ Formation

By the middle of the third week, the mesoderm cells of the central
axis have formed a notochord. The central nervous system begins
to develop above the notochord. Neural folds arise from the
ectoderm and fuse to form the neural tube, which swells anteriorly
to form the brain (fig. 44.11). The somites, which become the
dermis of the skin, the muscle tissue, and the vertebrae and other
bones, appear on both sides of the neural tube.

The heart and blood vessels also arise from the mesoderm.
At first, the heart is 2 tiny tubes. These fuse to form a single
chamber as the third week ends. The primitive heart bulges and is
easily observed as it begins to pump blood through simple arteries
and veins. These vessels take blood to the developing esophagus,
stomach, liver, pancreas, and intestine.

At 4 weeks, the embryo is barely larger than the height of
this print. There is a bridge of mesoderm called the body stalk,
which connects the caudal end of the embryo with the chorion (fig.
44.10*c*). The fourth extraembryonic membrane, the **allantois,** is
contained within this stalk, and its blood vessels become the
umbilical blood vessels. Then the head and the tail lift up and body
stalk moves anteriorly by constriction (fig. 44.10*d*). Once this
process is complete, the umbilical cord that connects the developing embryo to the placenta is fully formed (fig. 44.10*e*).

Figure 44.10

Stages showing the early appearance of the extraembryonic membranes and the formation of the umbilical cord in the human embryo. *a.* At 2 weeks, the amniotic cavity appears. *b.* At 3 weeks, the chorion and the yolk sac are apparent. *c.* At 4 weeks, the body stalk and the allantois form. *d.* At 5 weeks, the embryo begins to take shape as the umbilical cord forms. *e.* Eventually, the umbilical cord is formed fully.

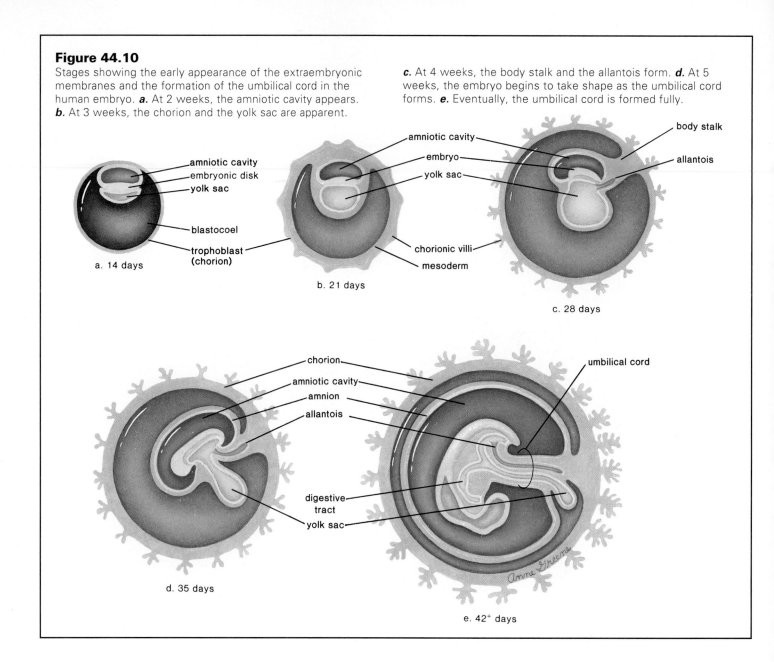

a. 14 days

amniotic cavity
embryonic disk
yolk sac
blastocoel
trophoblast (chorion)

b. 21 days

amniotic cavity
embryo
yolk sac
chorionic villi
mesoderm

c. 28 days

body stalk
allantois

d. 35 days

chorion
amniotic cavity
amnion
allantois
digestive tract
yolk sac

e. 42⁺ days

umbilical cord

Little flippers called *limb buds* appear (fig. 44.12); later, the arms and the legs develop from the limb buds, and even the hands and the feet become apparent. At the same time, during the fifth week, the head becomes larger, and the sense organs become more prominent. It is possible to make out the developing eyes, ears, and even nose.

At first the skeleton is composed of cartilage, but the cartilage begins to be replaced by bone during the sixth week. There is a remarkable change in external appearance during the sixth through eighth weeks of development from a form that is difficult to recognize as human to one that is easily recognizable as human. Concurrent with brain development, the head achieves its normal relationship with the body as a neck region develops. The nervous system is developed well enough to permit reflex actions, such as a startle response to being touched. At the end of this period, the embryo is about 38 mm long and weighs no more than an aspirin tablet, even though all organ and systems are established.

Placenta Formation

The placenta begins formation once the embryo is implanted fully. Treelike extensions of the chorion called chorionic villi project into the maternal tissues (fig. 44.10c–e). Later, these disappear in all areas except where the placenta develops. By the tenth week, the placenta is fully formed and begins to produce progesterone and estrogen (fig. 44.13). These maintain the corpus luteum even though production of HCG by the placenta drops off. There is no follicle production or menstruation during pregnancy.

The umbilical cord contains the blood vessels that take fetal blood to and from the **placenta.** Here the chorionic villi are surrounded by maternal blood sinuses, yet the blood of the mother

Animal Structure and Function

Figure 44.11

Human embryo at 21 days. The neural folds still need to close at the anterior and posterior ends of the embryo. The pericardial area contains the primitive heart, and the somites are the precursors of the muscles.

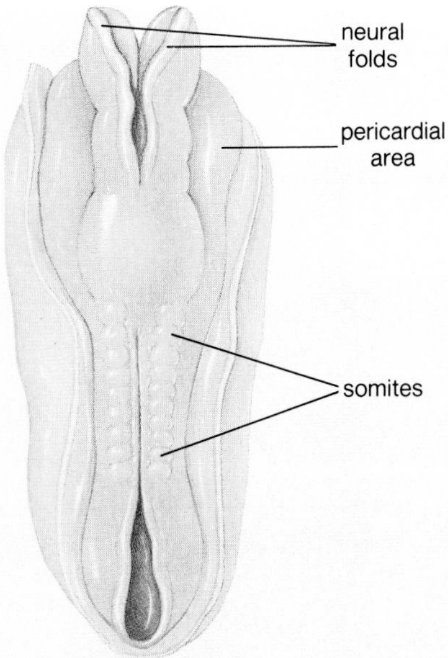

- neural folds
- pericardial area
- somites

Figure 44.12

Human embryo at beginning of fifth week. **a.** Scanning electron micrograph. **b.** The embryo is curled so that the head touches the heart, 2 organs whose development is further along than the rest of the body. The organs of the gastrointestinal tract are forming, and the arms and legs develop from the bulges that are called limb buds. The presence of the tail is an evolutionary remnant; its bones regress and become those of the coccyx. The pharyngeal pouches only become functioning gills in fishes and amphibians; in humans the first pair of pharyngeal pouches becomes the auditory cavity of the middle ear and the eustachian tube. The second pair becomes the tonsils, while the third and fourth become the thymus and the parathyroids.

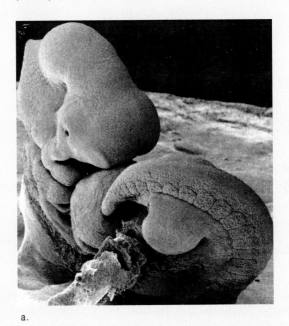

a.

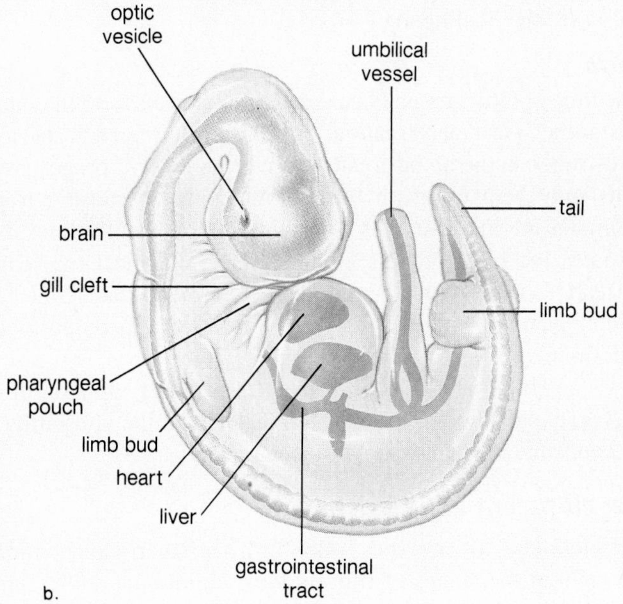

- optic vesicle
- umbilical vessel
- brain
- tail
- gill cleft
- limb bud
- pharyngeal pouch
- limb bud
- heart
- liver
- gastrointestinal tract

b.

does not mix with the blood of the fetus. Instead, exchange takes place across plasma membranes. Nitrogenous wastes and carbon dioxide move from the fetal side to the maternal side of the placenta, and nutrients and oxygen move from the maternal side to the fetal side of the placenta.

Harmful chemicals can also cross the placenta, and this is of particular concern during the embryonic period, when various structures are first forming. Each organ or part seems to have a sensitive period during which a substance can alter its normal function.

Fetal Development

Fetal development is marked by an extreme increase in size. Weight multiplies 600 times, going from less than 28 g to 3 kg. In this time, too, the fetus grows to about 50 cm in length. The genitalia finally make their appearance, and it is possible to tell if the fetus is male or female.

Soon, hair, eyebrows, and eyelashes add finishing touches to the face and head. In the same way, fingernails and toenails complete the hands and feet. A fine, downy hair (lanugo) covers the limbs and trunk, only to later disappear. The fetus looks like an old man because the skin is growing so fast that it wrinkles. A waxy, almost cheeselike substance (vernix caseosa) protects the wrinkly skin from the watery amniotic fluid.

The fetus at first only flexes its limbs and nods its head, but later it can move its limbs vigorously to avoid discomfort. The mother feels these movements from about the fourth month on. The other systems of the body also begin to function. The fetus begins to suck its thumb, swallow amniotic fluid, and urinate.

Development in Animals

Figure 44.13

This fetus is 6–7 months old, and the placenta is well developed. The placenta is the region of exchange between fetus and mother. Here, fetal circulation is separated from maternal circulation only by thin membranes.

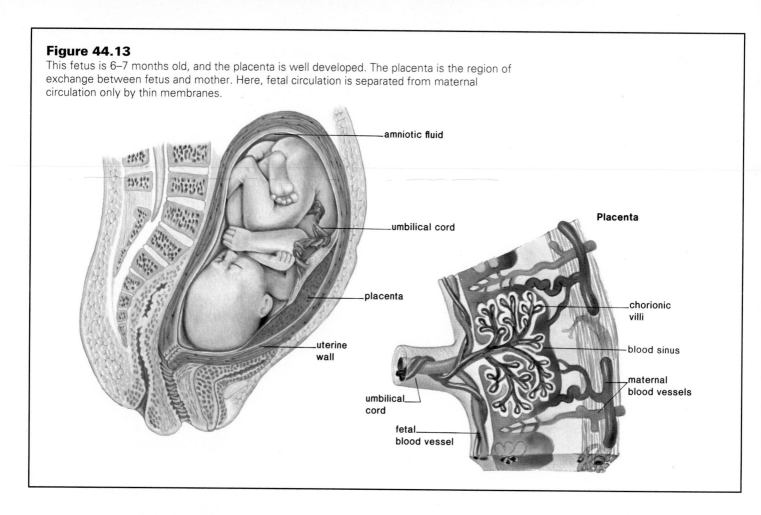

After 16 weeks, the fetal heartbeat is heard through a stethoscope. A fetus born at 24 weeks has a chance of surviving, although the lungs are still immature and often cannot capture oxygen adequately. Weight gain during the last couple of months increases the likelihood of survival.

Birth

The time of birth is usually calculated at 280 days from the start of the mother's last menstruation. As pregnancy progresses, the level of estrogen in the bloodstream exceeds the level of progesterone. This may help bring on birth of the fetus, because estrogen promotes uterine irritability. It is also possible that the fetus itself initiates the birth process by releasing a chemical messenger. Researchers have found in studies with sheep that if the fetal hypothalamus or pituitary is destroyed, birth can be postponed indefinitely.

The process of birth includes 3 stages: dilation of the cervix, expulsion of the fetus, and delivery of the afterbirth (the placenta and the extraembryonic membranes).

Development after Birth

Development does not cease once birth has occurred; it continues throughout the stages of life: infancy, childhood, adolescence, and adulthood.

Infancy lasts until about 2 years of age. It is characterized by tremendous growth and sensorimotor development. During *childhood,* the individual grows, and the body proportions change. *Adolescence* begins with *puberty,* when the secondary sex characteristics appear and the sexual organs become functional. At this time, there is an acceleration of growth leading to changes in height, weight, fat distribution, and body proportions. Males commonly experience a growth spurt later than females; therefore, they grow for a longer period of time. Males are generally taller than females and have broader shoulders and longer legs relative to their trunk length.

Young adults are at their physical peak in muscle strength, reaction time, and sensory perception. The organ systems at this time are best able to respond to altered circumstances in a homeostatic manner. From now on, however, there is an almost imperceptible, gradual loss in certain of the body's abilities.

Aging

Aging encompasses the progressive changes that contribute to an increased risk of infirmity, disease, and death. Figure 44.14 compares the percentage of organ function in a 75- to 80-year-old person to that of a 20-year-old person whose organs are assumed to function at 100% capacity. When making this comparison, we may note that the body has a vast functional reserve so that it can still perform well, even when not at 100% capacity.

Animal Structure and Function

Figure 44.14

Aging involves degenerative changes such as those illustrated on the left. On the right the percentage of the function/structure of various body processes/organs at 75–80 years of age compared to a person of 20.

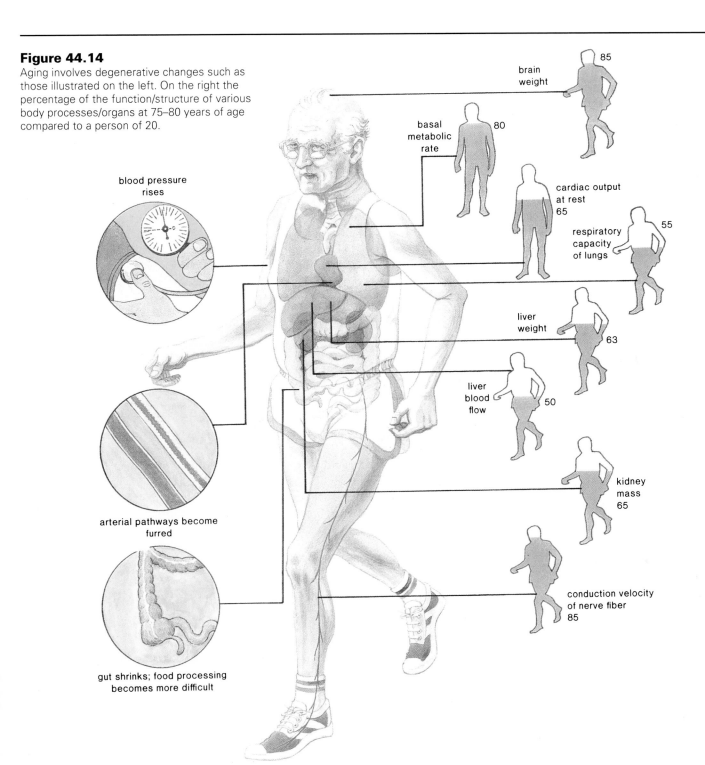

brain weight 85

basal metabolic rate 80

blood pressure rises

cardiac output at rest 65

respiratory capacity of lungs 55

liver weight 63

liver blood flow 50

arterial pathways become furred

kidney mass 65

conduction velocity of nerve fiber 85

gut shrinks; food processing becomes more difficult

There are many theories about what causes aging. We will consider 4 of them.

Genetic Origin Several lines of evidence indicate that we are genetically programmed to age. The maximum life span of animals is species specific; for humans it is about 110 years. But the children of long-lived parents tend to live longer than those of short-lived parents. Longevity may be related to the number of times the cells divide. A human cell divides a maximum of about 50 times. There could be a genetic off switch that causes cells to lose the ability to divide. David Danner, of the National Institutes of Aging, microinjected a mRNA for the protein prohibin into fibroblasts and they stopped dividing. The gene coding for this protein could be the off switch. It's also possible that other genes that normally block cell division are turned on in aging cells. Other researchers have found that the retinoblastoma (RB) gene is a tumor suppressor that growth factors fail to inactivate in older cells, for example.

Cellular Repair As cells degenerate, they become less efficient at self-repair. Ordinarily, whenever DNA replicates, any errors that occur are enzymatically repaired. Due to environmental insults, the DNA repair system may falter so that errors in replication are not corrected. Mutations may result that lead to the production of nonfunctional proteins. Eventually, the number of inadequately functioning cells may build up and contribute to an aging process.

It's also noted that as cells age, they begin to accumulate metabolic by-products that could damage DNA. For example, senescent cells accumulate free radicals, which are molecules that have an extra electron and can readily bond with cellular macromolecules. Housekeeping enzymes known as antioxidants ordinarily prevent the formation of free radicals. Longer-lived species are known to have more antioxidants than shorter-lived species, and it is hypothesized that older cells also have fewer copies of these beneficial enzymes.

Glucose is a destructive agent that attaches itself to protein. The protein-sugar complex starts a cross-linking process that produces a yellowish brown mass known as advanced glycosylation end products (AGEs). Like free radicals, AGEs may interact with DNA, causing mutations that interfere with the cell's ability to replicate and repair DNA molecules. The presence of AGE-derived cross-links explains why cataracts develop in older persons, and it may also help explain the development of atherosclerosis and the inefficiency of the kidneys in diabetics and older individuals. Researchers are presently experimenting with the drug aminoguanidine, which can prevent the development of AGEs.

Werner syndrome is a rare disorder in which those affected show signs of advanced aging in their 20s. Werner cells have an excess of collagen and fibronectin, as do aging cells in general. Collagen makes up the white fibers in connective tissues, and increasingly collagen cross-linking occurs as people age. Undoubtedly, this cross-linking contributes to the stiffening and loss of elasticity characteristic of aging tendons and ligaments. It may also account for the inability of organs such as the blood vessels, heart, and lungs to function as they once did.

Researchers have not decided what comes first: accumulation of metabolic by-products or a genetic defect that affects the housekeeping enzymes that prevent their formation. Ultimately, the hope is to identify the genetic and cellular events that lead to aging so that cells can live longer while maintaining their normal functions.

Whole Body Processes A decline in the hormonal system may affect many different organs of the body. For example, Type II diabetes is a common disorder in older individuals. The pancreas makes insulin, but the cells lack the receptors that enable them to respond. Menopause in women occurs for a similar reason. There is plenty of FSH in the bloodstream, but the ovaries do not respond. Perhaps aging results from the loss of hormonal activity and a decline in the functions they control.

With age, the immune system, too, no longer performs as it once did, and this may affect the body as a whole. The thymus gland gradually decreases in size, and eventually most of it is replaced by fat and connective tissue. The incidence of cancer increases among the elderly, which may signify that the immune system is no longer functioning as it should. This idea is further substantiated by the increased incidence of autoimmune diseases in older individuals.

Extrinsic Factors The present data on the effects of aging are often based on comparing the characteristics of the elderly with younger age groups. But today's elderly perhaps were not as aware of the importance of diet and exercise for general health. It's possible, then, that much of what we attribute to aging is instead due to years of poor health habits. For example, osteoporosis is associated with a progressive decline in bone density in both males and females. Fractures are more likely to occur after only minimal trauma. Osteoporosis is common in the elderly—by age 65, one-third of women will have had vertebral fractures, and by age 81, one-third of women and one-sixth of men will have suffered a hip fracture. While there is no denying that a decline in bone mass results from aging, certain extrinsic factors are also important. The occurrence of osteoporosis itself is associated with cigarette smoking, heavy alcohol intake, and inadequate calcium intake. Not only is it possible to eliminate these negative factors by personal choice, it is also possible to add a positive factor: a moderate exercise program has been found to slow down the progressive loss of bone mass.

Rather than collecting data on the average changes observed between different age groups, it may be more useful to note the differences within any particular age group so that any extrinsic factors contributing to a decline or promoting the health of an organ can be identified.

Summary

1. Any developmental change requires 3 processes: growth, differentiation, and morphogenesis.
2. The development of 3 types of animals (lancelet, frog, and chick) is compared. The first 3 stages (cleavage, blastulation, and gastrulation) differ according to the amount of yolk in the egg.
3. During cleavage, the zygote divides, but there is no overall growth. The result is a morula, which becomes the blastula when an internal cavity (the blastocoel) appears. During the gastrula stage, invagination of cells into the blastocoel results in formation of the germ layers: ectoderm, mesoderm, and endoderm. Later development of organs can be related to these layers.
4. During neurulation, the nervous system develops from mid-line ectoderm, just above the notochord. At this point, it is possible to draw a typical cross section of a vertebrate embryo (fig. 44.4).
5. Differentiation begins with cleavage, when the egg's cytoplasm is partitioned among the numerous cells. The cytoplasm is not uniform in content, and presumably each of the first few cells differ as to their cytoplasmic contents. Some probably contain substances that can influence gene activity—turning some genes on and others off.

Animal Structure and Function

6. After the first cleavage of a frog embryo, only a daughter cell that receives a portion of the gray crescent is able to develop into a complete embryo. This illustrates the importance of cytoplasmic inheritance to early development.
7. Morphogenesis involves the process of induction. The notochord induces the formation of the neural tube in frog embryos. The reciprocal induction that occurs between the lens and the optic vesicle is another good example of this phenomenon.
8. Today we envision induction as always present because cells are believed to constantly give off signals that influence the genetic activity of neighboring cells.
9. Studies of homeobox genes and pattern formation further indicate that morphogenesis involves genetic and environmental influences.
10. Human development can be divided into embryonic development (first 2 mon) and fetal development (3–9 mon). The early stages in human development resemble those of the chick. The similarities are probably due to their evolutionary relationship, not the amount of yolk the eggs contain, because the human egg has little yolk.
11. Fertilization occurs in the oviduct, and cleavage occurs as the embryo moves toward the uterus. The morula becomes the blastocyst before implanting in the uterine lining. Human gastrulation occurs as in the chick, and the three germ layers form.
12. The extraembryonic membranes appear early in human development. The trophoblast of the blastocyst is the first sign of the chorion, which goes on to become the fetal part of the placenta. The placenta is where exchange occurs between fetal and maternal blood. The amnion contains the amniotic fluid, which cushions and protects the embryo. The yolk sac and allantois are largely vestigial.
13. Organ development begins with neural tube and heart formation. There follows a steady progression of organ formation during embryonic development. During fetal development, refinement of features occurs, and the fetus adds weight. Birth occurs about 280 days after the start of the mother's last menstruation.
14. Development after birth consists of infancy, childhood, adolescence, and adulthood. Young adults are at their prime, and then the aging process begins. Aging may be genetic in origin, due to cellular repair changes, whole body processes, or extrinsic factors.

Writing Across the Curriculum

In order to practice writing skills, students should write out the answers to any or all of the study questions and the critical thinking questions. The study questions are sequenced in the same order as the text. Suggested answers to the critical thinking questions are in appendix D.

Study Questions

1. State, define, and give examples of the 3 processes that occur whenever a developmental change occurs.
2. Compare the process of cleavage and the formation of the blastula and gastrula in lancelets, frogs, and chicks.
3. State the germ layer theory, and tell which organs are derived from each of the germ layers.
4. Draw a cross section of a typical chordate embryo at the neurula stage, and label your drawing.
5. Give reasons for suggesting that differentiation and morphogenesis are dependent upon signals given off by neighboring cells. What do the signals bring about in the receiving cells?
6. Explain an experiment performed by Spemann that suggests the notochord induces formation of the neural tube. Give another well-known example of induction between tissues.
7. List the human extraembryonic membranes, give a function for each, and compare this function to that in the chick.
8. Tell where the stages of fertilization, cleavage, morula, and blastula occur in humans. What happens to the blastula in the uterus?
9. Describe the structure and function of the placenta in humans.
10. Outline the developmental changes that occur in humans from fetal development to adulthood.
11. Discuss 3 theories of aging.

Objective Questions

1. Which of these stages is the first one out of sequence?
   a. cleavage
   b. blastula
   c. morula
   d. gastrula
2. Which of these stages is mismatched?
   a. cleavage—cell division
   b. blastula—gut formation
   c. gastrula—three germ layers
   d. neurula—nervous system
3. Which of the germ layers is best associated with development of the heart?
   a. ectoderm
   b. mesoderm
   c. endoderm
   d. All of these.
4. Differentiation begins at what stage?
   a. cleavage
   b. blastula
   c. gastrula
   d. neurula
5. Morphogenesis is best associated with
   a. overall growth.
   b. induction of one tissue by another.
   c. genetic mutations.
   d. All of these.
6. In humans, the placenta develops from the chorion. This indicates that human development
   a. resembles that of the chick.
   b. is dependent upon extraembryonic membranes.
   c. cannot be compared to lower animals.
   d. only begins upon implantation.
7. In humans, the fetus
   a. is surrounded by 4 extraembryonic membranes.

b. has developed organs and is recognizably human.
  c. is dependent upon the placenta for excretion of wastes and acquisition of nutrients.
  d. Both b and c.
8. Developmental changes
  a. require growth, differentiation, and morphogenesis.
  b. stop occurring as soon as one is grown.
  c. are dependent upon a parceling out of genes into daughter cells.
  d. Both a and c.
9. Mesoderm forms by invagination of cells in the
  a. lancelet.
  b. frog.
  c. chick.
  d. All of these.
10. Which of these is mismatched?
  a. brain—ectoderm
  b. gut—endoderm
  c. bone—mesoderm
  d. lens—endoderm

11. Label this diagram illustrating the placement of the extraembryonic membranes, and give a function for each membrane in humans:

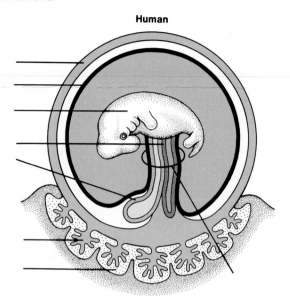

Human

Concepts and Critical Thinking

1. *Development occurs throughout the life of an animal.*

   Why is senescence considered a part of development?

2. *Chemical signals are involved in development.*

   Should these chemical signals be considered hormones? Why or why not?

3. *The genes control development.*

   Use the diagram on page 717 to substantiate a genetic theory of aging.

Selected Key Terms

differentiation (dif″er-en″she-a′shun) 712
morphogenesis (mor″fo-jen′ē-sis) 712
cleavage (klēv′ij) 713
blastocoel (blas′to-sēl) 713
blastula (blas′tu-lah) 713

gastrula (gas′troo-lah) 713
ectoderm (ek′to-derm) 713
endoderm (en′do-derm) 713
mesoderm (mes′o-derm) 714
germ layers (jerm la′erz) 714

chorion (kor′e-on) 718
amnion (am′ne-on) 719
yolk sac (yōk sak) 719
allantois (ah-lan′to-is) 719
placenta (plah-sen′tah) 720

PART 5 *Animal Structure and Function*

CRITICAL THINKING CASE STUDY

Skin Breathing

Most animals have specialized respiratory organs like gills or lungs for gas exchange. Many animals, especially amphibians, are also capable of skin breathing, or exchanging gases across the skin. Physiologists decided to test the hypothesis that *amphibians have physiological adaptations that favor skin breathing under appropriate conditions.*

Skin capillaries can be open (carrying blood) or closed (not carrying blood). The amount of open capillaries was determined in normal frogs and in frogs prevented from using their lungs. What do you predict about the amount of open capillaries in the skin of frogs prevented from using their lungs compared to those that use their lungs?

Prediction 1 There will be more open capillaries in the skin of frogs prevented from using their lungs.

Result 1 The predicted results were observed. Apparently amphibians are capable of regulating the amount of open capillaries in the skin to make more blood available for skin breathing.

I In another experiment, frogs that had been submerged and were skin breathing in the water were transferred to air. Give 2 reasons why you would expect the amount of open capillaries in the skin to decrease.

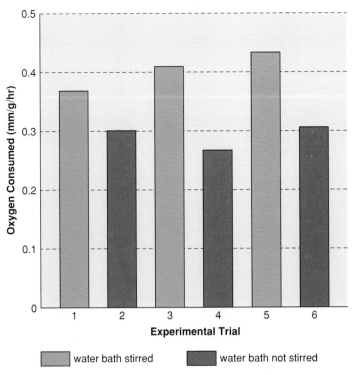

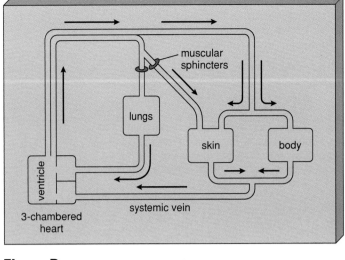

Figure B
Circulatory system in amphibian showing the 3-chambered heart and circulation to lungs, skin, and body. The muscle sphincters can shunt deoxygenated blood to skin or lungs depending on the environmental conditions.

Figure A
Oxygen consumed by frog via skin-breathing. Adapted from "Skin Breathing in Vertebrates" by Martin E. Feder and Warren W. Burggran. Copyright © 1985 Scientific American, Inc. All rights reserved.

Decreased distance between skin capillaries and the external medium speeds the process of diffusion, and this facilitates skin breathing. One group of tadpoles was raised in oxygen-rich water and another group was raised in oxygen-poor water. What do you predict about the distance of skin capillaries from the external environment in these 2 groups of tadpoles?

Prediction 2 Tadpoles raised in oxygen-poor water will have skin capillaries closer to the external environment than tadpoles raised in oxygen-rich water.

Result 2 Examination revealed that tadpoles raised in the oxygen-poor environment had skin capillaries 20 micrometers closer to the environment than tadpoles raised in an oxygen-rich environment. Moreover, these skin capillaries were finer and grew in a more dense network.

II What are the relative chances of effective skin breathing if tadpoles raised in an oxygen-poor environment are placed in an oxygen-rich environment? if tadpoles raised in an oxygen-rich environment are placed in an oxygen-poor environment?

Flowing water as opposed to stagnant water makes more oxygen available to the skin and therefore favors skin breathing. Amphibians are sometimes observed to have rhythmic body movements when submerged in water. Such movement may facilitate skin breathing. A frog immobilized in a wire cage was immersed in a water bath so that only its mouth and nostrils were exposed to air. Oxygen uptake from the water was determined when the water was stirred and when the water was not stirred. (Oxygen uptake was determined by periodically measuring the oxygen concentration in the water bath. A layer of mineral oil was placed on the surface of the water to prevent atmospheric oxygen from replacing the oxygen consumed.) What do you predict regarding oxygen consumption when the water was stirred and when it was not stirred?

Prediction 3 There will be more oxygen consumption when the water is stirred as opposed to when it is not stirred.

Result 3 The results of this experiment are shown in figure A. These data indicate that oxygen consumption is greater when the water bath is stirred than when it is not stirred. These results suggest that the body movements of amphibians are a behavioral adaptation to promote skin breathing.

III What would be the results given in figure A if mineral oil hadn't been placed on the water?
What would be the results in figure A if the experiment was repeated with no frog? Could this experiment serve as a control? Why?
Would a frog have more skin capillaries open when the water was stirred or not stirred? Why?

Figure B diagrams the circulatory system of an amphibian. Note the muscular sphincters, which apparently can control blood flow to the lungs or the skin. It is possible to measure blood flow in the pulmonary artery, which carries blood to the lungs, and in the cutaneous artery, which takes blood to the skin. Which blood vessel do you predict will carry more blood when a toad is prevented from using its lungs? when a toad is using its lungs?

Prediction 4 If a toad is prevented from using its lungs, the cutaneous artery will carry more blood. If a toad is using its lungs, the pulmonary artery will carry more blood.

Result 4 Subsequent observations were consistent with the prediction.

IV Suppose the data show that there is always more blood in the cutaneous artery even when a frog is using its lungs. What would be the implication regarding skin breathing in amphibians? Would this refute the hypothesis? Why?

A diversity of experiments indicates support for the hypothesis that amphibians have physiological adaptations that favor skin breathing under appropriate conditions.

Other Questions

1. Would you expect to find skin breathing in reptiles and mammals? Why or why not?
2. Effective gas exchange is dependent upon speed of diffusion. Order the following circumstances according to effective gas exchange. Explain your answer.
   a. capillaries close to external medium; oxygen-poor water
   b. capillaries close to external medium; oxygen-rich water
   c. capillaries far from external medium; oxygen-poor water
   d. capillaries far from external medium; oxygen-rich water

References

Feder, M. E., and Burggren, W. W. 1985. Cutaneous gas exchange in vertebrates: Design, patterns, control and implications. *Biological Reviews*. 60:1–45.
———. 1985. Skin breathing in vertebrates. *Scientific American*, 126–43.

Suggested Readings for Part 5

Alkon, D. L. July 1989. Memory storage and neural systems. *Scientific American*.

Aral, S. O., and Holmes, K. K. February 1991. Sexually transmitted diseases in the AIDS era. *Scientific American*.

Atkinson, M. A., and Maclaren, N. K. July 1990. What causes diabetes? *Scientific American*.

Barlow, R. B., Jr. April 1990. What the brain tells the eye. *Scientific American*.

Berns, M. B. June 1991. Laser surgery. *Scientific American*.

Cantin, M., and Genest, J. February 1986. The heart as an endocrine gland. *Scientific American*.

Cerami, A., et al. May 1987. Glucose and aging. *Scientific American*.

Cohen, I. R. April 1988. The self, the world and autoimmunity. *Scientific American*.

Crapo, L. 1985. *Hormones, the messengers of life*. New York: W. H. Freeman and Co.

DeRobertis, E. M., Oliver, G., and Wright, C. V. E. July 1990. Homeobox genes and the vertebrate body plan. *Scientific American*.

Freeman, W. J. February 1991. The physiology of perception. *Scientific American*.

Golde, D. W. December 1991. The stem cell. *Scientific American*.

Grey, H. M., Sette, A., and Buus, S. November 1989. How T cells see antigen. *Scientific American*.

Hole, J. W. 1990. *Human anatomy and physiology*. 5th ed. Dubuque, Iowa: Wm. C. Brown Publishers.

Holloway, M. March 1991. Rx for addiction. *Scientific American*.

Kalil, R. E. December 1989. Synapse formation in the developing brain. *Scientific American*.

Koretz, J. F., and Handelman, G. H. July 1988. How the human eye focuses. *Scientific American*.

Mills, J., and Masur, H. August 1990. AIDS-related infections. *Scientific American*.

Nathans, J. February 1989. The genes for color vision. *Scientific American*.

Orci, L., et al. September 1988. The insulin factor. *Scientific American*.

Powell, C. P. June 1991. Peering inward. *Scientific American*.

Rasmussen, H. October 1989. The cycling of calcium as an intracellular messenger. *Scientific American*.

Rennie, J. December 1990. The body against itself. *Scientific American*.

Scientific American. October 1988. Entire issue is devoted to articles on AIDS.

Scrimshaw, N. S. October 1991. Iron deficiency. *Scientific American*.

Selkoe, D. J. November 1991. Amyloid protein and Alzheimer's disease. *Scientific American*.

Smith, K. A. March 1990. Interleukin-2. *Scientific American*.

Ulmann, A., Teutsch, G., and Philibert, D. June 1990. RU 486. *Scientific American*.

Uvnas-Moberg, K. July 1989. The gastrointestinal tract in growth and reproduction. *Scientific American*.

von Boehmer, H., and Kisielow, P. October 1991. How the immune system learns about self. *Scientific American*.

Wassarman, P. M. December 1988. Fertilization in mammals. *Scientific American*.

Young, J., and Cohn, Z. January 1988. How killer cells kill. *Scientific American*.

Zivin, J. A., and Choi, D. W. July 1991. Stroke therapy. *Scientific American*.

Animal Structure and Function

APPENDIX A

Classification of Organisms

The classification system given here is a simplified one, containing all the major kingdoms, as well as the major divisions (called phyla in the kingdom Protista and the kingdom Animalia).

Kingdom Monera

Prokaryotic, unicellular organisms. Nutrition principally by absorption, but some are photosynthetic or chemosynthetic.
Division Archaebacteria: methanogens, halophiles, and thermoacidophiles
Division Eubacteria: all other bacteria, including cyanobacteria (formerly called blue-green algae)

Kingdom Protista

Eukaryotic, unicellular organisms (and the most closely related multicellular forms). Nutrition by photosynthesis, absorption, or ingestion.
Phylum Sarcodina: amoeboid protozoans
Phylum Ciliophora: ciliated protozoans
Phylum Zoomastigina: flagellated protozoans
Phylum Sporozoa: parasitic protozoans
Phylum Chlorophyta: green algae
Phylum Pyrrophyta: dinoflagellates
Phylum Euglenophyta: *Euglena* and relatives
Phylum Chrysophyta: diatoms
Phylum Rhodophyta: red algae
Phylum Phaeophyta: brown algae
Phylum Myxomycota: slime molds
Phylum Oomycota: water molds

Kingdom Fungi

Eukaryotic organisms, usually having haploid or multinucleated hyphal filaments. Spore formation during both asexual and sexual reproduction. Nutrition principally by absorption.
Division Zygomycota: black bread molds
Division Ascomycota: sac fungi
Division Basidiomycota: club fungi
Division Deuteromycota: imperfect fungi (means of sexual reproduction not known).

Kingdom Plantae

Eukaryotic, terrestrial, multicellular organisms with rigid cellulose cell walls and chlorophylls *a* and *b*. Nutrition principally by photosynthesis. Starch is the food reserve.
Division Bryophyta: mosses and liverworts
Division Psilophyta: whisk ferns
Division Lycophyta: club mosses
Division Sphenophyta: horsetails
Division Pterophyta: ferns
Division Cycadophyta: cycads
Division Ginkgophyta: ginkgo
Division Gnetophyta: gnetae
Division Coniferophyta: conifers
Division Anthophyta: flowering plants
Class Dicotyledonae: dicots
Class Monocotyledonae: monocots

Kingdom Animalia

Eukaryotic, usually motile, multicellular organisms without cell walls or chlorophyll. Nutrition principally ingestive, with digestion in an internal cavity.
Phylum Porifera: sponges
Phylum Cnidaria: radially symmetrical aquatic animals
Class Hydrozoa: hydras, Portuguese man-of-war
Class Scyphozoa: jellyfishes
Class Anthozoa: sea anemones and corals
Phylum Platyhelminthes: flatworms
Class Turbellaria: free-living flatworms
Class Trematoda: parasitic flukes
Class Cestoda: parasitic tapeworms
Phylum Nematoda: roundworms
Phylum Rotifera: rotifers
Phylum Mollusca: soft-bodied, unsegmented animals
Class Polyplacophora: chitons
Class Monoplacophora: *Neopilina*
Class Gastropoda: snails and slugs
Class Bivalvia: clams and mussels
Class Cephalopoda: squids and octopuses
Phylum Annelida: segmented worms
Class Polychaeta: sandworms
Class Oligochaeta: earthworms
Class Hirudinea: leeches

Phylum Arthropoda: chiton exoskeleton, jointed appendages
    Class Crustacea: lobsters, crabs, barnacles
    Class Arachnida: spiders, scorpions, ticks
    Class Chilopoda: centipedes
    Class Diplopoda: millipedes
    Class Insecta: grasshoppers, termites, beetles
Phylum Onychophora: small, sluglike animals with legs; internal features of both annelids and arthropods
Phylum Echinodermata: marine; spiny, radially symmetrical animals
    Class Crinoidea: sea lilies and feather stars
    Class Asteroidea: sea stars
    Class Ophiuroidea: brittle stars
    Class Echinoidea: sea urchins and sand dollars
    Class Holothuroidea: sea cucumbers
Phylum Chordata: dorsal supporting rod (notochord) at some stage; dorsal hollow nerve cord; pharyngeal pouches or gill slits
    Subphylum Urochordata: tunicates
    Subphylum Cephalochordata: lancelets
    Subphylum Vertebrata: vertebrates
        Class Agnatha: jawless fishes (lampreys, hagfishes)
        Class Chondrichthyes: cartilaginous fishes (sharks, rays)
        Class Osteichthyes: bony fishes
            Subclass Dipnoi: lungfishes
            Subclass Crossopterygii: lobe-finned fishes
            Subclass Actinopterygii: ray-finned fishes
        Class Amphibia: frogs, toads, salamanders
        Class Reptilia: snakes, lizards, turtles
        Class Aves: birds
        Class Mammalia: mammals

Subclass Prototheria: egg-laying mammals
    Order Monotremata: duckbilled platypus, spiny anteater
Subclass Metatheria: marsupial mammals
    Order Marsupialia: opossums, kangaroos
Subclass Eutheria: placental mammals
    Order Insectivora: shrews, moles
    Order Chiroptera: bats
    Order Edentata: anteaters, armadillos
    Order Rodentia: rats, mice, squirrels
    Order Lagomorpha: rabbits, hares
    Order Cetacea: whales, dolphins, porpoises
    Order Carnivora: dogs, bears, weasels, cats, skunks
    Order Proboscidea: elephants
    Order Sirenia: manatees
    Order Perissodactyla: horses, hippopotamuses, zebras
    Order Artiodactyla: pigs, deer, cattle
    Order Primates: lemurs, monkeys, apes, humans
        Suborder Prosimii: lemurs, tree shrews, tarsiers, lorises, pottos
        Suborder Anthropoidea: monkeys, apes, humans
            Superfamily Ceboidea: New World monkeys
            Superfamily Cercopithecoidea: Old World monkeys
            Superfamily Hominoidea: apes and humans
                Family Hylobatidae: gibbons
                Family Pongidae: chimpanzees, gorillas, orangutans
                Family Hominidae: *Australopithecus,* * *Homo erectus,* * *Homo sapiens sapiens*

*extinct

APPENDIX B

Table of Chemical Elements

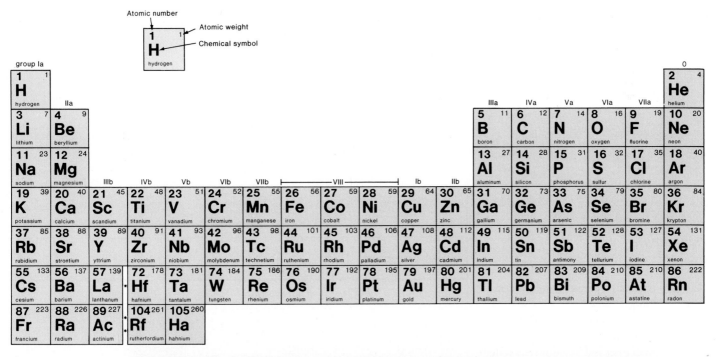

Atomic number → 1
Atomic weight → 1
Chemical symbol → H
hydrogen

group Ia																	0
1 1 **H** hydrogen	IIa											IIIa	IVa	Va	VIa	VIIa	**2** 4 **He** helium
3 7 **Li** lithium	**4** 9 **Be** beryllium											**5** 11 **B** boron	**6** 12 **C** carbon	**7** 14 **N** nitrogen	**8** 16 **O** oxygen	**9** 19 **F** fluorine	**10** 20 **Ne** neon
11 23 **Na** sodium	**12** 24 **Mg** magnesium	IIIb	IVb	Vb	VIb	VIIb		VIII		Ib	IIb	**13** 27 **Al** aluminum	**14** 28 **Si** silicon	**15** 31 **P** phosphorus	**16** 32 **S** sulfur	**17** 35 **Cl** chlorine	**18** 40 **Ar** argon
19 39 **K** potassium	**20** 40 **Ca** calcium	**21** 45 **Sc** scandium	**22** 48 **Ti** titanium	**23** 51 **V** vanadium	**24** 52 **Cr** chromium	**25** 55 **Mn** manganese	**26** 56 **Fe** iron	**27** 59 **Co** cobalt	**28** 59 **Ni** nickel	**29** 64 **Cu** copper	**30** 65 **Zn** zinc	**31** 70 **Ga** gallium	**32** 73 **Ge** germanium	**33** 75 **As** arsenic	**34** 79 **Se** selenium	**35** 80 **Br** bromine	**36** 84 **Kr** krypton
37 85 **Rb** rubidium	**38** 88 **Sr** strontium	**39** 89 **Y** yttrium	**40** 91 **Zr** zirconium	**41** 93 **Nb** niobium	**42** 96 **Mo** molybdenum	**43** 98 **Tc** technetium	**44** 101 **Ru** ruthenium	**45** 103 **Rh** rhodium	**46** 106 **Pd** palladium	**47** 108 **Ag** silver	**48** 112 **Cd** cadmium	**49** 115 **In** indium	**50** 119 **Sn** tin	**51** 122 **Sb** antimony	**52** 128 **Te** tellurium	**53** 127 **I** iodine	**54** 131 **Xe** xenon
55 133 **Cs** cesium	**56** 137 **Ba** barium	**57** 139 **La** lanthanum	**72** 178 **Hf** hafnium	**73** 181 **Ta** tantalum	**74** 184 **W** tungsten	**75** 186 **Re** rhenium	**76** 190 **Os** osmium	**77** 192 **Ir** iridium	**78** 195 **Pt** platinum	**79** 197 **Au** gold	**80** 201 **Hg** mercury	**81** 204 **Tl** thallium	**82** 207 **Pb** lead	**83** 209 **Bi** bismuth	**84** 210 **Po** polonium	**85** 210 **At** astatine	**86** 222 **Rn** radon
87 223 **Fr** francium	**88** 226 **Ra** radium	**89** 227 **Ac** actinium	**104** 261 **Rf** rutherfordium	**105** 260 **Ha** hahnium													

58 140 **Ce** cerium	**59** 141 **Pr** praseodymium	**60** 144 **Nd** neodymium	**61** 147 **Pm** promethium	**62** 150 **Sm** samarium	**63** 152 **Eu** europium	**64** 157 **Gd** gadolinium	**65** 159 **Tb** terbium	**66** 163 **Dy** dysprosium	**67** 165 **Ho** holmium	**68** 167 **Er** erbium	**69** 169 **Tm** thulium	**70** 173 **Yb** ytterbium	**71** 175 **Lu** lutetium
90 232 **Th** thorium	**91** 231 **Pa** protactinium	**92** 238 **U** uranium	**93** 237 **Np** neptunium	**94** 242 **Pu** plutonium	**95** 243 **Am** americium	**96** 247 **Cm** curium	**97** 247 **Bk** berkelium	**98** 249 **Cf** californium	**99** 254 **Es** einsteinium	**100** 253 **Fm** fermium	**101** 256 **Md** mendelevium	**102** 254 **No** nobelium	**103** 257 **Lr** lawrencium

APPENDIX C

Metric System

Metric System		
Standard Metric Units		**Abbreviations**
Standard unit of mass	gram	g
Standard unit of length	meter	m
Standard unit of volume	liter	l
Common Prefixes		**Examples**
kilo (k)	1,000	a kilogram is 1,000 grams
centi (c)	0.01	a centimeter is 0.01 of a meter
milli (m)	0.001	a milliliter is 0.001 of a liter
micro (μ)	one-millionth	a micrometer is 0.000001 (one-millionth) of a meter
nano (n)	one-billionth	a nanogram is 10^{-9} (one billionth) of a gram
pico (p)	one-trillionth	a picogram is 10^{-12} (one trillionth) of a gram

Think Metric
Length

1. The speed of a car is 60 miles/hr or 100 km/hr.
2. A man who is 6 feet tall is 180 centimeters tall.
3. A 6-inch ruler is 15 centimeters long.
4. One yard is almost a meter (0.9 m).

Units of Length		
Unit	**Abbreviation**	**Equivalent**
meter	m	approximately 39 in
centimeter	cm	10^{-2} m
millimeter	mm	10^{-3} m
micrometer	μm	10^{-6} m
nanometer	nm	10^{-9} m
angstrom	Å	10^{-10} m

Length Conversions

1 in = 2.5 cm	1 mm = 0.039 in	1 m = 1.094 yd
1 ft = 30 cm	1 cm = 0.39 in	1 km = 0.6 mi
1 yd = 0.9 m	1 m = 39 in	
1 mi = 1.6 km		

To Convert	**Multiply By**	**To Obtain**
inches	2.54	centimeters
feet	30	centimeters
centimeters	0.39	inches
millimeters	0.039	inches

Think Metric
Volume

1. One can of beer (12 oz) contains 360 milliliters.
2. The average human body contains between 10-12 pints of blood or between 4.7-5.6 liters.
3. One cubic foot of water (7.48 gal) is 28.426 liters.
4. If a gallon of unleaded gasoline costs $1.00, a liter costs 26¢.

Units of Volume

Unit	Abbreviation	Equivalent
liter	l	approximately 1.06 qt
milliliter	ml	10^{-3} l (1 ml = 1 cm^3 = 1 cc)
microliter	µl	10^{-6} l

Volume Conversions

1 tsp = 5 ml	1 pt = 0.47 l	1 ml = 0.03 fl oz
1 tbsp = 15 ml	1 qt = 0.95 l	1 l = 2.1 pt
1 fl oz = 30 ml	1 gal = 3.8 l	1 l = 1.06 qt
1 cup = 0.24 l		1 l = 0.26 gal

To Convert	Multiply by	To Obtain
fluid ounces	30	milliliters
quarts	0.95	liters
milliliters	0.03	fluid ounces
liters	1.06	quarts

Think Metric
Weight

1. One pound of hamburger is 448 grams.
2. The average human male brain weighs 1.4 kilograms (3 lb 1.7 oz).
3. A person who weighs 154 pounds weighs 70 kilograms.
4. Lucia Zarate weighed 5.85 kilograms (13 lbs) at age 20.

Units of Weight

Unit	Abbreviation	Equivalent
kilogram	kg	10^3 g (approximately 2.2 lb)
gram	g	approximately 0.035 oz
milligram	mg	10^{-3} g
microgram	µg	10^{-6} g
nanogram	ng	10^{-9} g
picogram	pg	10^{-12} g

Weight Conversions

1 oz = 28.3 g	1 g = 0.035 oz
1 lb = 453.6 g	1 kg = 2.2 lb
1 lb = 0.45 kg	

To Convert	Multiply By	To Obtain
ounces	28.3	grams
pounds	453.6	grams
pounds	0.45	kilograms
grams	0.035	ounces
kilograms	2.2	pounds

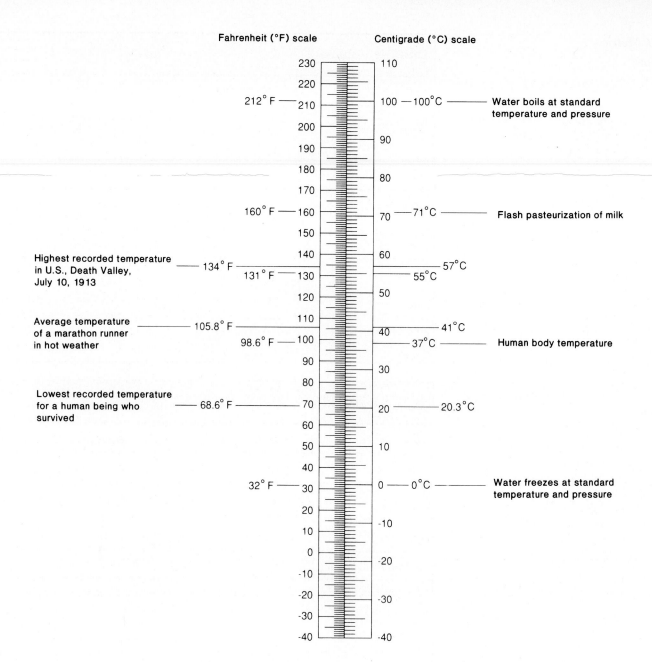

230 — 110

212° F — 210 — 100 — 100°C ——— Water boils at standard
temperature and pressure

220

200 — 90

190 — 80

180

170

160° F — 160 — 70 — 71°C ——— Flash pasteurization of milk

150

Highest recorded temperature
in U.S., Death Valley,
July 10, 1913 — 134° F — 140 — 60 — 57°C

131° F — 130 — 55°C

120 — 50

Average temperature
of a marathon runner
in hot weather — 105.8° F — 110 — 41°C

98.6° F — 100 — 40 — 37°C ——— Human body temperature

90 — 30

80

Lowest recorded temperature
for a human being who
survived — 68.6° F — 70 — 20 — 20.3°C

60

50 — 10

40

32° F — 30 — 0 — 0°C ——— Water freezes at standard
temperature and pressure

20 — -10

10

0 — -20

-10

-20 — -30

-30

-40 — -40

APPENDIX D

Answers

Chapter 1
Objective Questions

1. d
2. c
3. a
4. b
5. e
6. c
7. b
8. c
9. a. plants; b. animals; c. nutrients for plants; d. death and decay.

Concepts and Critical Thinking

1. A common ancestor has passed on to all living things common characteristics, such as using DNA genes.
2. Ways of life are diverse; therefore, organisms are diverse.
3. Nonliving things may have levels of organization, but they do not demonstrate emergent properties, where the whole is greater than the sum of the parts. For example, a cell is alive, but the structures, which make up a cell, are not alive.

Chapter 2
Objective Questions

1. d
2. a
3. d
4. a
5. b
6. c
7. b
8. d
9. a. Bacteria exposed to sunlight do not die if medium contains dye. b. Bacteria can be protected from UV light by dye. c. Experimental plate contains bacteria and dye; control contains only bacteria; exposure to UV light causes all bacteria to die. d. Hypothesis refuted.

Concepts and Critical Thinking

1. Even though the results of experimentation or observation support a hypothesis, it could be that the further studies will prove it false.
2. In the everyday sense, a theory means a supposition; in the scientific sense, a theory means a hypothesis that has been supported by many experiments and observations.
3. A scientist does studies to understand the natural world; citizens make decisions how these results should be used.

Chapter 3
Objective Questions

1. c
2. b
3. c
4. d
5. d
6. d
7. b
8. c
9. a
10. 7 p and 7 n in nucleus; 2 electrons in inner shell, and 5 electrons in outer shell. This means that nitrogen needs 3 more electrons to complete its outer shell; therefore the formula for ammonia is NH_3.

Concepts and Critical Thinking

1. For example, people take supplemental calcium to keep their bones strong.
2. Cells have a complex structure that is built upon the atoms and molecules in cells.
3. For example, if water was not slow to heat up and slow to cool down, living things might be subject to the killing effects of rapid heating and cooling.

Chapter 4
Objective Questions

1. c
2. a
3. d
4. b
5. c
6. b
7. c
8. c
9. d
10. c
11. a
12. a
13. a. monomer; b. condensation; c. polymer; d. hydrolysis. The diagram shows the manner in which macromolecules are synthesized and degraded in cells.
14. a. primary level; b. secondary level; c. tertiary level; d. quaternary level.
15. AGTTCGGCATGC

Concepts and Critical Thinking

1. They are unified in the use of carbohydrate as a structural molecule, but they are diversified as to the particular carbohydrate.
2. Butter is solid, and an oil is a liquid at room temperature because a fat containing saturated hydrocarbon chains melts at a higher temperature than one containing unsaturated chains.

3. Phospholipids have a polar head and nonpolar tails; they make up the plasma membrane, which is selectively permeable. Keratin, a fibrous protein that contains only helix polypeptides, is found in tough structures such as hair and nails.

Chapter 5
Objective Questions

1. c   2. c
3. d   4. a
5. c   6. c
7. a   8. d
9. d   10. d
11. a. Golgi apparatus further modifies; b. smooth ER modifies; c. rough ER produces; d. nucleolus—RNA helps; e. chromatin—DNA directs.
12. a. See text. b. Mitochondria and chloroplasts are a pair because they are both membranous structures involved in energy metabolism. c. Centrioles and flagella are a pair because they both contain microtubules; centrioles give rise to the basal bodies of flagella. d. Endoplasmic reticulum and ribosomes are a pair because together they are rough ER, which produces proteins.

Concepts and Critical Thinking

1. Show them microscopic slides of all sorts of tissues from different organisms. You would have to find an organism whose tissues did not contain cells.
2. The various organelles listed in table 5.1 show that the cell is compartmentalized; each organelle has a separate function.
3. All the structures labeled in question 11 plus mitochondria supply ATP energy. All the other parts of a cell assist to some degree also.

Chapter 6
Objective Questions

1. b   2. b
3. a   4. c
5. c   6. d
7. d   8. b
9. b
10. See figure 6.3. Phospholipid tails have no polar groups to interact with water's polar groups.
11. a. hypertonic—cell shrinks due to loss of water; b. hypotonic—cell has swelled due to gain of water.

Concepts and Critical Thinking

1. Structurally, the plasma membrane is the outer boundary of the cell. Functionally, the plasma membrane regulates what enters and leaves a cell.
2. A cell can die in either a severely hypertonic or severely hypotonic solution.
3. The plasma membrane secretes chemical messengers that communicate with distant cells; the plasma membrane participates in the structure of junctions between adjacently located cells.

Chapter 7
Objective Questions

1. a   2. d
3. c   4. a
5. d   6. c
7. d   8. c
9. a. macromolecules; b. degradative reactions; c. ADP + P → ATP; d. NADP → NADPH; e. small molecules; f. synthetic reactions; g. macromolecules.
10. See figure 7.13.

Concepts and Critical Thinking

1. Both the rough ER and Golgi apparatus produce molecules that are used for structural purposes. It takes energy to produce molecules.
2. When glucose energy is converted to ATP, there is a loss, and the energy released when ATP is broken down eventually becomes heat.

3. The temperature of the body is not high, and enzymes are needed to bring reactants together.

Chapter 8
Objective Questions

1. d   2. d
3. a   4. d
5. c   6. d
7. c   8. d
9. See figure 8.3.
10. a. water; b. oxygen; c. carbon dioxide; d. carbohydrate; e. ADP + P → ATP; f. NADP → NADPH.

Concepts and Critical Thinking

1. Show that solar energy is needed for photosynthesis, and photosynthetic organisms are the food for the biosphere, including humans. The bodies of plants became the fossil fuels, which we use to produce electricity and heat houses and convert to gasoline for cars.
2. Chlorophyll captures solar energy, which is converted to ATP energy and used to produce NADPH. ATP and NADPH are used to reduce CO_2 to a carbohydrate.
3. The thylakoid space is separated from the stroma by the thylakoid membrane; the build-up of hydrogen ions in the thylakoid space leads to ATP production.

Chapter 9
Objective Questions

1. b   2. c
3. a   4. c
5. c   6. c
7. a   8. b
9. c   10. d
11. c   12. a
13. b   14. d
15. b   16. a
17. d   18. a
19. c   20. b
21. See figure 9.11a.
22. a. 32; b. electron transport system; c. NADH; d. NADH; e. NADH; f. pyruvate; g. acetyl CoA; h. Krebs cycle; i. 2; j. lactate; k. CO_2; l. 2; m. CO_2.

Concepts and Critical Thinking
1. Flow of energy; photosynthesis converts the energy of the sun into carbohydrates that are converted to ATP energy by cellular respiration. Recycling of matter: Carbohydrates and oxygen from photosynthesis participate in cellular respiration, whose end product carbon dioxide reenters plants again.
2. Nutrient molecules are broken down to molecules that participate in cellular respiration, with the concomitant build-up of ATP. It is the ATP that is used by cells.
3. Glycolysis is a metabolic pathway that is almost universally found in organisms; it must have evolved in an ancestor common to all other organisms.

Chapter 10
Objective Questions

1.	c	2.	b
3.	d	4.	d
5.	a	6.	c
7.	b	8.	c
9.	b	10.	b
11.	See figure 10.5		

Concepts and Critical Thinking
1. As cells grow, they become ready for reproduction by producing more plasma membrane, cytoplasm, and organelles.
2. The daughter cells inherit DNA along with its inherent regulation from the parent cell.
3. Mitosis involves the use of a spindle, a structure that ensures that each daughter cell receives a copy of each chromosome.

Chapter 11
Objective Questions

1.	b	2.	c
3.	c	4.	b
5.	a	6.	c
7.	d	8.	d
9.	c		
10.	a—It shows bivalents at equators.		

Concepts and Critical Thinking
1. Asexual reproduction and sexual reproduction begin with cells, usually single cells. This shows how fundamental the cell is to the life of an organism.
2. Variation during asexual reproduction is limited to the occurrence of mutations. Variation during sexual reproduction is introduced due to mutations, crossing-over, and recombination (from independent assortment and fertilization).
3. Advantage: favorable mutations may be of immediate advantage. Disadvantage: unfavorable mutations may be of immediate disadvantage.

Chapter 12
Practice Problems 1
1. a. 100% *W;* b. 50% *W,* 50% *w;* c. 50% *T,* 50% *t;* d. 100% *T*
2. a. gamete; b. genotype; c. gamete

Practice Problems 2
1. *bb*
2. *Tt* ✕ *tt, tt*
3. 3/4 or 75%
4. *Yy* and *yy*

Practice Problems 3
1. a. 100% *tG;* b. 50% *TG,* 50% *tG;* c. 25% *TG,* 25% *Tg;* 25% *tG;* 25% *tg;* d. 50% *TG,* 50% *Tg*
2. a. genotype; b. gamete; c. genotype; d. gamete

Practice Problems 4
1. *BbTt* only
2. a. *LlGg* ✕ *llgg*
   b. *LlGg* ✕ *LlGg*
3. 9/16

Objective Questions

1.	b	2.	a
3.	c	4.	d
5.	c	6.	d
7.	a	8.	b
9.	b	10.	b
11.	c	12.	d

Additional Genetics Problems
1. 100% chance for widow's peak and 0% chance for continuous hairline
2. *Ee*

3. 50%
4. 210 gray body and 70 black body; 140 = heterozygous; cross fly with recessive
5. F_1 = all black with short hair; F_2 = 9:3:3:1; offspring would be 1 brown long: 1 brown short: 1 black long: 1 black short
6. *Bbtt* ✕ *bbTt* and *bbtt*
7. *GGLl*
8. 25%

Concepts and Critical Thinking
1. You inherit chromosomes containing the same types of genes from parents, but each parent contributes one-half of the particular genes inherited.
2. Plants and animals both have chromosomes containing the genetic material, DNA.
3. The individual's genes were not modified by the accident.

Chapter 13
Practice Problems 1
1. Cc^h (wild); $c^{ch}c^h$ (light gray); Cc^h (wild); c^hc^h (Himalaya)
2. 1 pink:1 white
3. 7
4. pleiotropy, epistasis

Practice Problems 2
1. females: X^RX^R, X^RX^r, X^rX^r, males: X^RY (gametes X^R, Y); X^rY (gametes X^r, Y)
2. b; 1:1
3. 100%, none, 100%
4. mother: $X^BX^bRr;$ father: $X^BYRr;$ son: X^bYrr

Practice Problems 3
1. 9:3:3:1, linkage
2. $54.5 - 13.0 = 41.5\%$
3. 12
4. bar eye, scalloped wings, garnet eye

Objective Questions

1.	b	2.	c
3.	e	4.	a
5.	b	6.	b
7.	d	8.	b
9.	c	10.	d

Additional Genetics Problems

1. pleiotropy
2. codominance, F^BF^B, F^WF^W, F^BF^W
3. 50%
4. 6, yes
5. linkage, 10
6. Both males and females are 1:1.
7. $X^BX^B \times X^bY$ = both males and females bar-eyed;
   $X^bX^b \times X^BY$ = females bar-eyed; males all normal
8. XbXbWw
9. females—1 short-haired tortoise shell: 1 short-haired yellow;
   males—1 short-haired black: 1 short-haired yellow

Concepts and Critical Thinking

1. Most human traits, such as height and color of skin, are examples of polygenic inheritance.
2. The genes behave similarly to the chromosomes (i.e., they come in pairs, they segregate during meiosis, and there is only one in the gametes).
3. Almost all males have an Y chromosome; XO individuals are females. This shows that most likely sex determining alleles on the Y chromosome have some degree of dominance to sex-determining alleles on the X chromosome.

Chapter 14

Practice Problems 1

1. 25%
2. 50%
3. heterozygous
4. heterozygous—unless a new mutation, which is a fairly common cause of NF.

Practice Problems 2

1. Hb^AHb^S; either Hb^SHb^S or Hb^AHb^S
2. light
3. white
4. Doe—baby #1
   Jones—baby #2

Practice Problems 3

1. mother; mother X^HX^h, father X^HY, son X^hY

2. 0%, 0%, 100%
3. rrX^bX^b and RrX^BX^b, rrX^BX^b, RrX^bX^b
4. colorblind

Objective Questions

1. c   2. a
3. b   4. c
5. All are consistent.
6. c
7. b
8. d
9. c
10. a
11. autosomal dominant condition

Additional Genetics Problems

1. 0%, 0%, 100%
2. 25%
3. 50%, 50%
4. AB, Yes, A, B, AB, or O
5. light, white
6. X^cYww, X^cX^cWw, X^cYWw; normal vision with widow's peak
7. **a.** recessive, Aa
   b. dominant, Aa
   c. sex-linked recessive, X^AX^a

Concepts and Critical Thinking

1. Yes, the concept still holds, because any differences don't negate the basic similarities.
2. Divide the cells of an early embryo to produce as large a number of identical children as possible. (This solves the nature question.) Raise each child under entirely different circumstances and then see how similar the children are.
3. This is a question that each person must decide on their own. Genetic counseling and genetic testing can sometimes tell parents their chances of having a child with a genetic disease.

Chapter 15

Objective Questions

1. b   2. d
3. a   4. c
5. d   6. c
7. a   8. a
9. b   10. d

11. The parental helix is heavy-heavy, and each daughter helix is heavy-light. (The cell was supplied with ^{14}N nucleotides as replication began.) This shows that each daughter helix is composed of one template strand and one new strand. This is consistent with semiconservative replication.

Concepts and Critical Thinking

1. ^{35}S would have been found with the bacterial cells, showing that it was needed for viral replication.
2. Replication is a part of duplication of the chromosomes. Replication before mitosis ensures that each daughter cell will have the diploid number of chromosomes, and replication before meiosis ensures that the sperm and egg will have the haploid number of chromosomes.
3. Mutation is the ultimate source of genetic variation. Without variation, new organisms would never evolve.

Chapter 16

Objective Questions

1. b   2. d
3. a   4. c
5. a   6. a
7. c   8. b
9. c   10. d

11. a. ACU CCU GAA UGC AAA;
    b. UGA GGA CUU ACG UUU;
    c. threonine-proline-glutamate-cysteine-lysine.

Concepts and Critical Thinking

1. The genetic information stored by DNA is the sequence of amino acids in a protein; this is stored in the sequence of DNA bases.
2. A universal code means that the ancestry of all types of organisms can be traced to a common origin.
3. The sequence of bases of DNA ultimately determines the sequence of bases in a protein. Figure 16.5 shows how the molecules are colinear.

Chapter 17

Objective Questions

1.	c	2.	a
3.	d	4.	a
5.	b	6.	b
7.	b	8.	d
9.	d	10.	d

11. a. See figure 17.2*a*. b. For the trp operon to be in the off position, a corepressor has to be attached to the repressor.

Concepts and Critical Thinking

1. Only certain mRNA molecules are present in cells at any particular time.
2. Gene expression necessitates a functioning protein product. Posttranslational control involves regulation of the activity of the protein product.
3. Checks and balances are often seen in biological systems. This is a safety feature that allows greater control.

Chapter 18

Objective Questions

1.	a	2.	c
3.	d	4.	a
5.	d	6.	d
7.	b	8.	c
9.	d	10.	a

11. a. AATT; b. TTAA.
12. 2. reverse transcription occurs; 4. most of the viral genes are removed, etc.; 5. transcription of DNA occurs; 6. recombinant RNA is repackaged; 8. reverse transcription occurs.

Concepts and Critical Thinking

1. It is possible to make recombinant DNA, to modify natural DNA in the laboratory, to perform PCR analysis, to do DNA fingerprinting, and to determine the sequence of the bases of DNA. This is disturbing to people who resist thinking of humans as physical and chemical machines.
2. Human genes can be placed in a bacterium, where they function normally.

Chapter 19

Objective Questions

1.	d	2.	b
3.	b	4.	d
5.	d	6.	d
7.	d	8.	b

9. a. All the continents of today were one continent during the Triassic period, and the reptiles spread throughout the land. b. All vertebrates share a common ancestor, who had pharyngeal pouches during development. c. Two different continents can have similar environments and therefore unrelated organisms that are similarly adapted. d. This demonstrates that diversification has occurred in the nightshade family.

Concepts and Critical Thinking

1. Both bacteria and humans use DNA as the hereditary material and have similar metabolic pathways.
2. If adaptation were purposeful, organisms wouldn't have structures that don't serve a current function.
3. It is possible to formulate a hypothesis about evolution and then make observations to see if the hypothesis is correct.

Chapter 20

Objective Questions

1.	c	2.	b
3.	c	4.	c
5.	c	6.	c
7.	d	8.	c
9.	b		

10. See figure 20.7*b*.

Population Genetics Problems

1. 99%
2. recessive allele = 0.2; dominant allele = 0.8; homozygous recessive = 0.04; homozygous dominant = 0.64; heterozygous = 0.32

Concepts and Critical Thinking

1. Traits are passed from one generation to the next via the gametes, and no mechanism has been found by which phenotype changes can affect the genes in the gametes.

2. The more fit organisms have more offspring, and this is the way by which adaptive traits accumulate in a population.
3. Increase variation between populations: genetic drift, natural selection, mutation, nonrandom mating. Decrease variation between populations: gene flow. These mechanisms alter gene frequencies in populations.

Chapter 21

Objective Questions

1.	d	2.	d
3.	c	4.	b
5.	b	6.	d
7.	a	8.	d
9.	b		

10. Do away with 5 middle arrows so that 2 sets of circles remain. The circles of each set are connected by arrows.

Concepts and Critical Thinking

1. The higher taxa show how species are related. Species in different kingdoms are distantly related, and those in the same genus are closely related, for example.
2. All new species have evolved from previous species; therefore all organisms have a phylogenetic history.
3. One group gives rise to many groups, each adapted to a particular environment.

Chapter 22

Objective Questions

1.	d	2.	c
3.	b	4.	b
5.	a	6.	c
7.	c	8.	b
9.	b	10.	d

11. The diagram suggests that the eukaryotic cell acquired mitochondria and, if present, chloroplasts by engulfing particular prokaryotic cells. This hypothesis is called the endosymbiotic theory.

Concepts and Critical Thinking

1. All 3 hypotheses suggest that a chemical evolution produced the first cell(s).
2. The increase in variability among the offspring of sexually reproducing organisms (as compared to asexual ones) provides more opportunity for change.
3. Under the conditions of the primitive earth, a chemical evolution produced the first cell(s). Under the conditions of today's earth, life comes only from life.

Chapter 23
Objective Questions

1.	b	2.	b
3.	c	4.	b
5.	c	6.	c
7.	d	8.	c
9.	b	10.	d
11.	d		

12. See figure 23.3.

Concepts and Critical Thinking

1. Viruses are unable to carry on reproduction outside a living cell. They reproduce by taking over the machinery of the cell; therefore, they are obligate parasites.
2. Monerans are metabolically diverse, as witnessed by the fact that there is probably no single type of organic compound that some Moneran cannot digest.
3. This statement refers to structure, not biochemistry.

Chapter 24
Objective Questions

1.	d	2.	a
3.	b	4.	c
5.	d	6.	a
7.	a	8.	d

9. a. Protista, Chlorophyta (green algae); b. Fungi, Ascomycota (sac fungi); c. Protista, Sarcodina (amoeboid protozoans).
10. a. locomotion; b. pigmentation.
11. See figure 24.9b. Sexual reproduction involves the union of gametes.

Concepts and Critical Thinking

1. The protozoans are heterotrophic; the algae are photosynthetic; and the slime molds are saprophytic.
2. In the haplontic cycle, the zygote undergoes meiosis, and therefore the adult is haploid. In alternation of generations, the sporophyte produces haploid spores that mature into the gametophyte. In the diplontic cycle, meiosis produces haploid gametes, the only part of the cycle that is haploid.
3. Fungi live on the organic material they digest. Plants produce their own organic food, and animals go out and find it.

Chapter 25
Objective Questions

1.	d	2.	c
3.	c	4.	c
5.	d	6.	b
7.	b	8.	b
9.	d		

10. See figure 25.3b.
11. a. leafy shoot; b. heart-shaped structure; c. stalk with sporangium; d. rhizome with fronds, roots. The fern has the sporophyte dominant. There are sporangia on the underside of the fronds, and this is the generation that persists longer.
12. a. underside of scales of male cone; b. in the anther of stamens; c. upper surface of scales of female cone; d. in ovary of pistil. The flowering plant produces fruit, which develops from the ovary.

Concepts and Critical Thinking

1. The 5-kingdom system of classification is based in part on mode of nutrition. From this standpoint, it would seem that all photosynthesizers should be in the same kingdom. On the other hand, it would seem that we need some other criteria to go by; for example, all organisms in the plant kingdom are multicellular with some degree of complexity. In this text, all organisms in the plant kingdom protect the zygote.
2. Green algae at the base could lead to nonvascular plants (bryophytes) and vascular plants. The primitive vascular plants as a group could lead to gymnosperms (4 divisions) and angiosperms (2 classes). The tree goes from the first evolved to the latest evolved according to the fossil record.
3. Roots enable a plant to take water out of the soil; stems hold the leaves aloft to catch solar energy, and the leaves also take in carbon dioxide at stomata.

Chapter 26
Objective Questions

1.	d	2.	c
3.	d	4.	d
5.	b	6.	b
7.	a		

8. See figure 26.4.
9. Sponges (no items)
   Cnidarians
     radial symmetry
     tissue level of organization
   Flatworms
     3 germ layers
   Roundworms
     pseudocoelom
     3 germ layers
     tube-within-a-tube
10. Sponges
     collar cells
   Planarians
     eyespots
     gastrovascular cavity
   Tapeworms
     proglottids
   Roundworms
     uterus

Concepts and Critical Thinking

1. For a contrast of the plant cell and the animal cell, see table 5.2. Note that animal cells cannot make their own food (lack chloroplasts) and are motile (have flagella), 2 distinct characteristics of animals.
2. A common ancestor would have to have those features shared by both types of organisms: bilateral symmetry, 3 germ layers, and organs.

3. Internal parasites need to have a means of protecting themselves from host attack, absorbing nutrients from the host, and reproducing that allows dispersal to new hosts.

Chapter 27

Objective Questions

1. a
2. a
3. c
4. c
5. a
6. d
7. d
8. b
9. d
10. Mollusks
    organ system level of
        organization
    coelom
    cephalization in some
        representatives
    complete gut
    Annelids
    organ system level of
        organization
    segmentation
    coelom
    cephalization in some
        representatives
    complete gut
    Arthropods
    organ system level of
        organization
    segmentation
    coelom
    cephalization in some
        representatives
    complete gut
    Chordates
    organ system level of
        organization
    segmentation
    coelom
    cephalization in some
        representatives
    complete gut
11. Clam
    Mollusk
    3 ganglia
    gills
    open circulatory system
    hatchet foot

Earthworm
    annelid
    ventral nerve cord
    closed circulatory system
    hydrostatic skeleton
    setae
12. Label wings (for flying in the air), spiracle (for breathing air), tympanum (for picking up sound waves), ovipositor (for laying eggs in earth), digestive system (for eating grass), Malpigian tubules (for excretion without loss of water), tracheae (for breathing air), vagina and seminal receptacle (for reception and storage of sperm so they do not dry out). See figure 27.12 for placement of labels.

Concepts and Critical Thinking

1. Body rings indicate that the earthworm's body is segmented and there are setae on each segment. In the grasshopper, segmentation is not so obvious because there is a head, thorax with legs, and abdomen—that is, specialization of parts.
2. See table 27.2, which lists all the features that arthropods and chordates have in common. The fact that both lineages have these features must mean that these features increase the fitness of organisms having them.
3. See answer to objective question 12.

Chapter 28

Objective Questions

1. a
2. d
3. a
4. b
5. b
6. d
7. d
8. b
9. b
10. c
11. a. mammals; b. birds; c. reptiles; d. amphibians; e. modern bony fishes; f. cartilagenous fishes; g. modern jawless fishes.
12. Primate (all), Prosimian (only lemurs), Anthropoid (all but lemurs), Hominoid (all but lemurs and monkeys), Hominid (only Australopithecines and humans).

13. Kingdom Animalia, phylum Chordata, subphylum Vertebrata, class Mammalia, order Primate, superfamily Hominoidea, family Hominidae, genus and species *Homo sapiens.*

Concepts and Critical Thinking

1. The number of species in varied habitats is a good criterion. Vertebrates are chordates; therefore table 27.2 lists the features they have in common. In addition, they both have a rigid, but jointed, skeleton.
2. Gametes are protected from drying out: reptiles pass sperm directly from male to female; angiosperms rely on the pollen grain. Embryo is protected from drying out: reptiles develop in a shelled egg; angiosperms produce seeds that protect the embryo until germination under favorable conditions occurs.
3. No living organism evolved from another living organism. Instead, 2 living organisms may have shared a common ancestor.

Chapter 29

Objective Questions

1. c
2. b
3. c
4. c
5. b
6. b
7. b
8. c
9. c
10. b
11. a. epidermis; b. cortex; c. endodermis; d. phloem; e. xylem.
12. a. cork; b. phloem; c. vascular cambium; d. xylem; e. pith; f. bark.
13. a. upper epidermis; b. palisade mesophyll; c. leaf vein; d. spongy mesophyll; e. lower epidermis.

Concepts and Critical Thinking

1. A leaf that is broad and flat is well shaped to catch the sun's rays; the cells contain chloroplasts; the leaf veins bring water from the roots; the stomata admit carbon dioxide into the air spaces of the spongy layer.
2. Figure 29.1 and table 29.1 show how a plant is organized.

3. Wood helps a plant resist the pull of gravity. It serves as an internal skeleton and allows the plant to lift the leaves, exposing them to the rays of the sun.

Chapter 30
Objective Questions

1.	d	2.	a
3.	c	4.	b
5.	c	6.	c
7.	c	8.	d

9. The diagram shows that air pressure pushing down on mercury in the pan can only raise a column of mercury to 76 cm. When water above the column is transpired, it pulls on the mercury and raises it higher. This suggests that transpiration would be able to raise water to the top of trees.
10. See figure 30.6b and c. After K^+ enters guard cells, water follows by osmosis and the stoma opens.
11. There is more solute in (1) than (2); therefore water enters (1); this creates a pressure that causes water, along with solute, to flow toward (2).

Concepts and Critical Thinking

1. A physical process accounts for the transport of water in xylem; causes a stoma to open; and accounts for the transport of organic substances in phloem.
2. Because vessel cells are hollow and nonliving, they form a continuous pipeline from the roots to the leaves. This allows plants to transport water and minerals from the roots to the leaves.

Chapter 31
Objective Questions

1.	c	2.	a
3.	b	4.	a
5.	a	6.	b
7.	a	8.	d
9.	c	10.	c

11. a. sporophyte; b. microsporangium; c. megasporangium (in ovule); d. meiosis; e. microspore; f. megaspore; g. male gametophyte (pollen grain); h. female gametophyte; i. egg and sperm; j. fertilization; k. zygote in seed.

Concepts and Critical Thinking

1. Flowering plants follow the alternation of generations life cycle, but the gametophyte is reduced to a microscopic size and is protected by the sporophyte. The gametes are also protected; the egg stays within the ovule within the ovary and awaits the sperm. A pollen grain is transported to the pistil by wind or animals and then germinates, allowing the sperm to move down the style to the egg. The zygote (new sporophyte) develops into an embryo within the ovule, which becomes the seed enclosed by fruit. Fruit is dispersed in various ways.
2. As stated, the gametophyte is protected by the sporophyte and the embryonic sporophyte is protected by the seed.
3. Coevolution results in one type of pollinator gathering food from a particular plant. This reduces competition and assures the pollinator of food from this particular source.

Chapter 32
Objective Questions

1.	d	2.	b
3.	e	4.	a
5.	c	6.	d
7.	b	8.	d
9.	d	10.	a

11. The arrows show the movement of auxin from the shady side to the sunny side of the stem. Auxin causes the cells on the shady side to elongate; therefore, the oat seedling bends toward the light.
12. The arrows (K^+ ions) leave guard cells under the influence of abscisic acid (ABA) when a plant is water stressed. Water osmotically follows the movement of K^+ and the stoma closes.

Concepts and Critical Thinking

1. The best example is that the ratio of auxin to cytokinin determines whether a plant tissue will form an undifferentiated mass or form roots, vegetative shoots and leaves, or flowers.
2. It is adaptive, for example, for leaves to unfold and stomata to open during the day, when photosynthesis can occur. A biological clock system needs a receptor that is sensitive to light and dark; a timekeeper, that is, a biological clock; and a means of communication within the body of the organism. In regard to flowering, phytochrome is the receptor and hormones are probably the means of communication. The possible timekeeper is not known.

Chapter 33
Objective Questions

1.	b	2.	b
3.	a	4.	d
5.	d	6.	d
7.	b	8.	d
9.	d	10.	c

11. a. smooth muscular tissue; b. blood cells—connective tissue; c. nervous tissue; d. ciliated columnar epithelium.

Concepts and Critical Thinking

1. The cell is alive, but the organelles making up a cell are not alive; a tissue performs functions that the individual cell in the tissue cannot perform and so forth from level to level.
2. Acquiring food is located to the right and left of the diagram. Cellular respiration is a degradative pathway requiring an exchange of gases. Excretion of waste accompanies degradation.
3. Acquiring food provides materials and energy; the musculoskeletal system helps us get food; exchange of gases gets rid of carbon dioxide and brings in oxygen; transporting materials brings nutrients to and takes wastes away from cells;

protecting the body from disease maintains its integrity; coordinating body activities makes sure that all the other functions are performed.

Chapter 34
Objective Questions

1.	b	2.	b
3.	a	4.	d
5.	d	6.	c
7.	d	8.	b
9.	b	10.	b
11.	d		
12.	a (See figure 34.6.)		

Concepts and Critical Thinking

1. Tissue fluid remains relatively constant because materials are constantly being added and removed at the capillaries.
2. The cells are far removed from the exterior, and therefore they have to have nutrients brought to them and wastes removed from them.
3. ATP energy keeps the heart pumping (arterial flow) and the skeletal muscles contracting (venous flow). ATP energy is produced by mitochondria, where chemical energy is transformed into usable energy.

Chapter 35
Objective Questions

1.	d	2.	d
3.	a	4.	b
5.	c	6.	b
7.	a	8.	a
9.	b	10.	d

Concepts and Critical Thinking

1. Death occurs when immunity fails to prevent microorganisms from taking over the body.
2. Immunity consists of immediate defense mechanisms and specific defense mechanisms, which work more slowly.
3. Bones are of course part of the skeletal system. Red bone marrow is the site of red and white blood cell formation; therefore, it is part of the circulatory and lymphatic systems.

Chapter 36
Objective Questions

1.	d	2.	b
3.	d	4.	b
5.	c	6.	a
7.	c	8.	c
9.	d	10.	c
11.	d	12.	c
13.	See figure 36.6.		
14.	Test tube 1: No digestion		

14. Test tube 1: No digestion
No enzyme and no HCl
Test tube 2: Some digestion
No HCl
Test tube 3: No digestion
No enzyme
Test tube 4: Digestion
Both enzyme and HCl are present

Concepts and Critical Thinking

1. In humans, the digestive system takes in nutrients, excretes certain metabolites like heavy metals, and prevents bacterial invasion by the low pH of the stomach.
2. See figure 36.3.
3. Only autotrophs are able to produce their own food. Heterotrophs feed on autotrophs and acquire material and energy. Autotrophs use wastes from heterotrophs (fertilizer, CO_2).

Chapter 37
Objective Questions

1.	a	2.	b
3.	b	4.	d
5.	c	6.	b
7.	c	8.	b
9.	d	10.	b
11.	See figure 37.7.		

Concepts and Critical Thinking

1. The respiratory system carries out exchange of gases and helps maintain the pH of the blood.
2. Gills are external extensions and lungs are internal cavities. Both are minutely divided and highly vascularized for the exchange of gases.
3. In humans, the lungs expand when the chest moves up and out and deflate when the chest moves down and in. The lungs contain many alveoli (air sacs) and are highly vascularized. Oxygen enters the blood following inhalation, and carbon dioxide leaves the body upon exhalation.

Chapter 38
Objective Questions

1.	d	2.	a
3.	c	4.	b
5.	a	6.	c
7.	b	8.	d
9.	a	10.	b
11.	See figure 38.10.		

Concepts and Critical Thinking

1. When molecules undergo degradation, wastes result. For example, following glucose metabolism, carbon dioxide must be excreted. Following amino acid breakdown, urea must be excreted.
2. Water contains a lower amount of oxygen than air. The countercurrent mechanism in the gills of fishes helps them extract oxygen out of water. Mammals evolved on land; a countercurrent mechanism in the kidneys helps them conserve water.
3. Reabsorption, in part by active transport, takes place in the proximal convoluted tubule. An increased surface area (microvilli) helps reabsorption, and mitochondria supply the ATP energy for active transport.

Chapter 39
Objective Questions

1.	c	2.	b
3.	d	4.	c
5.	a	6.	a
7.	b	8.	d
9.	b	10.	b

11. a. receptor (initiates nerve impulse); b. sensory neuron (takes impulses to cord); c. interneuron (passes impulses to motor neuron and other interneurons in the cord); d. white matter (contains tracts that take impulses up and down the cord); e. motor neuron (takes impulses to effector); f. effector (brings about adaptive response).

Concepts and Critical Thinking

1. Controls internal organs like the beating of the heart and skeletal muscles, which allow an animal to seek environments compatible with a constant internal environment.
2. There is a spinal nerve in each segment, but the brain is an obvious specialization of part of the spinal cord.
3. A neuron has long processes that conduct nerve impulses. The postsynaptic membrane is part of a dendrite, and the presynaptic membrane is part of an axon. Impulses always flow in this direction because only an axon contains synaptic vesicles.

Chapter 40
Objective Questions

1.	d	2.	c
3.	c	4.	d
5.	d	6.	c
7.	c	8.	b
9.	d	10.	c

11. See figure 40.5 and table 40.1.

Concepts and Critical Thinking

1. The sense organs help animals find favorable environments and avoid unfavorable ones; they help animals find food and avoid being eaten; they help animals communicate with others for the purpose of cooperation.
2. The sense organs are receptors for external and internal stimuli; they generate nerve impulses, which supply information to the central nervous system.
3. Relative fitness determines what structures characterize a species. Therefore, animals have those sense organs that are most adaptive.

Chapter 41
Objective Questions

1.	b	2.	f
3.	c	4.	e
5.	b	6.	b
7.	b	8.	c
9.	a	10.	d

11. See table 41.3.
12. See figure 41.10a.

Concepts and Critical Thinking

1. Consult, for example, figure 21.19 and note how the skeleton of a horse was adapted at first to a forest and then later to a grassland habitat.
2. Animals have to go out and get their food; therefore, locomotion is essential to them.
3. Nutrient molecules from the digestive system travel in the blood to muscle cells. Here, glucose energy is converted to ATP energy, largely within mitochondria. ATP energy is used by myosin filaments to pull actin filaments, and muscle contraction occurs.

Chapter 42
Objective Questions

1.	f	2.	b
3.	c	4.	a
5.	e	6.	c
7.	a	8.	d
9.	b	10.	d
11.	b	12.	b

13. See figure 42.7. In a negative feedback system, the output (i.e., thyroxin) cancels the input (e.g., thyroid growth hormone from the pituitary); therefore, the production of thyroxin is shut off when it gets too high.

Concepts and Critical Thinking

1. Neurotransmitters are released when needed and then reabsorbed by the presynaptic membrane or broken down within the synaptic cleft by an enzyme, for example. The level of a hormone is controlled typically by using negative feedback to regulate its production.
2. Specialized structures (i.e., neurons) deliver messages in the nervous system, whereas the endocrine system uses the blood system.
3. The human body undergoes a dramatic change following puberty.

Chapter 43
Objective Questions

1.	b	2.	d
3.	c	4.	d
5.	c	6.	c
7.	c	8.	c
9.	a	10.	a

11. See figure 43.3 and table 43.1. Testis, epididymis, vas deferens, urethra (in penis).

Concepts and Critical Thinking

1. They are similarly adapted, except the embryos of reptiles develop in shelled eggs while the embryos of placental mammals develop within the uterus of the female. Even so, the same extraembryonic membranes surround the embryo.
2. Reproduction is a process that takes some time; it is not like movement, which takes place quickly.

Chapter 44
Objective Questions

1.	b	2.	b
3.	b	4.	a
5.	b	6.	b
7.	d	8.	a
9.	d	10.	d

11. See figure 44.8. Chorion exchanges wastes for nutrients with the mother; amnion protects and prevents desiccation; blood vessels of allantois become umbilical blood vessels; yolk sac is the first site of blood cell formation.

Concepts and Critical Thinking

1. Senescence is a change that is part of the normal life cycle from fertilization to death.
2. They could be considered local hormones because they are messengers that are produced by one cell and act on another cell.
3. Perhaps as individuals age, genes are turned off (instead of on), and this leads to degenerative changes.

Chapter 45
Objective Questions
1.	d	**2.**	a
3.	d	**4.**	a
5.	c	**6.**	d
7.	d	**8.**	c
9.	d	**10.**	a

11. a. communication, chemical; b. learning, imprinting; c. communication, sound; d. learning, operant conditioning; e. communication, touch; f. communication, visual.

Concepts and Critical Thinking
1. The genes control the development of the brain and therefore even insight learning has a genetic basis.
2. Feeding behavior patterns affect fitness because an animal needs a body in good working order for reproduction to be successful. Territoriality affects fitness because an animal needs enough food to nourish its offspring. Reproductive patterns affect fitness because an animal needs to have a mate to reproduce.
3. Altruistic acts can be shown to increase inclusive fitness.

Chapter 46
Objective Questions
1.	b	**2.**	c
3.	d	**4.**	d
5.	d	**6.**	c
7.	b	**8.**	c

9. See figure 46.3*b*. The *r* stands for rate of natural increase.

Concepts and Critical Thinking
1. When there are unlimited resources, such as space, food, and shelter, and where there are no deaths due to natural disasters, population growth could be infinite.
2. Human life strategy is only like that of a *K*-strategist. See table 46.4.

Chapter 47
Objective Questions
1.	b	**2.**	d
3.	c	**4.**	d
5.	c	**6.**	b
7.	d	**8.**	c
9.	d		

10. a. mutualism; b. parasitism; c. commensalism.

Concepts and Critical Thinking
1. The organisms themselves provide niches for other organisms. For example, several types of warblers find niches in spruce trees (fig. 47.3).
2. As prey populations decrease in size, it is less likely that a predator will find that type of prey, and prey populations have defenses against predation.
3. Parasites keep down population sizes when they weaken or directly kill off their hosts. Parasites increases diversity because, for example, even parasites have parasites. As mentioned in number 1, organisms themselves provide niches for other organisms.

Chapter 48
Objective Questions
1.	a	**2.**	a
3.	d	**4.**	a
5.	a	**6.**	c
7.	d	**8.**	c
9.	a		

10. a. 4+ for both; b. 3+ for both; c. 2+ for both; d. 1+ for both.

Concepts and Critical Thinking
1. The neritic province is more productive because it receives both nutrients and sunlight. In contrast, only the epipelagic zone of the oceanic province receives adequate sunlight. This zone is not as nutrient rich as the neritic province.
2. Succession does not require a definite sequence, but there is a sequence of communities from the simple to the complex.
3. It gets colder as you go from the base of a mountain to the top; therefore, the communities will change as noted.

Chapter 49
Objective Questions
1.	b	**2.**	a
3.	c	**4.**	d
5.	b	**6.**	d
7.	c	**8.**	c
9.	a	**10.**	b

11. a. algae; b. zooplankton; c. small fishes; d. large fishes; e. humans.

Concepts and Critical Thinking
1. For example, the atmosphere is the abiotic component of the nitrogen cycle, and the passage of nitrogen compounds from producer to consumers is the biotic component of the nitrogen cycle.
2. It is impossible to create energy to keep an ecosystem going (first law), and as energy is passed from one trophic level to the next, there is always some loss of energy (second law). Therefore, there is a flow of energy from the sun through an ecosystem.

Chapter 50
Objective Questions
1.	a	**2.**	d
3.	c	**4.**	b
5.	b	**6.**	c
7.	d	**8.**	c
9.	a	**10.**	c

Concepts and Critical Thinking
1. The primary productivity of the other terrestrial biomes might increase due to carbon dioxide buildup following tropical rain forest destruction.
2. The figure shows, for example, that nitrous dioxide (NO_2), nitrous oxide, carbon dioxide, methane, CFCs, and halons all contribute to the greenhouse effect, and of these CFCs and halons also caused ozone shield destruction.
3. Fossil fuel energy contributes to agricultural yield and therefore helps support the increased human population size of today.

GLOSSARY

A

abscisic acid (ABA) (ab-sis'ik) a plant hormone causing stomata to close and initiating and maintaining dormancy *528*

abscission (ab-sizh'un) the dropping of leaves, fruits, or flowers from a plant body *528*

acetylcholine (as"ĕ-til-ko'lēn) a neurotransmitter used within both the peripheral and central nervous systems *639*

acetyl CoA (as"ĕ-til-ko-A) molecule made up of a 2-carbon acetyl group attached to coenzyme A, which enters the Krebs cycle for further oxidation *143*

acid a compound tending to dissociate and yield hydrogen ions in a solution and to lower its pH numerically *36*

acid deposition the return to earth as rain or snow of the sulfate or nitrate salts of acids produced by commercial and industrial activities on earth *800*

acoelomate (a-sēl'o-māt) condition where an organism does not develop a coelom or true body cavity *424*

ACTH adrenocorticotrophic hormone, which is secreted by the anterior pituitary and stimulates activity in the adrenal cortex *683*

actin a muscle protein making up the thin filaments in a sarcomere; its movement shortens the sarcomere, yielding muscle contraction *671*

action potential change in the polarity of a neuron caused by opening and closing of ion channels in the axon plasma membrane *639*

active site that part of an enzyme molecule where the substrate fits and the chemical reaction occurs *109*

active transport use of a plasma membrane carrier molecule to move particles from a region of lower to higher concentration; it opposes an equilibrium and requires energy *94*

adaptation an organism's modification in structure, function, or behavior to increase the likelihood of continued existence *8, 307*

adaptive radiation formation of a large number of species from a common ancestor *336*

adenine (A) (ad'ē-nīn) one of 4 organic bases in the nucleotides composing the structure of DNA and RNA *54, 238*

adenosine diphosphate (ADP) (ah-den'o-sēn di-fos'fāt) one of the products of the hydrolysis of ATP, a process that liberates energy *116*

adenosine triphosphate (ATP) (ah-den'o-sēn tri-fos'fāt) a compound containing adenine, ribose, and 3 phosphates, 2 of which are high-energy phosphates. The breakdown of ATP to ADP makes energy available for energy-requiring processes in cells *115*

adipose tissue type of loose connective tissue in which fibroblasts enlarge and store fat *541*

ADP *see* adenosine diphosphate

adrenal gland lies atop a kidney and secretes the stress hormones epinephrine, norepinephrine, and the corticoid hormones *685*

aerobic process in which oxygen is required *145*

age structure diagram a representation of the number of individuals in each age group in a population *748*

agglutination (ah-glūt-en-ā'shun) clumping of red blood cells, as when incompatible red cell antigens and antibodies mix *568*

AIDS acquired immunodeficiency syndrome, caused by the HIV virus in humans; the virus attacks T4 lymphocytes, making them ineffective and leading to the development of other infections *707*

aldosterone (al"do-ster'on) hormone secreted by the adrenal gland that maintains the sodium and potassium balance of the blood *630, 686*

alga (pl., algae) aquatic organisms carrying out photosynthesis and belonging to the kingdom Protista *383*

allantois (ah-lan'to-is) an extraembryonic membrane that accumulates nitrogenous wastes in the embryo of birds and reptiles and contributes to the formation of umbilical blood vessels in mammals *719*

allele (ah-lēl') alternative forms of a gene, for example, an allele for long wings and an allele for short wings in fruit flies *186*

allopatric speciation origin of new species in populations that are separated geographically *334*

alternation of generations cycle life cycle in which a haploid generation alternates with a diploid generation *384*

alveolus (al-ve'o-lus) terminal, microscopic, grapelike air sacs found in vertebrate lungs *612*

altruism form of social behavior found in insect societies, exemplified by the willingness of the daughters to help the queen rather than reproduce their own offspring *740*

amino acid organic subunits each having an amino group and an acid group, that covalently bond to produce protein molecules *50*

amnion an extraembryonic membrane of the bird, reptile, and mammal embryo that forms an enclosing fluid-filled sac *719*

amphibian class of terrestrial vertebrates that includes frogs, toads, and salamanders; still tied to a watery environment for reproduction *452*

anaerobic a process that does not require oxygen *142*

angiosperm flowering plant having double fertilization that results in development of seed-bearing fruit *404*

annelid (ah'nel-id) a phylum of segmented worms characterized by a tube-within-a-tube body plan, bilateral symmetry, and segmentation of some organ systems *434*

anther part of stamen where pollen grains develop *407, 510*

antheridium (an"ther-id'i-um) male structures in nonseed plants where swimming sperm are produced *398*

antibody a protein molecule usually formed naturally in an organism to combat antigens or foreign substances *563, 577*

anticodon 3 nucleotides on a tRNA molecule attracted to a complementary codon on mRNA *257*

antidiuretic hormone (ADH) (an"tī-di"u-ret'ik) hormone secreted by the posterior pituitary that increases the permeability of the collecting duct in a nephron *628, 681*

antigen substances, usually proteins, that are not normally found in the body *563, 577*

aorta largest systemic artery, which transports blood from the heart to all other systemic arteries *560*

appendicular skeleton part of the skeleton forming the upper appendages, shoulder girdle, lower appendages, and hip girdle *666*

archaebacterium (ar"ke-bak-te're-um) probably the earliest prokaryote; the cell wall and plasma membrane differ from other bacteria, and many live in extreme environments *374*

archegonium (ar"kē-go'ne-um) female structure in nonseed plants where an egg is produced *398*

artery a blood vessel that transports blood away from the heart toward arterioles *555*

arthropod a diverse phylum of animals that have jointed appendages; includes lobsters, insects, and spiders *437*

asexual reproduction reproduction involving only mitosis which produces offspring genetically identical to the parent *166*

aster short, radiating fibers produced by the centrioles; important during mitosis and meiosis *160*

atom the smallest particle of an element that displays its properties *26*

ATP *see* adenosine triphosphate

autonomic nervous system a branch of the peripheral nervous system administering motor control over internal organs *641*

autosome any chromosome other than the sex-determining pair *204*

autotroph self-nourishing organism; referring to producers starting food chains that make organic molecules from inorganic nutrients *356*

auxin plant hormone regulating growth, particularly cell elongation; also called indoleacetic acid *523*

axial skeleton part of the skeleton forming the vertical support or axis, including the skull, rib cage, and vertebral column *666*

axon the part of a neuron that conducts the impulse from cell body to synapse *635*

B

B lymphocyte white blood cell that produces and secretes antibodies that can combine with antigens *577*

bacteriophage a virus that parasitizes a bacterial cell as its host, destroying it by lytic action *237*

bacterium (pl., bacteria) one-celled organism that lacks a nucleus and cytoplasmic organelles other than ribosomes; reproduces by binary fission and occurs in one of 3 shapes (rod, sphere, spiral) *66, 370*

Barr body X chromosome in females that is inactive *269*

basal body structure in the cell cytoplasm, located at the base of a cilium or flagellum *78*

base a compound tending to lower the hydrogen ion concentration in a solution and raise its pH numerically *36*

behavior all responses made by an organism to changes in the environment *5*

benthic division the ocean floor with a unique set of organisms in contrast to the pelagic division or open waters *768*

bicarbonate ion form in which most of the carbon dioxide is transported in the bloodstream *614*

bilaterally symmetrical a body having 2 corresponding or complementary halves *419*

bile a substance released from the gallbladder to small intestine to emulsify fats prior to chemical digestion *594*

binary fission splitting of a parent cell into 2 daughter cells; serves as an asexual form of reproduction in single-celled organisms *156, 371*

biogeography the study of the geographical distribution of organisms *303*

biological clock internal mechanism that maintains a biological rhythm in the absence of environmental stimuli *530, 732*

biological magnification process by which substances become more concentrated in organisms in the higher trophic levels of the food chain *784*

biome a major biotic community having well-recognized life forms and a typical climax species *766*

biosphere a thin shell of air, land, and water around the earth that supports life *8, 766*

biotic potential maximum rate of natural increase of a population that can occur under ideal circumstances *745*

bivalent (tetrad) (bi-va'lent) homologous chromosomes each having sister chromatids that are joined by a nucleoprotein lattice during meiosis *172*

blastocoel (blas'to-sēl) the fluid-filled cavity of a blastula in the early embryo *713*

blastula a hollow, fluid-filled ball of cells prior to gastrula formation of the early embryo *713*

blood type of connective tissue in which cells are separated by a liquid called plasma *541*

blood pressure force of flowing blood pushing against the inside wall of an artery *561*

bone type of connective tissue in which a matrix of calcium salts is deposited around protein fibers *541*

bronchus (pl., bronchi) one of 2 main branches of the trachea in vertebrates that have lungs *612*

bronchiole small tube that conducts air from a bronchus to the alveoli *612*

bryophyte a plant of a group that includes mosses and liverworts *398*

budding in animals an asexual form of reproduction whereby a new organism develops as an outgrowth of the body of the parent *695*

buffer a substance or group of substances that tend to resist pH changes in a solution, thus stabilizing its relative acidity *36*

bundle sheath cells that surround the xylem and phloem of leaf veins *131, 491*

C

calorie 1/1,000 of a kilocalorie; a unit used to express the quantity of energy of carbohydrates, fats, and proteins *34, 109*

C₃ plant plant using the enzyme rubisco to fix carbon dioxide to RuBP; the first detected molecule after fixation is PGAL, a 3-carbon molecule *131*

C₄ plant plant that does not directly use the Calvin cycle and produces an immediate 4-carbon molecule during photosynthesis *131*

Calvin cycle series of photosynthetic reactions in which carbon dioxide is reduced in the chloroplast *129*

cambium meristematic, or growth, tissue of a plant; for example, vascular cambium between the xylem and phloem in a stem or the cork cambium beneath the epidermis of the stem *484*

CAM plant plant that uses PEP carboxylase to fix carbon dioxide at night; CAM stands for crassulacean acid metabolism *133*

cancer condition by which cells grow and divide uncontrollably, detaching from tumors and spreading throughout the body *271*

capillary microscopic blood vessel for gas and nutrient exchange with body cells *555*

carbohydrate a family of organic compounds consisting of carbon, hydrogen, and oxygen atoms; includes subfamilies monosaccharides, disaccharides, and polysaccharides *44*

carbon cycle circulation of carbon between the biotic and abiotic component of both terrestrial and aquatic ecosystems *787*

carbon dioxide (CO_2) fixation photosynthetic reaction in which carbon dioxide is attached to an organic compound *130*

carbonic anhydrase an enzyme in red blood cells that speeds up the formation of carbonic acid from water and carbon dioxide *614*

carcinogen agent that contributes to the development of cancer *273*

cardiovascular system animal organ system consisting of the blood, heart, and a series of blood vessels that distribute blood under the pumping action of the heart *555*

carnivore a secondary consumer in a food chain that eats other animals *781*

carpel simple plant reproductive unit of a simple pistil; consisting of 3 parts, the stigma, style, and ovary *509*

carrier an individual who is capable of transmitting an infectious or genetic disease; a protein that combines with and transports a molecule across the plasma membrane *94, 219*

carrying capacity the maximum size of a population that can be supported by the environment in a particular locale *747*

cartilage a type of connective tissue having a matrix of protein and protein-aceous fibers that is found in animal skeletal systems *541*

Casparian strip a waxy layer bordering 4 sides of root endodermal cells; prevents water and solute transport between adjacent cells *481*

catastrophism (kah-tas'tro-fizm) belief espoused by Cuvier that periods of catastrophic extinctions occurred, after which repopulation of surviving species took place and gave the appearance of change through time *301*

cell the smallest unit that displays properties of life; composed of cytoplasmic regions, possibly organelles, and surrounded by a plasma membrane *3, 61*

cell plate (sel plāt) structure across a dividing plant cell that signals the location of new plasma membranes and cell walls *164*

cell theory statement that all organisms are made of cells and that cells are produced only from preexisting cells *61*

cellular respiration metabolic reactions that provide energy to cells by the step-by-step oxidation of carbohydrates *114, 138*

cellulose polysaccharide consisting of covalently bonded glucose chains, hydrogen-bonded within fibrils; important in plant cell walls *45*

cell wall a relatively rigid structure composed mostly of polysaccharides that surrounds the plasma membrane of plants, fungi, and bacteria *64*

central nervous system the brain and nerve (spinal) cord in animals *634*

centriole cell organelle, existing in pairs, that possibly organizes a mitotic spindle for chromosome movement during cell division *78*

centromere a constriction where duplicates (chromatids) of chromosomes are held together *157*

cephalization (sef'al-i-za'shun) having a well-recognized anterior head with concentrated nerve masses and receptors *421*

cerebellum portion of the brain that is dorsal to the pons and medulla oblongata; functioning in muscle coordination to produce smooth graceful motions *645*

cerebrum foremost part of the brain consisting of 2 large masses—cerebral hemispheres; the largest part of the brain in humans *644*

chemical evolution an increase in the complexity of chemicals that could have led to first cells *355*

chemiosmotic phosphorylation (kem"ĭ-os-mot'ik fos"for-ĭ-la'shun) production of ATP by utilizing the energy released when H^+ ions flow through an ATP synthase complex in mitochondria and chloroplasts *117*

chemoreceptor a receptor that is sensitive to chemical stimulation, for example, receptors for taste and smell *653*

chemoautotroph bacterium capable of oxidizing inorganic compounds, such as ammonia, nitrites, and sulfides, to gain energy to synthesize carbohydrates *373*

chiasma during synapsis in meiosis, the region of attachment between nonsister chromatids of a bivalent due to crossing-over *173*

chitin (ki'tin) a strong but flexible, nitrogenous polysaccharide that is found in the exoskeleton of arthropods *46, 437*

chlorophyll green pigment that absorbs sunlight energy and is important in photosynthesis *125*

chloroplast a membrane-bounded organelle with membranous grana that contain chlorophyll and where photosynthesis takes place *74, 125*

cholesterol one of the major lipids found in animal plasma membranes; makes the membrane impermeable to many molecules *50, 88*

chordate phylum of animals that includes lancelets, tunicates, fishes, amphibians, reptiles, birds, and mammals; characterized by a notochord, dorsal nerve cord, and gill pouches *443*

chorion (kor'e-on) an extraembryonic membrane functioning for respiratory exchange in the eggs of birds and reptiles; contributes to placenta formation in mammals *718*

chromatid one of 2 identical parts of a chromosome, produced through DNA replication; within the same chromosome, one chromatid is the sister chromatid of the other *157*

chromatin the mass of DNA and associated proteins observed within a nucleus that is not dividing *68, 157, 269*

chromosome association of DNA and proteins arranged linearly in genes and visible only during cell division *68, 157*

chromosome theory of inheritance theory stating that the genes are on chromosomes accounting for their similar behavior *202*

cilium (pl., cilia) short, hairlike projection from the cell membrane, found as one of a group (cilia) *78*

circadian rhythm (ser"kah-de'an) a biological rhythm with a 24-hour cycle *530, 731*

citric acid cycle a cycle of reactions in mitochondria, that begins and ends with citric acid; produces CO_2, ATP, NADH, and $FADH_2$; also called the Krebs cycle *144*

cladistics (clah-dis'tiks) means of classification that places a common ancestor, and all organisms evolved from it, into one taxon; based on studying homologous structures and other similarities among organisms *342*

classical conditioning one type of learning in which an animal learns to give a response to an irrelevant stimulus; also called associative learning *735*

cleavage earliest division of the zygote developmentally, without cytoplasmic addition or enlargement *163, 713*

cleavage furrowing indenting of the plasma membrane that leads to division of the cytoplasm during cell division *163*

climax community mature stage of a community that can sustain itself indefinitely as long as it is not stressed *771*

cloned (klōnd) the production of identical copies; in genetic engineering, the production of many, identical copies of a gene 279

cnidarian (ni-dah're-an) phylum of animals with a gastrovascular cavity; includes sea anemones, corals, hydras, and jellyfishes 419

cochlea (kok'le-ah) spiral-shaped structure of the inner ear containing the receptor hair cells for hearing 660

codominance a condition in heredity in which both alleles in the gene pair of an organism are expressed 197

codon 3 nucleotides of DNA or mRNA, codes for a particular amino acid 253

coelomate (sēl'o-māt) characterizes animals with a true coelom which is completely lined with mesoderm 424

coenzyme a nonprotein organic part of an enzyme structure, often with a vitamin as a subpart 112

cohesion-tension model explanation for the transport of water to great heights in a plant by water molecules clinging together as transpiration occurs 497

collecting duct the final portion of a nephron where reabsorption of water can be controlled 625

commensalism a symbiotic relationship in which one species is benefited and the other is neither harmed nor benefited 760

communication an action by a sender that influences the behavior of the recipient 738

community many different populations of organisms that interact with each other 755

competitive exclusion principle theory that no 2 species can occupy the same niche 756

complementary base pairing bonding between particular purines and pyrimidines in DNA 55, 240

complement system series of proteins, produced by the liver and present in the plasma, that produces a cascade of reactions as part of the immune system 576

compound (kom'pownd) a chemical substance having 2 or more different elements in fixed ratio 29

compound eye a type of eye found in arthropods composed of many independent visual units 437, 654

condensation chemical change producing the covalent bonding of 2 monomers with the accompanying loss of a water molecule 44

cone photoreceptors in vertebrate eyes that respond to bright light and allow color vision 655

conifer conifers are one of the 4 groups of gymnosperm plants; cone-bearing trees that include pine, cedar, and spruce 405

conjugation the transfer of genetic material from one cell to another through a cytoplasmic bridge 385

connective tissue animal tissue type that binds structures together, provides support and protection, fills spaces, stores fat, and forms blood cells 540

consumer an organism that feeds on another organism in a food chain; primary consumers eat plants, and secondary consumers eat animals 781

control group a sample that goes through all the steps of an experiment except the one being tested; a standard against which results of an experiment are checked 15

corepressor molecule that binds to a repressor, allowing the repressor to bind to an operator in a repressible operon 268

cork outer covering of bark of trees; made up of dead cells that may be sloughed off 477

corpus luteum (kor'pus lu'te-um) a follicle that has released an egg and increases its secretion of progesterone 701

cortex layer of the young plant root or stem beneath the epidermis; consisting of large, thin-walled parenchyma cells 481

cotyledon (kot-el-ēd'en) seed leaf for embryonic plant, providing nutrient molecules for the developing plant before its mature leaves begin photosynthesis 478

covalent bond (ko-va'lent) a chemical bond in which the atoms share one pair of electrons 31

cristae shelflike folds of the inner membrane of the mitochondrion, projecting into the inner space of the organelle 75

Cro-Magnon hominid who lived 40,000 years ago; accomplished hunters, made compound stone tools, and possibly had language 467

crossing-over an exchange of segments between nonsister·chromatids of a bivalent during meiosis 172

cultural eutrophication human activities result in an acceleration of the process by which nutrient poor lakes and ponds become nutrient rich; speeds up aquatic succession, in which a pond or lake eventually fills in and disappears 796

cyanobacterium (si"ah-no-bak-te're-um) formerly called blue-green alga; photosynthetic prokaryote that contains chlorophyll and releases oxygen 66, 375

cyclic AMP an ATP-related compound that promotes chemical reactions in the body; the "second messenger" in peptide hormone activity, it initiates activity of the metabolic machinery 679

cyclic photophosphorylation (fo"to-fos"for-ĭ-la'shun) photosynthetic pathway in which electrons from photosystem P700 pass through the electron transport system, produce ATP, and eventually return to P700 128

cystic fibrosis a lethal genetic disease affecting the mucus in the lungs and digestive tract 221

cytochrome system a series of electron carrier molecules, part of the electron transport systems of mitochondria and chloroplasts 128, 144

cytokinesis (si"to-ki-ne'sis) the division of the cytoplasm following mitosis and meiosis 157

cytokinin (si"to-ki'nin) plant hormone that promotes cell division; often works in combination with auxin during organ development in plant embryos 526

cytoplasm contents of a cell between the nucleus (nucleoid) and the plasma membrane 64

cytosine (C) (si'to-sin) one of 4 organic bases in the nucleotides composing structure of DNA and RNA 238

cytoskeleton internal framework of the cell, consisting of microtubules, actin filaments, and intermediate filaments 76

cytosol fluid medium that bathes the contents of the cytoplasm 64

D

datum (pl., data) fact or information collected through observation and/or experimentation 14

decomposer organism, usually a bacterial or fungal species, that breaks down large organic molecules into elements that can be recycled in the environment 373

deductive reasoning a process of logic and reasoning, using "if . . . then" statements 14

demographic transition a decline in death rate, followed shortly by a decline in birthrate, resulting in slower population growth 750

denatured condition of an enzyme when its shape is changed so that its active site cannot bind substrate molecules 111

dendrite the part of a neuron that sends impulses toward the cell body 635

denitrification (de-ni"trĭ-fi-ka'shun) the process of converting nitrogen compounds to atmospheric nitrogen; is a part of the nitrogen cycle 781

deoxyribonucleic acid (DNA) an organic molecule produced from covalent bonding of nucleotide subunits; the chemical composition of genes on chromosomes 54, 235

dependent variable result or change that occurs when the experimental variable is manipulated *15*

dermis deeper, thicker layer of the skin that consists of fibrous connective tissue and contains various structures such as sense organs *544*

desert treeless biome where the annual rainfall is less than 25 cm; the rain that does fall is subject to rapid runoff and evaporation *773*

desertification (dez"ert-ĭ-fi-ka'shun) transformation of marginal lands to desert *793*

detritus (de-tri'tus) falling, settled remains of plants and animals on the land-floor or water bed *781*

deuterostome (du'ter-o-stōm") a group of coelomate animals in which the second embryonic opening becomes the mouth; the first embryonic opening, the blasto-pore, becomes the anus *430*

diabetes mellitus (di"ah-be'tēz mě-lī'tus) a disease caused by a lack of insulin production by the pancreas *686*

diaphragm a dome-shaped muscle separating the thoracic cavity from the abdominal cavity in the animal body and involved in respiration *609*

diastole (di-as'to-le) relaxation period of a heart during the cardiac cycle *558*

dicotyledon a flowering plant group; members show 2 embryonic leaves, net-veined leaves, cylindrical arrangement of vascular bundles, and other characteristics *407*

dictyosome organelle in a plant cell that is called a Golgi apparatus in animal cells *72*

differentiation specialization of early embryonic cells with regard to structure and function *712*

diffusion the movement of molecules from a region of high to low concentration, requiring no energy and tending toward an equal distribution *89*

dihybrid a genetic cross involving 2 traits *190*

diploid (2N) number (dip'loid) the condition in which cells have 2 of each type of chromosome *157*

diplontic cycle (dip-lon'tik) life cycle in which the adult is diploid *384*

directional selection outcome of natural selection in which an extreme phenotype is favored, usually in a changing environment *326*

disruptive selection outcome of natural selection in which extreme phenotypes are favored over the average phenotype and can lead to polymorphic forms *325*

distal convoluted tubule the portion of a nephron between the loop of the nephron and the collecting duct, where tubular secretion occurs *625*

DNA *see* deoxyribonucleic acid

DNA ligase (li'gas) an enzyme that links DNA fragments; used in genetic engineering to join foreign DNA to the vector DNA *280*

DNA marker genetic variation found within highly repetitive DNA and inherited in a Mendelian fashion; the distinctive pattern of repetition in a person serves as a DNA fingerprint or means of identification *290*

DNA polymerase (pol-im'er-ās) enzyme that joins complementary nucleotides to form double-stranded DNA *241*

DNA probe known sequences of DNA that are used to find complementary DNA strands; can be used diagnostically to determine the presence of particular genes *283*

dominance hierarchy organization of animals in a group that determines the order in which the animals have access to resources *736*

dominant allele (ah-lēl') allele of a given gene that hides the effect of the recessive allele in a heterozygous condition *186*

Down syndrome a genetic disease caused by trisomy 21 that is characterized by mental retardation and a specific set of physical features *216*

dryopithecine (dri"o-pith'e-sin) a possible hominoid ancestor; a forest-dwelling primate with characteristics somewhat like those of living apes *461*

duodenum (dū-ah-dē'-num) first part of the small intestine where chyme enters from the stomach *593*

Duchenne muscular dystrophy (du-shen') an X-linked, recessive disorder that causes muscular weakness and eventually death at a young age *230*

E

echinoderm (e-kin'o-derm) a phylum of marine animals that includes sea stars, sea urchins, and sand dollars; characterized by radial symmetry and a water vascular system *440*

ecological pyramid pictorial graph representing the biomass, organism number, or energy content of each trophic level in a food web from the producer to the final consumer populations *782*

ecology the study of the interactions of organisms with their living and physical environment *745*

ecosystem a biological community together with the associated abiotic environment *8, 781*

ectoderm outermost of an animal's primary germ layers *713*

effector organ that makes a response when signaled by a motor neuron *641*

electromagnetic spectrum solar radiation divided on the basis of wavelength, with gamma rays having the shortest wavelength and radio waves having the longest wavelength *122*

electron negative subatomic particle, moving about in energy levels around the nucleus of the atom *26*

electron transport system a mechanism whereby electrons are passed along a series of carrier molecules to produce energy for the synthesis of ATP *114, 144*

element the simplest of substances consisting of only one type of atom; for example, carbon, hydrogen, oxygen *26*

embryo early developmental stage of a plant or animal, produced from a zygote *510*

emigrate (n., emigration) to move from a geographical area *748*

endocrine system one of the major systems involved in the coordination of body activities; uses messengers called hormones which are secreted into the bloodstream *677*

endocytosis (en"do-si-to'sis) moving particles or debris into the cell from the environment by phagocytosis (cellular eating) or pinocytosis (cellular drinking) *97*

endoderm innermost of an animal's primary germ layers *713*

endodermis internal plant root tissue forming a boundary between the cortex and vascular cylinder *481*

endometrium (en-do-me'tre-um) a mucous membrane lining the inside free surface of the uterus *700*

endoplasmic reticulum (en"do-plas'mik rě-tik'u-lum) a system of membranous saccules and channels in the cytoplasm *71*

endosperm nutritive tissue in a seed; often triploid from a fusion of sperm cell with 2 polar nuclei *510*

endospore a bacterium that has shrunk its cell, rounded up within the former plasma membrane, and secreted a new and thicker cell wall in the face of unfavorable environmental conditions *371*

endosymbiotic theory statement that chloroplasts and mitochondria (and perhaps cilia) were originally prokaryotes that evolved a mutualistic relationship with the eukaryotic cell *75*

energy capacity to do work and bring about change; occurs in a variety of forms *28, 105*

environment surroundings with which a cell or organism interacts *8*

environmental resistance the opposing force of the environment on the biotic potential of a population *747*

enzyme a protein that acts as an organic catalyst to speed up reaction rates in living systems *50, 108*

epicotyl (ep″ĭ-kot′il) plant embryo portion above the cotyledon that contributes to shoot development *515*

epidermis covering tissue of roots and leaves of plants, plus stems of nonwoody organisms; outer, protective layer of the skin *476, 544*

epiglottis a flaplike covering, hinged to the back of the larynx and capable of covering the glottis or air-tract opening *593, 612*

epiphyte (ep′ĭ-fīt) a plant that takes its nourishment from the air because its attachment to other plants gives it an aerial position *503, 778*

epistatic gene (epistasis) (ĕp″ĭ-stat′ik jēn) a gene that interferes with the expression of alleles that are at a different locus *198*

epithelial tissue (ep-ah-thē′lē-al) animal tissue type forming a continuous layer over most body surfaces (i.e., skin) and inner cavities *539*

erythrocyte (e-rith′ro-sit) red blood cell; important for oxygen transport in blood *563*

esophagus a muscular tube for moving swallowed food from pharynx to stomach *593*

essential amino acid an amino acid that is required by the body but must be obtained in the diet because the body is unable to produce it *149, 598*

estrogen female ovarian sex hormone that has numerous effects on the endometrium of the uterus throughout the ovarian cycle *700*

estuary the end of a river where fresh water and salt water mix as they meet *766*

ethylene plant hormone that causes ripening of fruit and is also involved in abscission *527*

eubacterium most common type of bacteria, including the photosynthetic bacteria *374*

euchromatin diffuse chromatin, which is being transcribed *269*

eukaryotic cell typical of organisms, except bacteria and cyanobacteria, having organelles and a well-defined nucleus *64*

evolution genetic and phenotypic changes that occur in populations of organisms with the passage of time, often resulting in increased adaptation of organisms to the prevailing environment *300*

exhalation stage during respiration when air is pushed out of the lungs *609*

exocytosis a process by which cells expel particles or debris in vesicles passing through a plasma membrane to the extracellular environment *97*

exon in a gene, the portion of the DNA code that is expressed as the result of polypeptide formation *254*

exoskeleton a protective, external skeleton, as in arthropods *437*

experimental variable condition that is tested in an experiment by varying it and observing the results *15*

exponential referring to a geometrically multiplying, rapid population growth rate *745*

F

facilitated transport process whereby molecules pass through a plasma membrane, utilizing a carrier molecule but without the expenditure of energy *94*

FAD (flavin adenine dinucleotide) (fla′vin ad′ĕ-nīn di-nu′kle-o-tīd) coenzyme that functions as an electron acceptor in cellular oxidation-reduction reactions *114*

fat organic compound consisting of 3 fatty acids covalently bonded to one glycerol molecule *47*

fatty acid subunit of a fat molecule, characterized by a carboxyl acid group at one end of a long hydrocarbon chain *47*

feedback inhibition process by which a substance, often an end product of a reaction or a metabolic pathway, controls its own continued production by binding with the enzyme which produced it *112*

fermentation anaerobic breakdown of carbohydrates that results in products such as alcohol and lactate *139*

fibroblast cell type of loose and fibrous connective tissue with cells at some distance from one another and separated by a jellylike substance *540*

filament the elongated stalk of a stamen bearing the anther at the tip *510*

fitness ability of an organism to survive and reproduce in its local environment *305*

fixed-action pattern (FAP) a stereotyped behavior pattern that occurs automatically *733*

flower reproductive organ of a flowering plant, consisting of several kinds of modified leaves arranged in concentric rings and attached to a modified stem called the receptacle *407*

fluid-mosaic model model for the plasma membrane based on the changing location and pattern of protein molecules in a fluid phospholipid bilayer *86*

follicle structures in the ovary of animals that contain oocytes; site of egg production *700*

follicle-stimulating-hormone (FSH) gonadotrophic hormone secreted by anterior pituitary that controls the events of the ovarian cycle by being secreted at different rates throughout the cycle *701*

food chain a succession of organisms in an ecosystem that are linked by an energy flow and the order of who eats whom *782*

food web a complex pattern of interlocking and crisscrossing food chains *782*

founder effect the tendency for a new, small population to experience genetic drift after it has separated from an original, larger population *323*

fruit flowering plant structure consisting of one or more ripened ovaries that usually contain seeds *407, 511*

fruiting body a spore-bearing structure found in certain types of fungi, such as mushrooms *391*

FSH *see* follicle-stimulating hormone

fungus saprophytic decomposer; body is made up of filaments called hyphae that form a mass called a mycelium *391*

G

gallbladder attached to the liver; serves as a storage organ for bile *594*

gamete (gam′et) haploid sex cell *157*

gametophyte haploid generation of the alternation of generations life cycle of a plant; it produces gametes that unite to form a diploid zygote *398*

ganglion (sing., ganglia) (gang′gle-on) a knot or bundle of neuron cell bodies outside the central nervous system *640*

gastrovascular cavity a blind, branched digestive cavity that also serves a circulatory (transport) function in animals that lack a circulatory system *419*

gastrula (gas′troo-lah) cup-shaped early embryo with 2 primary germ layers, ectoderm and endoderm, enclosing a primitive digestive tract *713*

gel electrophoresis (ĭ-lek-tra-fa-rē′sus) technique that allows DNA fragments or proteins to be separated according to differences in the charge and size *290*

gene the unit of heredity occupying a particular locus on the chromosome and passed on to offspring *5*

gene flow sharing of genes between 2 populations through interbreeding *320*

gene locus the specific location of a particular gene on a chromosome *186*

gene mutation (jēn mu-ta'shun) a change in the base sequence of DNA such that the sequence of amino acids is changed in a protein *316*

gene pool the total of all the genes of all the individuals in a population *318*

gene therapy use of transplanted genes to overcome an inborn error in metabolism *288*

genetic drift change in the genetic make-up of a population due to chance (random) events; important in small populations or when only a few individuals mate *322*

genetic engineering alteration of the genome of an organism by technological processes *279*

genome (jē'nōm) all of the genes of an organism *244*

genotype (jen'o-tīp) the genes of an organism for a particular trait or traits; for example, *BB* or *Aa 187*

germination resumption of growth by a seed or any other reproductive structure of a plant or protist *515*

germ layer developmental layer of body; that is, ectoderm, mesoderm, and endoderm *714*

gibberellin (gib-ber-el'in) plant hormone producing increased stem growth by cell division and enlargement; also involved in flowering and seed germination *526*

gill respiratory organ in most aquatic animals; most common in fish as an outward extension of the pharynx *607*

girdling removing a strip of bark from around a tree *503*

glomerular capsule a cuplike structure that is the initial portion of a nephron; where pressure filtration occurs *625*

glomerulus (glo-mer'u-lus) a capillary network within a glomerular capsule of a nephron *625*

glottis opening for airflow in the larynx *612*

glucagon (glū'kah-gon) hormone secreted by the pancreas that stimulates the breakdown of stored nutrients and increases the level of glucose in the blood *686*

glucose monosaccharide with the molecular formula of $C_6H_{12}O_6$ found in the blood of animals, serving as an energy source in most organisms and transported in the blood of animals *45*

glycogen polysaccharide consisting of covalently bonded glucose molecules, characterized by many side branches, energy storage product in animals *45*

glycolysis (gli-kol'i-sis) pathway of metabolism converting a sugar, usually glucose, to pyruvate; resulting in a net gain of 2 ATP and 2 NADH molecules *139*

goiter an enlargement of the thyroid gland caused by a lack of iodine in the diet *683*

Golgi apparatus an organelle consisting of a central region of saccules and vesicles that modifies and/or activates proteins and produces lysosomes *72*

gonadotrophic hormone (GNRH) (gō-nad-ah-trō'fik) either FSH or LH, a substance secreted by the anterior pituitary that stimulates the gonads *482*

granum (pl., grana) (gra'num) in a chloroplast, a stack of flattened membrane sacs, thylakoids, that contain chlorophyll *74, 123*

gravitropism (grav"ĭ-tro'pizm) directional growth in response to the earth's gravity; roots demonstrate positive gravitropism, stems demonstrate negative gravitropism *529*

greenhouse effect the reradiation of heat toward the earth caused by gases in the atmosphere *800*

growth hormone (GH) substance stored and secreted by the anterior pituitary; promotes cell division, protein synthesis, and bone growth *683*

guanine (G) (gwan'in) one of 4 organic bases in the structure of nucleotides composing the structure of DNA and RNA *54, 238*

guard cell plant cell type, found in pairs, with one on each side of a leaf stoma; changes in the turgor pressure of these cells regulate the size and passage of gases through the stoma *498*

guttation (gut-ta'shun) liberation of water droplets from the edges and tips of leaves *497*

gymnosperm a vascular plant producing naked seeds, as in conifers *405*

H

habitat region of an ecosystem occupied by an organism *756*

haploid (N) number a cell condition in which only one of each type of chromosome is present *157*

haplontic cycle life cycle in which the adult is haploid *384*

Hardy-Weinberg law a law stating that the frequency of an allele in a population remains stable under certain assumptions, such as random mating; therefore, no change or evolution occurs *318*

helper T cell cell that secretes lymphokines, which stimulate all kinds of immune cells *579*

hemoglobin red respiratory pigment of erythrocytes for transport of oxygen *554, 563, 614*

hemophilia a disease in which the blood fails to clot; caused by insufficient clotting factor VIII *229*

herbaceous plant a plant that lacks persistent woody tissue *407*

herbivore a primary consumer in a food chain; a plant eater *781*

hermaphroditic animal (her'maf"ro-dit'ik) characterizes an animal having both male and female sex organs *422*

heterochromatin highly compacted chromatin during interphase *269*

heterogamete (het"er-o-gam'ēt) a nonidentical gamete, for example, large, nonmotile egg and small, flagellated sperm *385*

heterotroph (het'er-o-trōf") an organism that cannot synthesize organic compounds from inorganic substances and therefore must take in preformed food *356*

heterozygous possessing 2 different alleles for a particular trait *187*

homeostasis the maintenance of internal conditions in cell or organisms, for example, relatively constant temperature, pH, blood sugar *546*

hominid referring to humans; ancestors of humans who had diverged from the ape lineage *461*

Homo erectus hominid who lived during the Pleistocene; with a posture and locomotion similar to modern humans *466*

Homo habilis (ho'mo hah'bĭ-lis) fossil hominid of 2 million years ago; possible first direct ancestor of modern humans *465*

homologous chromosome a pair of chromosomes that carry genes for the same traits and synapse during prophase of the first meiotic division *171*

homologous structure in evolution, structure derived from a common ancestor *310*

homologue (hom'o-log) member of homologous pair of chromosomes *171*

homozygous possessing 2 identical alleles for a particular trait *186*

hormone a chemical secreted into the bloodstream in one part of the body that controls the activity of other parts *523*

Huntington disease (HD) a lethal, genetic disease affecting one in 10,000 people in the United States; affects neurological functioning and generally does not manifest itself until middle age 223

hydra a freshwater cnidarian with only a polyp stage that reproduces both sexually and asexually 420

hydrogen bond a weak bond that arises between a partially positive hydrogen and a partially negative oxygen, often on different molecules or separated by some distance 33

hydrolysis splitting of a compound into parts in a reaction that involves addition of water, with the H^+ ion being incorporated in one fragment and the OH^- ion in the other 44

hydrophilic a type of molecule that interacts with water, such as polar molecules, which dissolve in water or form hydrogen bonds with water molecules 34, 43

hydrophobic a type of molecule that does not interact with water, such as nonpolar molecules, which do not dissolve in water or form hydrogen bonds with water molecules 43

hydroponics water culture method of growing plants that allows an experimenter to vary the nutrients and minerals provided so as to determine specific growth requirements 500

hydrostatic skeleton characteristic of animals in which muscular contractions are applied to fluid-filled body compartments 434

hypertonic solution lesser water concentration, greater solute concentration than the cytoplasm of a particular cell; situation in which cell tends to lose water by osmosis 92

hypha a filament of the vegetative body of a fungus 391

hypocotyl (hi"po-kot'il) plant embryo portion below the cotyledon that contributes to development of stem 515

hypothalamus portion of the brain that regulates the internal environment of the body; involved in control of heart rate, body temperature, water balance, and glandular secretions of the stomach and pituitary gland 644

hypothesis a supposition that is established by reasoning after consideration of available evidence and that can be tested by obtaining more data, often by experimentation 14

hypotonic solution greater water concentration, lesser solute concentration than the cytoplasm of a particular cell; situation in which a cell tends to gain water by osmosis 92

I

immune system includes many leukocytes and various organs and provides specific defense against foreign antigens 577

immunity the ability of the body to protect itself from foreign substances and cells, including infectious microbes 575

imprinting a behavior pattern with both learned and innate components that is acquired during a specific and limited period, usually when very young 733

incomplete dominance a pattern of inheritance in which the offspring shows characteristics intermediate between 2 extreme parental characteristics, for example, a red and a white flower producing pink offspring 197

inducer a molecule that brings about activity of an operon by joining with a repressor and preventing it from binding to the operator 268

inducible operon an operon that is normally inactive but is turned on by an inducer 268

induction the ability of a chemical or tissue to influence the development of another tissue 715

inductive reasoning a process of logic and reasoning, using specific observations to arrive at a hypothesis 14

inhalation stage during respiration when air is drawn into the lungs 609

inheritance of acquired characteristics Lamarckian belief that organisms become adapted to their environment during their lifetime and pass on these adaptations to their offspring 301

inorganic referring to molecules that are not compounds of carbon and hydrogen 42

insight learning ability to respond correctly to a new, different situation the first time it is encountered by using reason 735

insulin hormone secreted by the pancreas that stimulates fat, liver, and muscle cells to take up and metabolize glucose, stimulates the conversion of glucose into glycogen in muscle and liver cells, and promotes the buildup of fats and proteins 686

interferon an antiviral agent produced by an infected cell that blocks the infection of another cell 577

interneuron neuron, within the central nervous system, conveying messages within parts of the central nervous system 636

interphase stage of cell cycle during which DNA synthesis occurs and the nucleus is not actively dividing 159

intron noncoding segments of DNA that are transcribed but removed before mRNA leaves nucleus 254

invertebrate referring to an animal without a serial arrangement of vertebrae or a backbone 416

ion charged derivative of an atom, positive if the atom loses electrons and negative if the atom gains electrons 30

ionic bond an attraction between charged particles (ions) through their opposite charge that involves a transfer of electrons from one atom to another 31

isomer molecules with the same molecular formula but different structure and, therefore, shape 43

isotonic solution a solution that is equal in solute and water concentration to that of the cytoplasm of a particular cell; situation in which cell neither loses nor gains water by osmosis 91

isotope forms of an element having the same atomic number but a different atomic weight due to a different number of neutrons 27

J

jointed appendage freely moving appendages of arthropods 437

K

karyotype (kar'e-o-tīp) chromosomes are cut from a photograph of the nucleus just prior to cell division and arranged in groups for display 216

kidney bean-shaped, reddish brown organ in humans that regulates the chemical composition of the blood and produces a waste product called urine 624

killer T cell (kil'er te sel) cell that attacks and destroys cells that bear a foreign antigen 579

kingdom a taxonomic category grouping related phyla (animals) or divisions (plants) 3

kin selection an explanation for the evolution of altruistic behavior that benefits one's relatives 740

Klinefelter syndrome a condition caused by the inheritance of a chromosome abnormality in number; an XXY (trisomy) condition where normally a chromosome pair exists 218

Krebs cycle *see* citric acid cycle

***K*-strategist** a species that has evolved characteristics that keep its population size near carrying capacity, for example, few offspring, longer generation time 749

L

lacteal (lak'te-al) a lymphatic vessel in an intestinal villus; aids in the absorption of fats 596

larynx voicebox, cartilage-made box for airflow between the glottis and trachea 612

leaf usually broad, flat structure of a plant shoot system, containing cells that carry out photosynthesis 482

learning change in behavior as a result of experience; there are 3 kinds of learning—imprinting, operant conditioning, and insight learning 733

leukocyte (lu'ko-sit) white blood cell; important for immune response 563

LH *see* luteinizing hormone

lichen (li'ken) a symbiotic relationship between certain fungi and algae which has long been thought to be mutualistic, the fungi providing inorganic food and the algae providing organic food 375, 392

life cycle entire sequence of developmental phases from zygote formation to gamete formation 384

ligament structure consisting of fibrous connective tissue that connects bones at joints 541, 668

light-dependent reaction a reaction of photosynthesis that requires light energy to proceed; involved in the production of ATP and NADH 125

light-independent reaction photosynthetic reaction that does not directly require light energy; involved primarily in the conversion of carbon dioxide to carbohydrate utilizing the products of the light-dependent reaction 125

limbic system part of the brain beneath the cortex that contains pathways that connect various portions of the brain; allows the experience of many emotions 646

linkage group genes that are located on the same chromosome and tend to be inherited together 206

lipid (lip'id) a family of organic compounds that tend to be soluble in nonpolar solvents such as alcohol; includes fats and oils 46

liposome (lip'o-som) droplet of phospholipid molecules formed in a liquid environment 356

loop of the nephron the portion of a nephron between the proximal and distal convoluted tubules where some water reabsorption occurs 625

lung respiratory organ mainly found in terrestrial vertebrates, evolving as a vascularized outgrowth of the lower pharyngeal region 608

luteinizing hormone (LH) (lūt'ē-ah-nīz-ing) gonadotrophic hormone that controls events of the ovarian cycle by being secreted at different rates throughout the cycle 701

lymph tissue fluid collected by the lymphatic system and returned to the general systemic circuit 574

lymphatic system mammalian organ system consisting of lymphatic vessels and lymphoid organs 753

lymphocyte specialized white blood cell, produced by lymphoid tissue, that fights infection; occurs in 2 forms—T lymphocyte and B lymphocyte 563

lymphokine (lim'fo-kin) chemicals secreted by T cells that stimulate immune cells 579

lysogenic cycle one of the bacteriophage life cycles in which the virus incorporates its DNA into that of the bacterium; only later does it begin a lytic cycle which ends with the destruction of the bacterium 367

lysosome membrane-bounded vesicles that contain hydrolytic enzymes to digest macromolecules 73

lytic cycle (lit'ik) one of the bacteriophage life cycles in which the virus takes over the operation of the bacterium immediately upon entering it and subsequently destroys the bacterium 367

M

macroevolution large-scale evolutionary change, for example, the formation of new species 340

macrophage (mak'ro-faj) large, phagocytotic white blood cell 563

Malpighian tubule (mal-pig'i-an) blind, threadlike excretory tubule near anterior end of insect hindgut 438, 623

mammal a class of vertebrates characterized especially by the presence of hair and mammary glands 455

marsupial a mammal bearing immature young nursed in a marsupium, or pouch; for example, kangaroo, opossum 457

mass extinction recorded in the fossil record as a very large number of extinctions over a relatively short geologic time 343

matrix inner space of the mitochondrion, filled with a gellike fluid 75

mechanoreceptor a cell that is sensitive to mechanical stimulation, such as that from pressure, sound waves, and gravity 658

medulla oblongata (mě-dul'ah ob"long-ga'tah) a brain region at the base of the brainstem controlling heartbeat, blood pressure, breathing, and other vital functions 644

megaspore reproductive structure produced by the megasporangium of seed plants; of the 4 produced, one develops into a female gametophyte 405

meiosis (mi-o'sis) type of nuclear division that occurs as part of sexual reproduction in which the daughter cells receive the haploid number of chromosomes 157, 171

meiosis I (mi-o'sis) first division of meiosis, which ends when duplicated homologous chromosomes separate and move into different cells 173

meiosis II (mi-o'sis) second division of meiosis in which sister chromatids separate and move into different cells, resulting in the formation of haploid cells 175

melanocyte-stimulating hormone (mah-lan'ah-sīt) substance stored and secreted by the anterior pituitary; causes color changes in many fish, amphibians, and reptiles 683

memory B cell a B lymphocyte that remains in the bloodstream after an immune response ceases and is capable of producing antibodies to an antigen 578

memory T cell a T lymphocyte that remains in the bloodstream after an immune response ceases 579

menstruation periodic shedding of tissue and blood from the inner lining of the uterus 703

meristem tissue undifferentiated, embryonic tissue in the active growth regions of plants 475

mesoderm middle layer of an animal's primary germ layers 714

mesoglea (mes"-o-gle'ah) a jellylike layer between the epidermis and endodermis of cnidarians 419

mesophyll inner, thickest layer of a leaf consisting of a palisade layer of elongated cells and a spongy layer of irregularly spaced cells 491

messenger RNA (mRNA) a nucleic acid (ribonucleic acid) complementary to genetic DNA and bearing a message to direct cell protein synthesis at the ribosome 251

metabolic pool metabolites that are the products of and/or the substrates for key reactions in cells allowing one type of molecule to be changed into another type, such as the conversion of carbohydrates to fats 148

metabolism all of the chemical reactions that occur in a cell during growth and repair involve and produce metabolites 4, 107

metafemale a female with 3 X chromosomes; most show no obvious physical abnormalities 218

metamorphosis change in shape and form that some animals, such as insects, undergo during development 438

MHC protein (major histocompatibility complex) protein on the plasma membrane of a macrophage that aids in the stimulation of T cells 580

microbody membrane-bounded vesicle containing specific enzymes involved in lipid and alcohol metabolism, photosynthesis, and germination 73

microsphere formed from proteinoids exposed to water; has properties similar to today's cells 356

microspore reproductive structure produced by a microsporangium of a seed plant; it develops into a pollen grain 510

microtubule small cylindrical organelle that is believed to be involved in maintaining the shape of the cell and directing various cytoplasmic structures 76

mimicry superficial resemblance of an organism to one of another species; often used to avoid predation 758

mitochondrion membrane-bounded organelle of the cell known as the powerhouse because it transforms the energy of carbohydrates and fats to ATP energy 74

mitosis a process in which a parent nucleus reproduces 2 daughter nuclei, each identical to the parent nucleus; this division is necessary for growth and development 157

model a suggested explanation for experimental results that can help direct future research 17

molecule a unit of a chemical substance formed by the union of 2 or more atoms by covalent or ionic bonding; smallest part of a compound that retains the properties of the compound 29

mollusk a invertebrate phylum that includes squids, clams, snails, and chitons and is characterized by a visceral mass, a mantle, and a foot 431

molt periodic shedding of the exoskeleton in arthropods 437

Monera kingdom that includes bacteria, which are microscopic single-celled prokaryotes 370

monocotyledon a flowering plant group; members show one embryonic leaf, parallel leaf veins, scattered vascular bundles, and other characteristics 407

monohybrid a genetic cross involving one trait 185

monotreme an egg-laying mammal; for example, duckbill platypus or spiny anteater 457

morphogenesis (mor″fo-jen′ĕ-sis) movement of early embryonic cells to establish body outline and form 712

multiple allele more than 2 alleles for a particular trait 200

muscle fiber a cell with myofibrils containing actin and myosin filaments arranged within sarcomeres; a group of muscle fibers is a muscle 671

muscular (contractile) tissue animal tissue type, composed of fibers with actin and myosin filaments, that can shorten its length and produce movements 542

mutagen an agent, such as radiation or a chemical, that brings about a mutation in DNA 260

mutation a change in the composition of DNA, due to either a chromosomal or a genetic alteration 7, 208, 235

mutualism a symbiotic relationship in which both species benefit 761

mycelium (mi-se′le-um) a tangled mass of hyphal filaments composing the vegetative body of a fungus 391

mycorrhiza (mi″ko-ri′zah) a "fungus root" composed of a fungus growing around a plant root, which assists the growth and development of the plant 502

myelin sheath (mi′ĕ-lin) covering on neurons that is formed from Schwann cell membrane and aids in transmission of nervous impulses 636

myofibril a specific muscle cell organelle containing a linear arrangement of sarcomeres which shorten to produce muscle contraction 671

myosin (mi′o-sin) a muscle protein making up the thick filaments in a sarcomere; it pulls actin to shorten the sarcomere, yielding muscle contraction 671

N

NAD⁺ (nicotinamide-adenine dinucleotide) (nik″o-tin′ah-mīd ad′ĕ-nīn di-nu′kle-o-tīd) a coenzyme that functions as an electron and hydrogen ion carrier in cellular oxidation-reduction reactions of glycolysis and cellular respiration 113

NADP⁺ (nicotinamide-adenine dinucleotide phosphate) (nik″o-tin′ah-mīd ad′ĕ-nīn di-nu′kle-o-tīd fos′fāt) a coenzyme that functions as an electron and hydrogen ion carrier in cellular oxidation-reduction reactions of photosynthesis 114

natural selection the guiding force of evolution caused by environmental selection of organisms most fit to reproduce, resulting in adaptation 305

Neanderthal hominid who lived during the last ice age in Europe and the Middle East; made stone tools, hunted large game, and lived together in a kind of society 466

negative feedback loop pattern of homeostatic response in which the output is counter to and cancels the input 548

nematocyst (nem′ah-to-sist) in the cnidaria, a threadlike fiber enclosed within the capsule of a stinging cell; when released, aids in the capture of prey 419

nephridium (nĕ-frid′e-um) for invertebrates, a tubular structure for excretion; its contents are released outside through a nephridiopore 435

nephron microscopic kidney unit that regulates blood composition by filtration and reabsorption; over one million per human kidney 625

nerve bundle of axons and/or dendrites 640

neurofibromatosis (nu″ro-fi-bro′mah-to-sus) disease caused by an autosomal dominant allele, characterized by the proliferation of nerve cells into benign tumors under the skin 223

neuromuscular junction region where a nerve end comes into contact with a muscle fiber; contains a presynaptic membrane, synaptic cleft, and postsynaptic membrane 673

neuron a nerve cell; composed of dendrite(s), a cell body, and axon 543, 635

neurotransmitter a chemical made at the ends of axons that is responsible for transmission across a synapse or neuromuscular junction 639

neutron neutral subatomic particle, located in the nucleus of the atom 26

neutrophil (nu′tro-fil) white blood cell with granules 563

niche total description of an organism's functional role in an ecosystem, from activities to reproduction 756

nitrogen fixation process whereby free atmospheric nitrogen is converted into compounds, such as ammonia and nitrates, usually by soil bacteria 373, 788

nitrogen-fixing bacterium a bacterium that can convert atmospheric nitrogen into compounds such as ammonia and nitrates; free living in the soil or symbiotic with plants such as legumes 373

nodule (nod'ul) structure on plant roots that contains nitrogen-fixing bacteria *502*

noncyclic photophosphorylation (fo"to-fos"for-ĭ-la'shun) photosynthetic pathway in which electrons from photosystem P680 pass through the electron transport system and move on to another photosystem (P700) *128*

nonrenewable resource resource found in limited supplies that can be used up *803*

norepinephrine (nor"ep-ah-nef'ren) a hormone released by the adrenal medulla as a reaction to stress; also a neurotransmitter released by neurons of the central and peripheral nervous systems *639*

notochord (no'to-kord) a dorsal, elongated, supporting structure in chordates beneath the neural tube or spinal cord; present in the embryo of all vertebrates *714*

nuclear envelope double membrane that separates the nucleus from the cytoplasm *69*

nuclear power energy source generation by atomic reactions *803*

nucleic acid polymer of nucleotides; both DNA and RNA are nucleic acids *54, 235*

nucleoid area in prokaryotic cell where DNA is found *64, 156*

nucleolus (pl., nucleoli) dark-staining, spherical body in the cell nucleus that contains ribosomal RNA *68*

nucleotide the building block subunit of DNA and RNA consisting of a 5-carbon sugar bonded to a nitrogenous base and phosphorus group *54*

nucleus (nu'kle-us) region of a eukaryotic cell, containing chromosomes, that controls the metabolic function of the cell *66*

O

oil lipid containing triglycerides having unsaturated hydrocarbon chains; liquid at room temperature *47*

olfactory cell modified neuron that is a receptor for the sense of smell *653*

omnivore a food chain organism feeding on both plants and animals *781*

oncogene (ong'ko-jen) a cancer-causing gene *273*

oogenesis (o"o-jen'ĕ-sis) production of eggs in females by the process of meiosis and maturation *175*

operant conditioning trial-and-error learning in which behavior is reinforced with a reward or a punishment *734*

operator the sequence of DNA in an operon to which the repressor protein binds *266*

operon a group of structural and regulating genes that functions as a single unit *267*

optic nerve nerve that carries impulses from the retina of the eye to the brain *655*

orbital volume of space around a nucleus where electrons can be found most of the time *28*

organ combination of 2 or more different tissues performing a common function *544*

organ of Corti specialized region of the cochlea containing the hair cells for sound detection and discrimination *661*

organelle small, membranous structure in the cytoplasm having a specific function *64*

organic containing carbon and hydrogen; such molecules usually also have oxygen attached to the carbon(s) *42*

organic soup accumulation of simple organic compounds in the early oceans *355*

osmosis the diffusion of water through a selectively permeable membrane *89*

osmotic pressure measure of the tendency of water to move across a selectively permeable membrane into a solution; visible as an increase in liquid on the side of the membrane with higher solute concentration *91*

osteocyte bone cell found within the Haversian system of bone tissue *665*

ovary in flowering plants, enlarged, base portion of the pistil that eventually develops into the fruit; in animals, an egg-producing organ *407, 509, 700*

oviduct tubular portion of the female reproductive tract from ovary to the uterus *700*

ovulation bursting of a follicle at the release of an egg from the ovary *700*

ovule (o'vul) in seed plants, a structure that contains the megasporangium, where meiosis occurs and the female gametophyte is produced; develops into the seed *407, 510*

oxidation a chemical reaction that results in removal of one or more electrons from an atom, ion, or compound; oxidation of one substance occurs simultaneously with reduction of another *33*

oxidative phosphorylation (fos"for-ĭ-la'shun) the process of building ATP by an electron transport system that uses oxygen as the final receptor; occurs in mitochondria *145*

oxidizing atmosphere an atmosphere with free oxygen; for example, the atmosphere of today *355, 358*

oxygen debt use of oxygen to convert lactate, which builds up due to anaerobic conditions, to pyruvate *142*

oxytocin hormone made by the hypothalamus and stored in the posterior pituitary; causes the uterus to contract and stimulates the release of milk from mammary glands *681*

ozone shield formed from oxygen in the upper atmosphere; protects the earth from ultraviolet radiation *358, 801*

P

paleontology study of fossils that results in knowledge about the history of life *301*

palisade layer layer in a plant leaf containing elongated cells with many chloroplasts; along with the spongy layer, it is the site of most of photosynthesis *491*

pancreas an abdominal organ that produces digestive enzymes and the hormones insulin and glucagon *686*

parasitism (par'ah-si"tizm) a symbiotic relationship in which one species (parasite) benefits for growth and reproduction to the harm of the other species (host) *759*

parasympathetic nervous system a division of the autonomic nervous system that is involved in normal activities; uses acetylcholine as a neurotransmitter *641*

parathyroid gland embedded in the posterior surface of the thyroid gland; produces parathyroid hormone, which is involved in the regulation of calcium and phosphorus balance *685*

parathyroid hormone (PTH) substance secreted from the 4 parathyroid glands that increases the calcium level and decreases the phosphate level in the blood *685*

parenchyma (pah-ren'kah-mah) least specialized of all plant cell or tissue types; found in all organs of a plant with many of the cells containing chloroplasts *477*

parthenogenesis (par"thĕ-no-jen'-ĕ-sis) development of an egg cell into a whole organism without fertilization *695*

pelagic division (pe-laj'ik) the open portion of the sea *768*

penis male copulatory organ *697*

peptide 2 or more amino acids joined together by covalent bonding *50*

peptide bond a covalent bond between two amino acids resulting from a condensation reaction between the carboxyl group of one and the amino group of another *50*

pericycle external layer of cells in the vascular cylinder of a plant root, producing branch and secondary roots *481*

peripheral nervous system (PNS) made up of the nerves that branch off of the central nervous system *634*

peristalsis the rhythmic, wavelike contraction that moves food through the digestive tract *593*

petal plant structure belonging to an inner whorl of the flower; colored and internal to the outer whorl of green sepals *509*

phagocytosis (fag-ah-si-to'sis) transport process by which amoeboid-type cells engulf large material, forming an intracellular vacuole 97, 381

pharynx a common passageway (throat) for both food intake and air movement 612

phenotype the visible expression of a genotype; for example, brown eyes, height 187

phenylketonuria (PKU) (fen"il-ke"to-nu're-ah) a genetic disorder characterized by the absence of an enzyme that metabolizes phenylalanine; results in severe mental retardation 222

pheromone (fer'o-mōn) substance produced and discharged into the environment by an organism that influences the behavior of another organism 738

phloem (flo'em) vascular tissue conducting organic solutes in plants; contains sieve-tube cells and companion cells 399, 477

phospholipid a molecule having the same structure as a neutral fat except one of the bonded fatty acids is replaced by a phosphorous-containing group; an important component of plasma membranes 48, 87

photochemical smog air pollution that contains nitrogen oxides and hydrocarbons which react to produce ozone and peroxylacetyl nitrate 798

photon discrete packet of light energy; the amount of energy in a photon is inversely related to the wavelength of light in the photon 122

photoperiodism relative lengths of daylight and darkness that affect the physiology and behavior of an organism 530

photoreceptor light sensitive receptor cell 654

photosynthesis a process whereby chlorophyll-containing organisms trap sunlight energy to build a sugar from carbon dioxide and water 114

photosystem a photosynthetic unit where light is absorbed; contains an antenna (photosynthetic pigments) and an electron acceptor 125

phototropism (fo-tot'ro-pizm) movement in response to light; in plants, growth toward or away from light 529

pH scale measurement scale for the relative concentrations of hydrogen (H^+) and hydroxyl (OH^-) ions in a solution 36

phylogenetic tree (fī-lō-jah-net'ik) diagram with branches illustrating common evolutionary ancestors and the descendants produced from them through evolutionary divergence 340

phytochrome a plant photoreversible pigment; this reversion rate of 2 molecular forms is believed to measure the photoperiod 531

pinocytosis (pi"no-si-to'sis) process by which vesicles form around and bring macromolecules into the cell 97

pistil a female flower structure consisting of an ovary, style, and stigma 407, 509

pituitary gland a small gland that lies just below the hypothalamus and is important for its hormone storage and production activities 681

placenta a structure formed from an extraembryonic membrane, the chorion, and uterine tissue through which nutrient and waste exchange occur for the embryo and fetus 720

placental mammal a mammalian subclass that is characterized by a placenta, which is an organ of exchange between maternal and fetal blood that supplies nutrients to the growing offspring 458

plankton fresh- and saltwater organisms that float on or near the surface of the water 766

plasma liquid portion of blood; contains nutrients, wastes, salts, and proteins 562

plasma cell an activated B cell that is currently secreting antibodies 577

plasma membrane membrane that separates the contents of a cell from the environment and regulates the passage of molecules into and out of the cell 64

plasmid a self-duplicating ring of accessory DNA in the cytoplasm of bacteria 279

plasmodesmata (plaz-mah-dez'mah-tah) in plants, cytoplasmic channels bounded by the plasma membrane and connecting the cytoplasm of adjacent cells 101

platelet cell-like disks formed from fragmentation of megakaryocytes that are involved in blood clotting in vertebrate blood 563

plumule (ploō'mūl) embryonic plant shoot that bears young leaves 515

polar body a nonfunctional product of oogenesis; 3 of the 4 meiotic products are of this type 176

polar covalent bond (ko-va'lent) a bond in which the sharing of electrons between atoms is unequal 33

pollen grain a male gametophyte in seed plants 404

pollination transfer of pollen from an anther to a stigma 404, 510

pollutant substance that is added to the environment and leads to undesirable effects for living organisms 793

polygenic inheritance a pattern of inheritance in which a trait is controlled by several gene pairs that segregate independently 200

polymer macromolecule consisting of covalently bonded monomers; for example, a protein is a polymer of monomers called amino acids 43

polypeptide chain of many amino acids, covalently bonded together by peptide bonds 50

polyploid (polyploidy) a condition in which an organism has more than 2 sets of chromosomes 208

polysaccharide carbohydrate macromolecule consisting of many monosaccharides covalently bonded together 45

polysome a string of ribosomes, simultaneously translating different regions of the same mRNA strand during protein synthesis 257

population a group of organisms of the same species occupying a certain area and sharing a common gene pool 8, 745

portal system circulatory pathway that begins and ends in capillaries, such as the one found between the small intestine and liver 560

positive feedback loop pattern of homeostatic response in which the output intensifies and increases the likelihood of response instead of countering it and canceling it 548

prairie terrestrial biome that is a temperate grassland, changing from tall-grass prairies to short-grass prairies when traveling east to west across the Midwest of the United States 775

pressure-flow model model explaining transport through sieve-tube cells of phloem, with leaves serving as a source and roots as a sink in the summer, and vice versa in the spring 503

primate an animal order of mammals having hands and feet with 5 distinct digits and some having an opposable thumb 458

primitive atmosphere earliest atmosphere of the earth after the planet's formation 355

producer organism at the start of a food chain that makes its own food (e.g., green plants on land and algae in water) 781

progesterone female ovarian sex hormone that causes the endometrium of the uterus to become secretory during the ovarian cycle; along with estrogen maintains secondary sex characteristics in females 700

prokaryote first type of cell to evolve; found in the kingdom Monera and lacking a well-defined nucleus and organelles 64, 370

prokaryotic cell the first, primitive cells on earth, exemplified today by bacteria, which lack a defined nucleus and most organelles 64

promoter a sequence of DNA in an operon where RNA polymerase begins transcription *266*

prostaglandins (pros-tah-glan'dins) class of chemical messengers, acting locally in the body, with widespread effects; derived from fatty acids and stored in the plasma membrane *690*

protein macromolecule having, as its primary structure, a sequence of amino acids united through covalent bonding; many kinds, important in structure and metabolism *50*

proteinoid (pro'te-in-oid") abiotically polymerized amino acids that are joined in a preferred manner; possible early step in cell evolution *356*

protist organism belonging to the kingdom Protista; for example, protozoans, algae, or slime molds *380*

protocell a cell forerunner developed from cell-like microspheres *356*

proton positive subatomic particle, located in the nucleus of the atom *26*

protostome a group of coelomate animals in which the blastopore (the first embryonic opening) becomes the mouth *430*

protozoan animal-like, heterotrophic, unicellular organisms *381*

proximal convoluted tubule the portion of a nephron following the Bowman's capsule where selective reabsorption of filtrate occurs *625*

pseudocoelom (su"do-se'lom) a coelom that is not completely lined by mesoderm *423*

pseudopodia (sud-ah-pod-e-ah) cytoplas-mic extensions of amoeboid protists; used for locomotion by these organisms *381*

pulmonary circuit pathway of bloodflow between the lungs and heart *560*

Punnett square a gridlike graph that enables one to calculate the results of simple genetic crosses by lining gametic genotypes of 2 parents on the outside margin and their recombination in boxes inside the grid *187*

purine (pu'rin) a type of nitrogenous base, such as adenine and guanine, having a double-ring structure *54*

pyrimidine (pi-rim'i-din) a type of nitro-genous base, such as cytosine, thymine, and uracil, having a single-ring structure *54*

pyruvate the end product of glycolysis; its further fate involving fermentation or Krebs cycle incorporation depending on oxygen availability *139*

R

radially symmetrical body plan in which similar parts are arranged around a central axis like spokes of a wheel *419*

radicle part of the plant embryo that contains the root apical meristem and becomes the first root of the seedling *515*

receptor a cell, often in groups, that detects change or stimulus and initiates a nerve impulse *641, 653*

recessive allele (ah-lēl') form of a gene whose effect is hidden by the dominant allele in a heterozygote *186*

recombinant DNA (rDNA) DNA that contains genes from more than one source *279*

reducing atmosphere an atmosphere with little if any free oxygen, for example, the primitive atmosphere *355*

reduction a chemical reaction that results in addition of one or more electrons to an atom, ion, or compound. Reduction of one substance occurs simultaneously with oxidation of another *33*

reflex an automatic, involuntary response of an organism to a stimulus *641*

regeneration reforming a lost body part from the remaining body mass by active cell division and growth *695*

regulator gene in an operon, a gene that codes for a repressor *266*

renewable resource resource that can be replenished by physical or biological means *803*

repetitive DNA DNA with the same set of base pairs repeated many times, separating genes that direct the synthesis of cytoplas-mic proteins *244*

replication fork V-shaped region of a eukaryotic chromosome wherever DNA is being replicated *243*

repressible operon an operon that is normally active because the repressor must combine with a corepressor before the complex can bind to the operator *268*

repressor protein molecule that binds to an operator site, preventing RNA polymerase from binding to the promotor site of an operon *266*

reproduce make a copy similar to oneself; for example, bacteria dividing to produce more bacteria, or egg and sperm joining to produce offspring in more advanced organisms *5*

reptile class of terrestrial vertebrates with internal fertilization, scaly skin, and an egg with a leathery shell; includes snakes, lizards, turtles, and crocodiles *453*

resting potential state of an axon when the membrane potential is about -65 mV and no impulse is being conducted *637*

restriction enzyme enzyme that stops viral reproduction by cutting viral DNA; used in genetic engineering to cut DNA at specific points *279*

retina innermost layer of the eyeball containing the light receptors—rods and cones *655*

retrovirus RNA virus containing the enzyme reverse transcriptase that carries out RNA/DNA transcription; for example, viruses that carry cancer-causing oncogenes and the AIDS virus *368*

rhizoid rootlike hairs that anchor a plant and absorb minerals and water from the soil *398*

rhizome a rootlike, underground stem *485*

rhodopsin (ro-dop'sin) a light-absorbing molecule in rods and cones that contains a pigment and its attached protein *658*

ribonucleic acid (RNA) a nucleic acid with a sequence of subunits dictated by DNA; involved in protein synthesis *54, 235*

ribosomal RNA (rRNA) type of RNA found in ribosomes; sometimes called structural RNA *251*

ribosome cytoplasmic organelle; site of protein synthesis *70*

RNA *see* ribonucleic acid

RNA polymerase (pol-im'er-ās) enzyme that links ribonucleotides together during transcription *254*

rod photoreceptor in vertebrate eyes that responds to dim light *655*

root structure of a plant root system; serving to anchor the plant, absorb water and minerals from the soil, and store products of photosynthesis *479*

root pressure a force generated by an osmotic gradient that serves to elevate sap through xylem a short distance *497*

rough ER complex organelle that is continuous with the nuclear envelope, consisting of membranous channels and studded with ribosomes *71*

r-strategist a species that has evolved characteristics that maximize its rate of natural increase, for example, high birthrate *749*

RuBP (ribulose biphosphate) 5-carbon compound that combines with carbon dioxide during the Calvin cycle and is later regenerated by the same cycle *130*

S

SA node (es a nod) mass of specialized tissue in right atrium wall of heart initiates a heartbeat; the "pacemaker" 559

sac body plan a body with a digestive cavity that has only one opening, as in cnidarians and flatworms 421

saltatory conduction (sal'tah-to"re) movement of a nervous impulse from one node of Ranvier to another along a myelinated axon 639

saprophyte organism that carries out external digestion and absorbs the resulting nutrients across the plasma membrane of its cells; indicative of fungi and slime molds 373, 380

sarcolemma (sar"ko-lem'ah) plasma membrane of a muscle fiber that forms the tubules of the T system involved in muscular contraction 671

sarcomere (sar"ko-mer) a unit of a myofibril, many of which are arranged linearly along its length, that shortens to give muscle contraction 671

sarcoplasmic reticulum (sar"ko-plaz'mik rĕ-tik'u-lum) the modified endoplasmic reticulum of a muscle fiber that stores calcium ions, whose release initiates muscle contraction 671

savanna terrestrial biome that is a tropical grassland in Africa, characterized by a few trees and a severe dry season 775

scientific method a step-by-step process for discovery and generation of knowledge, ranging from observation and hypothesis to theory and principle 14

sclera outer, white, fibrous layer of the eye that surrounds the transparent cornea 655

secondary oocyte (sek'on-dary o'o-sīt) the largest functional product of meiosis; becomes the egg 176

seed a mature ovule that contains an embryo with stored food enclosed in a protective coat 404, 511

seed coat covering around the embryonic sporophyte and stored foods in a mature seed; derived from the integument of the ovule 513

segmented having repeating body units as is seen in the earthworm 434

selectively permeable ability of plasma membranes to regulate the passage of substances into and out of the cell, allowing some to pass through the membrane and preventing the passage of others 89

semicircular canal one of 3 half-circle-shaped canals of the inner ear that are fluid filled for registering changes in motion 660

semiconservative replication duplication of DNA resulting in a double helix having one parental and one new strand 241

seminal fluid (sem'in-al) thick, whitish fluid consisting of sperm and secretions from several glands of the male reproductive tract 697

sepal protective leaflike structure enclosing the flower when in bud 509

sessile filter feeder an organism that stays in one place and filters its food from the water 417

sex chromosome the chromosome that determines the biological sex of an organism 204

sex-influenced trait a genetic trait that is expressed differently in the 2 sexes but is not controlled by alleles on the sex chromosomes, for example, pattern baldness 231

sex-linked gene unit of heredity, a gene, located on a sex chromosome 204

sexual reproduction reproduction involving meiosis and gamete formation; produces offspring with genes inherited from each parent 166, 170

sexual selection type of natural selection by which one organism specifically prefers another organism for mating 327

sickle-cell disease a genetic disorder in which sickling of red blood cells clogs blood vessels; individuals who are heterozygous for the allele have a greater resistance to malaria; more common among Blacks, but even then quite rare 226

sister chromatid one of 2 genetically identical chromosomal units that are the result of DNA replication and are attached to each other at the centromere 157

slash-and-burn agriculture practice of cutting and burning forest to use the land for agriculture 795

smooth ER complex organelle that is continuous with the nuclear envelope; consists of membranous channels and saccules but not studded with ribosomes 71

society a group of individuals of the same species that shows cooperative behavior 740

sociobiology the biological study of social behavior 740

sodium-potassium pump a transport protein in the plasma membrane that moves sodium out of and potassium into animal cells; important in nerve and muscle cells 94

solar energy energy from the sun; light energy 803

speciation (spe"se-a'shun) the process whereby a new species is produced or originates 334

species a taxonomic category that is the subdivision of a genus; its members can breed successfully with each other but not with organisms of another species 8

sperm male sex cell with 3 distinct parts at maturity: head, middle piece, and tail 698

spermatogenesis (sper"mah-to-jen'ĕ-sis) production of sperm in males by the process of meiosis and maturation 175

sphincter circular muscle in the wall of a tubular structure, such as an artery or the digestive tract, that can open and close the vessel, therefore regulating the amount of material moving through it 555

spicule (spik'ūl) skeletal structures of sponges composed of calcium carbonate or silicate 418

spinal cord part of the central nervous system, continuous with the base of the brain and housed within the vertebral column 640

spindle microtubule structure that brings about chromosome movement during cell division 160

spongy layer layer of loosely packed cells in a plant leaf that increases the amount of surface area for gas exchange; along with the palisade layer, it is the site of most of photosynthesis 491

spore usually a haploid reproductive structure that develops into a haploid generation; characteristic product of meiosis in plants 384

sporophyte diploid generation of the alternation of generations life cycle of a plant; it produces haploid spores, by meiosis, that develop into the haploid generation 398

stabilizing selection outcome of natural selection in which extreme phenotypes are eliminated and the average phenotype is conserved 325

stamen a pollen-producing flower structure consisting of an anther on a filament tip 407, 510

starch polysaccharide consisting of covalently bonded glucose molecules, characterized by few side branches; typical storage product in plants 45

steady-state society a society with no yearly increase in population or resource consumption 804

stem upright, vertical portion of a plant shoot system, transporting substances to and from the leaves 482

steroid biologically active lipid molecule having 4 interlocking rings; examples are cholesterol, progesterone, testosterone 48

stigma enlarged, sticky knob at one end of the pistil where pollen grains are received during pollination 407, 509

stolon modified, horizontal stem of a plant that is aboveground and produces new plants where its nodes touch the ground 485

stoma (stō-ma) small opening with 2 guard cells on the underside of leaf epidermis; their opening controls the rate of gas exchange 476

stroma (stro′mah) large, central space in a chloroplast that is fluid filled and contains enzymes used in photosynthesis 74, 123

structural gene gene that codes for proteins in metabolic pathways 267

style the tubular part of the pistil of a flower where a pollen tube develops from a transferred pollen grain 407, 509

substrate the reactant in an enzymatic reaction; each enzyme has a specific substrate 108

substrate-level phosphorylation (fos″fōr-ĭ-la′shun) process in which ATP is formed by transferring a phosphate from a metabolic substrate to ADP 117

succession an orderly sequence of community replacement, one following the other, to an eventual climax community 769

suppressor T cell T lymphocyte that increases in number more slowly than other T cells and eventually brings about an end to the immune response 579

survivorship usually shown graphically to depict death rates or percentage of remaining survivors of a population over time 748

symbiosis (sim″bi-o′sis) a relationship that occurs when 2 different species live together in a unique way; may be beneficial, neutral, or detrimental to one and/or the other species 759

symbiotic (sim″bi-ot′ik) one species having a close relationship with another species; includes parasitism, mutualism, and commensalism 373

sympathetic nervous system a division of the autonomic nervous system that is involved in fight or flight responses; uses norepinephrine as a neurotransmitter 641

sympatric speciation (spe″se-a′shun) origin of new species in populations that overlap geographically 334

synapse a junction between neurons consisting of presynaptic (axon) membrane, the synaptic cleft, and the postsynaptic (usually dendrite) membrane 639

synapsis pairing of homologous chromosomes during meiosis I 173

systematics means to classify organisms based on their phylogeny or evolutionary history 340

systemic circuit pathway of bloodflow ranging from the heart to the tissues throughout the body and back to the heart 560

systole (sis′to-le) contraction period of a heart during the cardiac cycle 558

T

taiga (tī′gah) terrestrial biome that is a coniferous forest extending in a broad belt across northern Eurasia and North America 775

taste bud an elongated cell that functions as a taste receptor 654

taxis movement of an organism toward or away from a stimulus 736

taxonomy a system to classify organisms meaningfully into groups based on similarities and differences and using morphology and evolution 340

Tay-Sachs disease (ta saks′) a lethal, genetic lysosomal storage disease, best known among the U.S. Jewish population, which results from the lack of a particular enzyme; nervous system damage leads to early death 222

tendon structure consisting of fibrous connective tissue that connects skeletal muscles to bones 541, 669

territoriality behavior used to guarantee exclusive use of a given space for reproduction, feeding, etc. 736

testcross a genetic mating in which a possible heterozygote is crossed with an organism homozygous recessive for the characteristic(s) in question in order to determine its genotype 189

testosterone male sex hormone produced from interstitial cells in the testis 698

tetrad the set of 4 chromatids of a synapsed homologous chromosome pair, visible during prophase of meiosis I; also called bivalent 172

thalamus (thal′ah-mus) a lower forebrain region in vertebrates involved in crude sensory perception and screening messages intended for the higher forebrain or cerebrum 645

theory a conceptual scheme arrived at by the scientific method and supported by innumerable observations and experimentations 17

thermal inversion temperature inversion that traps cold air and its pollutants near the earth with the warm air above it 800

thigmotropism (thig-mot′ro-pizm) movement in response to touch; in plants, the coiling of tendrils 529

thoracic cavity (tho-ras′ik) internal body space of some animals that contains the lungs, protecting them from desiccation; the chest 610

thrombocyte cell fragment in the blood that initiates the process of blood clotting 563

thylakoid (thi′lah-koid) flattened disk of a granum; its membrane contains the photosynthetic pigments 74, 125

thymine (T) (thi′min) one of 4 organic bases in the nucleotides composing the structure of DNA 54, 238

thyroid gland large gland in the neck that produces several important hormones, including thyroxin and calcitonin 683

thyroxin a substance, T4, secreted from the thyroid gland, that promotes growth and development in vertebrates; in general, it increases the metabolic rate in cells 683

tissue group of similar cells combined to perform a common function 539

tissue fluid derivative of the blood plasma from capillaries that bathes the cells throughout the body 566

T lymphocyte white blood cell that directly attacks antigen-bearing cells 577

tonicity (tō-nis′ĭ-tē) referring to the solute concentration of a solution 91

trachea 1. an air tube (windpipe) of the respiratory tract in vertebrates; 2. an air tube in insects 438, 593, 612

transcription the process whereby the DNA code determines (is transcribed into) the sequence of codons in mRNA 252

transfer RNA (tRNA) active in protein synthesis, and transfers a particular amino acid to a ribosome; at one end it binds to the amino acid and at the other end it has an anticodon that binds to an mRNA codon 251

transgenic organism a multicellular organism that contains a transplanted gene 284

transition reaction a reaction involving the removal of CO_2 from pyruvate; connects glycolysis to the Krebs cycle 143

translation the process whereby the sequence of codons in mRNA determines (is translated into) the sequence of amino acids in a polypeptide 252

transpiration the plant's loss of water to the atmosphere, mainly through evaporation at leaf stomata 497

transposon (trans-po′zun) movable segment of DNA 245

triplet code genetic code (mRNA, tRNA) in which sets of 3 bases call for specific amino acids in the formation of polypeptides 253

trophic level (trof′ik) feeding level of one or more populations in a food web *782*

tropical rain forest forest found in warm climates with plentiful rainfall *777*

tropism (tro′pizm) in plants, a growth response toward or away from an external stimulus *529*

TSH thyroid stimulating hormone, secreted by the anterior pituitary; stimulates activity in the thyroid gland *683*

tube-within-a-tube body plan a body with a digestive tract that has both a mouth and an anus *421*

tumor suppressor gene unit of heredity (gene) that codes for proteins that ordinarily suppress cell division, thereby promoting organized cell growth *273*

tundra treeless terrestrial biome of cold climates; occurs on high mountains and in polar regions *773*

turgor pressure the pressure of the plasma membrane against the cell wall, determined by the water content of the plant cell vacuole; gives internal support to the cell *92*

Turner syndrome a condition that results from the inheritance of a single X chromosome; a second sex chromosome is absent: XO *218*

tympanic membrane eardrum; membranous region that receives air vibrations in an auditory organ *660*

U

uracil one of 4 nucleotides composing the structure of RNA *54*

urea main nitrogenous waste of terrestrial amphibians and mammals *621*

ureter (u-re′ter) a tubular structure conducting urine from kidney to urinary bladder *624*

urethra (u-re′thrah) tubular structure that receives urine from the bladder and carries it to the outside of the body *624*

uric acid main nitrogenous waste of insects, reptiles, birds, and some dogs *621*

urinary bladder organ where urine is stored *624*

urine liquid waste product made by the nephrons of the kidney through the processes of pressure filtration and selective reabsorption *624*

uterine cycle a cycle that runs concurrently with the ovarian cycle; prepares the uterus to receive a developing zygote *703*

uterus pear-shaped portion of female reproductive tract that lies between the oviducts and the vagina; site of embryo development *700*

V

vaccine a substance that wakes the immune response without causing illness *580*

vacuole organelle that is a membranous sac, particulary prominent in plant cells *74*

vagina muscular tube leading from the uterus; the female copulatory organ and the birth canal *700*

vas deferentia part of the male reproductive tract that transports sperm cells from the epididymis to the penis *697*

vascular plant any organism of the plant kingdom that contains the vascular tissues xylem and phloem as part of its structure *399*

vector in genetic engineering, a means to transfer foreign genetic material into a cell, for example, a plasmid *279*

vein a blood vessel that arises from venules and transports blood toward the heart *555*

vena cava (pl., venae cavae) (ve′nah ka′vah) one of 2 largest veins in the body; returns blood to the right atrium in a 4-chambered heart *560*

vertebra (pl., vertebrae) one of many bones of the vertebral column; held to other vertebrae by bony facets, muscles, and strong ligaments *666*

vertebrate referring to a chordate animal with a serial arrangement of vertebrae, or backbone *416*

villus (pl., villi) (vil′us) small, fingerlike projection of the inner small intestine wall *594*

viroid unusual infectious particle consisting of a short chain of naked RNA *368*

visible light a portion of the electromagnetic spectrum of light that is visible to the human eye *122*

vitamin an organic molecule that is required in small quantities for various biological processes and must be in an organism's diet because it cannot be synthesized by the organism; becomes part of enzyme structure *112, 600*

W

white matter regions of the brain and spinal cord, consisting of myelinated nerve fibers *642*

woody angiosperm characterizes angiosperm trees with hardwood *407*

X

X-linked gene gene located on the X chromosome that does not control a sexual feature of the organism *204*

xylem a vascular tissue that transports water and mineral solutes upward through the plant body *399, 477*

XYY male a genetic condition in which affected males have 2 Y chromosomes, are taller than average, and may exhibit learning difficulties *218*

Y

yolk sac one of 4 extraembryonic membranes in vertebrates; important in the development of fishes, amphibians, reptiles, and birds, it is largely vestigial in mammals *719*

Z

zero population growth no growth in population size *751*

zygote the diploid (2N) cell formed by the union of 2 gametes, the product of fertilization *510*

CREDITS

Conservation/Dr. Charles McDonald

Chapter 26
Opener: © E. Robinson/Tom Stack & Associates; **26.1a:** © BioMedia Associates; **26.1b:** © Stephen Krasemann/Peter Arnold, Inc.; **26.6a:** © Bill Cartsinger/Photo Researchers, Inc.; **26.6b:** © Ron Taylor/Bruce Coleman, Inc.; **26.6c:** © Carolina Biological Supply Company; **26.8a:** © Dr. Fred Whittaker; **26.10b:** © Fred Marsik/Visuals Unlimited; **26.11:** © Jim Solliday/Biological Photo Service; **26.12:** © Markell, E.K. and Voge, M.: *Medical Paristology*, 4th ed., W.B. Saunders Co. 1981

Chapter 27
Opener: © J. Alcock/Visuals Unlimited; **27.3b:** © William Jorgenson/Visuals Unlimited; **27.4a:** © William Ferguson; **27.4c:** © William Jorgenson/Visuals Unlimited; **27.6a:** © Michael DiSpezio/Images; **27.6c:** © Geral Corsi/Tom Stack & Associates; **27.7a:** © Michael DiSpezio/Images; **27.7c:** © Gary Milburn/Tom Stack & Associates; **27.9:** © Michael DiSpezio/Images; **27.10a:** © Robert Evans/Peter Arnold, Inc.; **27.10b:** © Tom McHugh/Photo Researchers, Inc.; **27.10c:** © Dwight Kuhn; **27.10d:** © John McGregor/Peter Arnold, Inc.; **27.10e, f:** © John Fowler/Valan Photos; **27.11a:** © James H. Carmichael/Bruce Coleman, Inc.; **27.11c:** © Fred Bavendam/Peter Arnold, Inc.; **27.Ab:** © Edward S. Ross; **27.14a:** © Michael DiSpezio/Images; **27.15b:** © Oxford Scientific Films, Ltd.

Chapter 28
Opener: © John Cancalosi/Peter Arnold, Inc.; **28.3:** © Field Museum of Natural History, Chicago; **28.4a:** © Tom Stack/Tom Stack & Associates; **28.4b:** © Douglas Faulkner/Sally Faulkner Collection; **28.5a:** © Paul L. Janosi/Valan Photos; **28.5b:** © Thomas Kitchin/Valan Photos; **28.6a:** © Tom McHugh/Steinhart Aquarium; **28.7a-d:** © Jane Burton/Bruce Coleman, Inc.; **28.8a:** © American Museum of Natural History; **28.8b:** © John Cunningham/Visuals Unlimited; **28.9a:** © Andrew Odum/Peter Arnold, Inc.; **28.9b:** © E.R. Degginger/Bruce Coleman, Inc.; **28.10a:** © Wolfgang Bayer Productions, Inc.; **28.11:** © American Museum of Natural History; **28.12:** © R. Austing/Photo Researchers, Inc.; **28.13:** © Jim David/Photo Researchers, Inc.; **28.14:** © Renee Stockdale/Animals Animals; **28.15:** © Stouffer Enterprises, Inc./Animals Animals; **28.18a:** © Martha Reeves/Photo Researchers, Inc.; **28.18b:** © Bios/Peter Arnold, Inc.; **28.18c:** © George Holton/Photo Researchers, Inc.; **28.18d:** © Tom McHugh/Photo Researchers, Inc.; **28.23:** © American Museum of Natural History

Chapter 29
Opener: © M.I. Walker/Science Source/Photo Researchers, Inc.; **29.2a:** © Ed Reschke/Peter Arnold, Inc.; **29.2b:** © John Cunningham/Visuals Unlimited; **29.2c:** © Biophoto Associates/Photo Researchers, Inc.; **29.3a:** © Biological Photo Service; **29.3b, c:** © Biophoto Associates/Photo Researchers, Inc.; **29.4b:** © Biological Photo Service; **29.5b:** © George Wilder/Visuals Unlimited; **29.8b:** © Carolina Biological Supply Company; **29.9:** © Dwight Kuhn; **29.10:** © John Cunningham/Visuals Unlimited; **29.11a:** © G.R. Roberts; **29.11b:** © E.R. Degginger/Color-Pic; **29.11c:** © David Newman/Visuals Unlimited; **29.12:** © J.R. Waaland, University of Washington/Biological Photo Service; **29.13a:** © Carolina Biological Supply Company; **29.13b, c:** © Runk/Schoenberger/Grant Heilman; **29.14:** © Carolina Biological Supply Company; **29.16b:** © Carolina Biological Supply Company; **29.B:** © Earl Roberge/Photo Researchers, Inc.; **29.19b:** © J.H. Troughton and F.B. Sampson; **29.21a:** © Martha Cooper/Peter Arnold, Inc.; **29.21b:** © Larry Mellichamp/Visuals Unlimited; **29.21c:** © Carolina Biological Supply Company

Chapter 30
Opener: © Cabisco/Visuals Unlimited; **30.6a:** © David Phillips/Visuals Unlimited; **30.8a-c:** © Dr. Mary E. Doohan; **30.10a:** © Gordon Leedale/BioPhoto Associates; **30.10b:** © Donald Marx/USDA Forest Service

Chapter 31
Opener: © D. Newman/Visuals Unlimited; **31.3:** © J. Robert Waaland/Biological Photo Service; **31.4a:** © BioPhoto Associates/Photo Researchers, Inc.; **31.4b:** © Ed Reschke/Peter Arnold, Inc.; **31.4c:** © David Scharf/Peter Arnold, Inc.; **31.Aa:** © Nicholas Smythe/Photo Researchers, Inc.; **31.Ab:** © H. Eisenbeiss/Photo Researchers, Inc.; **31.Ac:** © Anthony Mercieca/Photo Researchers, Inc.; **31.Ad:** © Donna Howell; **31.6a (1):** © Ralph A. Reinhold/Animals Animals; **31.6a (2):** © W. Ormerod/Visuals Unlimited; **31.6b (1):** © C.S. Lobban/Biological Photo Service; **31.6b (2):** © Biological Photo Service; **31.6c** (both): © Dwight Kuhn; **31.7:** © Ted Levin/Earth Scenes

Chapter 32
Opener: © D. Newman/Visuals Unlimited; **32.8:** Courtesy R.J. Weaver; **32.4:** © Tom McHugh/Photo Researchers, Inc.; **32.5:** © Robert E. Lyons/Visuals Unlimited; **32.6a-d:** © Kiem Tran Thanh Van and her colleagues; **32.7:** © Runk/Schoenberger; **32.10:** © John Cunningham/Visuals Unlimited; **32.11a, b:** © Tom McHugh/Photo Researchers, Inc.; **32.14:** © Frank B. Salisbury

Chapter 33
Opener: © Michael Gabridge/Visuals Unlimited; **33.2a-e:** © Ed Reschke; **33.2f:** © Ed Reschke/Peter Arnold, Inc.; **33.3a,b, 33.4b,c, 33.6a-c, 33.7:** © Ed Reschke

Chapter 34
Opener: © Manfred Kage/Peter Arnold, Inc.; **34.1a:** © Eric Grave/Photo Researchers, Inc.; **34.1b:** © Carolina Biological Supply Company; **34.1c:** © Michael DiSpezio/Images; **34.A:** The Bettmann Archive; **34.Ba:** © Lewis Lainey; **34.12b:** © Science Photo Library/Photo Researchers, Inc.; **34.15a:** © Stuart I. Fox

Chapter 35
Opener: © Manfred Kage/Peter Arnold, Inc.; **35.2:** © Keetin, Biological Science, Norton Publishing; **35.4:** © Boehringer Ingleheim International GmbH, Courtesy of Lennart Nilsson; **35.7:** © R. Feldman/Dan McCoy/Rainbow; **35.8:** © Boehringer Ingleheim International GmbH, courtesy Lennart Nilsson

Chapter 36
Opener: © Matt Meadows/Peter Arnold, Inc.; **36.1a:** © David Dennis/Tom Stack & Associates; **36.1b:** © Harry Rogers/Photo Researchers, Inc.; **36.1c:** © Grant Heilman Photography; **36.14a,b:** Courtesy of the World Health Organizations; **36.14c,d:** Courtesy of the Atlanta Centers for Disease Control

Chapter 37
Opener: © CNRI/Science Photo Library/Photo Researchers, Inc.; **37.10a:** © Lennart Nilsson: Behold Man, Little Brown and Company, Boston; **37.A, 37.B:** © Martin Rotker

Chapter 38
Opener: © CNRI/Science Photo Library/Photo Researchers, Inc.; **38.2b:** © Robert Myers/Visuals Unlimited

Chapter 39
Opener: © John Allison/Peter Arnold, Inc.; **39.13:** © Dan McCoy/Rainbow

Chapter 40
Opener: © Runk/Schoenberger/Grant Heilman; **40.4:** © T. Norman Tait/BioPhoto Associates; **40.6d:** © Dr. Frank Werblin, University of California at Berkeley; **40.A** (both): Courtesy Professor J.E. Hawkins, Robert S. Preston, Kresage

Chapter 41
Opener: © SIU/Visuals Unlimited; **41.1:** © Joe McDonald/Bruce Coleman, Inc.; **41.9b:** © H.E. Huxley; **41.11a:** © Victor Eichler/Bio Art

Chapter 42
Opener: © Manfred Kage/Peter Arnold, Inc.; **42.6:** © Bettina Cirone/Photo Researchers, Inc.; **42.9:** © Lester Bergman and Associates

Chapter 43
Opener: © John Mitchell/Photo Researchers, Inc.; **43.1b:** © John Shaw/Bruce Coleman, Inc.; **43.2a:** © Manfred Kage/Peter Arnold, Inc.; **43.2b:** © Matt Meadows/Peter Arnold, Inc.; **43.5:** © BioPhoto Associates/Photo Researchers, Inc.; **43.8b:** © Baganvandoss/Photo Researchers, Inc.

Chapter 44
Opener: © Karl Switak/Photo Researchers, Inc.; **44.12a:** © Edelmann/First Days of Life/Black Star

Chapter 45
Opener: © Gregory Scott/Photo Researchers, Inc.; **45.1a-d:** © Dr. Rae Silver; **45.5:** © C.C. Lockwood/Animals Animals; **45.6a:** © Steve Kaufman/Peter Arnold, Inc.; **45.6b:** Nina Leen, Life Magazine © 1964 Time, Inc.; **45.11:** © Susan Kuklin/Photo Researchers, Inc.

Chapter 45
Opener: © Gregory Scott/Photo Researchers, Inc.; **45.1a-d:** © Dr. Rae Silver; **45.5:** © C.C. Lockwood/Animals Animals; **45.6a:** © Steve Kaufman/Peter Arnold, Inc.; **45.6b:** © Nina Leen, Life Magazine © 1964 Time, Inc.; **45.11:** © Susan Kuklin/Photo Researchers, Inc.

Chapter 46
Opener: © Bruce Berg/Visuals Unlimited; **46.4a:** © Paul Janosi/Valan Photos

Chapter 47
Opener: © Stephen Dalton/Photo Researchers, Inc.; **47.4a:** © Dr. Gregory Antipa and H.S. Wesenbergand; **47.5a:** © Alan Carey/Photo Researchers, Inc.; **47.6a:** © Runk/Schoenberger/Grant Heilman; **47.6b:** © National Audubon Society/Photo Researchers, Inc.; **47.6c:** © Z. Leszczynski/Animals Animals; **47.7:** © Hans Pfletschinger/Peter Arnold, Inc.; **47.8:** © Cliff B. Frith/Bruce Coleman, Inc.' **47.9a-c:** © Dr. Daniel Jantzen

Chapter 48
Opener: © S.J. Krasemann/Peter Arnold, Inc.; **48.3:** © Douglas Faulkner/Sally Faulkner Collection; **48.7a:** © Stephen Krasemann/Peter Arnold, Inc.; **48.7b:** © Richard Ferguson/William Ferguson; **48.7c:** © Mary Thatcher/Photo Researchers, Inc.; **48.7d:** © Karlene Schwartz; **48.9:** © W.H. Hodge/Peter Arnold, Inc.; **48.10:** © Stephen Krasemann/Peter Arnold, Inc.; **48.12:** © R.S. Virdee/Grant Heilman; **48.13:** © Norman Owen Tomalin/Bruce Coleman, Inc.; **48.Aa:** © Stephen Dalton/Photo Researchers, Inc.; **48.Ab:** © Erwin and Peggy Bauer/Bruce Coleman, Inc.; **48.Ac:** © Bruce Coleman/Bruce Coleman, Inc.

Chapter 49
Opener: © Wyman Meinzer/Peter Arnold, Inc.; **49.1:** © Gallbridge/Visuals Unlimited; **49.8:** © Bob Coyle; **49.11:** © Jacques Jangoux/Peter Arnold, Inc.

Chapter 50
Opener: © Larry Brock/Tom Stack & Associates; **50.1:** © Link/Visuals Unlimited; **50.2a:** © Sydney Thomson/Animals Animals; **50.2b:** © Nichols/Magnum; **50.2c:** © G. Prance/Visuals Unlimited; **50.3:** © Gary Milburn/Tom Stack & Associates; **50.7a,b:** © Dr. John Skelly; **50.10:** NASA; **50.12:** © Thomas Kitchin/Tom Stack & Associates

Line Art

Chapter 2
2.3: Copyright © Mark Lefkowitz.

Chapter 3
3.1: Copyright © Mark Lefkowitz.

Chapter 4
4.C: Figure from *Cell Biology* by L. J. Kleinsmith and V. M. Kish. Copyright © 1988 by Harper & Row, Publishers, Inc. Reprinted by permission of HarperCollins Publishers.

Chapter 7
7.12: From Stuart Ira Fox, *Human Physiology*, 3d ed. Copyright © 1990 Wm. C. Brown Communications, Inc., Dubuque, Iowa. All Rights Reserved. Reprinted by permission.

Chapter 10
10.A, 10.8b, 10.12b-d: Copyright © Mark Lefkowitz.

Chapter 13
13.6: Copyright © Mark Lefkowitz.
13.14: From Robert F. Weaver and Philip W. Hedrick, *Genetics*. Copyright © 1989 Wm. C. Brown Communications, Inc., Dubuque, Iowa. All Rights Reserved. Reprinted by permission.

Chapter 14
14.3: From Robert F. Weaver and Philip W. Hedrick, *Genetics*. Copyright © 1989 Wm. C. Brown Communications, Inc., Dubuque, Iowa. All Rights Reserved. Reprinted by permission.
14.9b: From E. Peter Volpe, *Biology and Human Concerns*, 3d ed. Copyright © 1983 Wm. C. Brown Communications, Inc., Dubuque, Iowa. All Rights Reserved. Reprinted by permission.

Chapter 15
15.8: From A. H. Wang, et al., *Nature*, Vol. 282:684. Copyright © 1979 Macmillan Magazines, Ltd., London, England. Reprinted by permission.

Chapter 17
17.4b,c: Copyright © Mark Lefkowitz.

Chapter 18
18.10: From J. C. Stephens, "Mapping the Human Genome: Current Status" in *Science*, 250:237, 12 October 1990. Copyright 1990 by the AAAS.

Chapter 19
19.6, 19.13: Copyright © Mark Lefkowitz.

Chapter 20
20.5: From D. Hartl, *A Primer of Population Genetics*. Copyright © 1981 Sinauer Associates, Inc., Sunderland, MA. Reprinted by permission. Data from P. Buri, "Gene Frequency in Small Populations of Mutant *Drosophila*" in *Evolution* 10:367-402, 1956.
20.7a: From E. Peter Volpe, *Biology and Human Concerns*, 3d ed. Copyright © 1983 Wm. C. Brown Communications, Inc., Dubuque, Iowa. All Rights Reserved. Reprinted by permission.

Chapter 21
21.17, 21.19: Copyright © Mark Lefkowitz.

Chapter 23
23.5 (left): From Helena Curtis and N. Sue Barnes, *Biology*, 5th ed. Copyright © 1989 Worth Publishers, New York, NY. Reprinted by permission.

Chapter 24
24.10b: Copyright © 1974 Kendall/Hunt Publishing Company.

Chapter 26
26.2, 26.3, 26.5: Copyright © Mark Lefkowitz.

Chapter 27
27.1, 27.5: Copyright © Mark Lefkowitz.

Chapter 28
28.10: Adapted from figure in *Introduction to Embryology*, Fifth Edition, by B. I. Balinsky and B. C. Fabian, copyright © 1981 by Saunders College Publishing, reprinted by permission of the publisher.
28.19, 28.20, 28.21, 28.22: Copyright © Mark Lefkowitz.

Chapter 29
29.7: From Kingsley R. Stern, *Introductory Plant Biology*, 4th ed. Copyright © 1988 Wm. C. Brown Communications, Inc., Dubuque, Iowa. All Rights Reserved. Reprinted by permission.

Chapter 30
30.7: From Kingsley R. Stern, *Introductory Plant Biology*, 4th ed. Copyright © 1988 Wm. C. Brown Communications, Inc., Dubuque, Iowa. All Rights Reserved. Reprinted by permission.

Chapter 31
31.9, 31.10: From Kingsley R. Stern, *Introductory Plant Biology*, 4th ed. Copyright © 1988 Wm. C. Brown Communications, Inc., Dubuque, Iowa. All Rights Reserved. Reprinted by permission.
A.1a: Reproduced by permission of the National Research Council of Canada from the *Canadian Journal of Botany*, Vol. 39, pages 891-900, 1961.
A.1b: From H. W. Woolhouse, *Symposia Society for Experimental Biology*, 21:179. Copyright © 1967 Cambridge University Press, New York, NY. Reprinted by permission.

Chapter 33
33.4a: From John W. Hole, Jr., *Human Anatomy and Physiology*, 5th ed. Copyright © 1990 Wm. C. Brown Communications, Inc., Dubuque, Iowa. All Rights Reserved. Reprinted by permission.
33.9: From Kent M. Van De Graaff and Stuart Ira Fox, *Concepts of Human Anatomy and Physiology*, 2d ed. Copyright © 1989 Wm. C. Brown Communications, Inc., Dubuque, Iowa. All Rights Reserved. Reprinted by permission.

Chapter 34
34.5, 34.6: Copyright © Mark Lefkowitz.
34.B (b): From Kent M. Van De Graaff and Stuart Ira Fox, *Concepts of Human Anatomy and Physiology*, 2d ed. Copyright © 1989 Wm. C. Brown Communications, Inc., Dubuque, Iowa. All Rights Reserved. Reprinted by permission.

Chapter 36
36.6, page 605: Copyright © Mark Lefkowitz.

Illustrators

Part 1

Kathleen Hagelston:
5.16a.

Illustrious, Inc.:
text art, pages 20, 39, 47; 4.9, 4.B, 4.C.a, text art, page 50; 4.16a-d, text art, pages 58, 72; 5.18b-e, 5.19a-b, 6.11, 6.12, 7.9, 7.13, 7.14, text art, page 119; 8.9, text art, pages 130, 133, 136, 140, 142; 9.7, text art, page 144; 9.10, text art, page 151.

Carlyn Iverson:
2.7a, text art, page 82; 8.2, 8.3a, 8.7, 9.1b, 9.2.

Laurie O'Keefe:
5.10.

Mark Lefkowitz:
2.3, 3.1.

Rolin Graphics:
text art, page 15; 2.2, 2.4, 2.6, text art, pages 33, 34, 36; 3.2, 3.5, 3.6, 3.7, 3.8a-b, 3.A, 3.10, 3.11, 3.13, 3.14, 3.15, text art, page 43; 4.2, 4.3, 4.4, text art, page 48; 4.5, 4.6, text art, pages 54, 58; 4.10, 4.13, 4.14, 4.15, 4.17, 4.18, 4.20, text art, page 65; 5.2a-b, 5.3b, text art, page 74; 5.4b, 5.5a, 5.7a, 5.7c, 5.9b, 5.15, 5.17a, text art, page 85; 6.1a-b, text art, page 86; 6.2b-c, 6.3, text art, page 87; 6.4, text art, page 88; 6.5, 6.6, 6.8, 6.9, text art, page 97; 6.A.a-f, 6.17b, text art, pages 102, 103, 106; 7.3a-b, text art, page 109; 7.6a-b, 7.7, text art, pages 110, 113; 7.10, text art, page 114; 7.11, text art, pages 117, 122; 8.4, text art, pages 127, 128; 8.8b, text art, page 131; 8.10, 8.12, text art, pages 138, 140; 9.3, 9.4, text art, page 142; 9.5, 9.6, 9.8, text art, page 145; 9.9, text art, pages 146, 150; 9.11, 9.12.

Rolin/Iverson:
4.19.

Part 2

Molly Babich:
text art, page 184

Kathleen Hagelston:
18.3

Hagelston/Margorie Leggitt:
14.2B

Illustrious, Inc.:
11.1, text art, page 171; 13.3, 13.5, 13.8, 13.11, text art, page 209; 14.12A, 15.5B, 15.6, 15.7, 15.8, 16.2, 16.3B&C, 16.4, text art, page 253; 16.7, 16.11A, 16.15, text art pages 269, 274; 18.2, text art pages 280, 289, 18.10, 18.11, text art, page 294.

Illustrious, Inc./Leonard Morgan:
10.5A, 10.6A, 10.7A, 10.9A, 11.3, 11.7, text art, page 167.

Carlyn Iverson:
14.1, 16.10A, 18.B.

Laurie O'Keefe:
14.8A-C, 17.7, 17.8

Mark Lefkowitz:
10.8B, 10.A, 10.12B, 12.2, 13.6, 15.2A, 17.4B&C

Precision Graphics:
14.9B

Rolin Graphics:
text art, pages 155, 157; 10.2A, text art, page 160; 10.4, 11.2, text art, page 174; 11.4B-D, 11.6, 12.3, text art, page 187; 12.4, 12.5, text art, page 189; 12.6, 12.7, 12.8, 12.9, text art, page 192; 13.1, text art, page 201; 13.9, 13.12, 13.16, 14.3, 14.6, 14.7, 14.B.A-C, 14.14, text art, page 236; 15.3, text art, page 237; 15.4A, text art, page 240; 15.A, text art, pages 243, 249; 16.6, 16.8B, 16.9A-E, text art, page 266; 16.14, 17.1, text art, pages 258, 266; 17.A, 17.B, 17.2, 17.5, 18A, 18.8, 18.12B.

Rolin/Iverson:
16.1

Part 3

Molly Babich:
22.6, 24.14, text art, page 401; 27.A, 28.17

Illustrious, Inc.:
text art, page 309; 20.2A&B, text art, page 318; 20.5, 20.8, 20.12, 21.3, 21.5, text art, page 340; Tbl. 21.3, text art, pages 364, 446, 470; fig. A.

Carlyn Iverson:
21.6A&B, 21.7, 21.16B

Margorie Leggitt:
28.16

Mark Lefkowitz:
19.6A&B, 19.13, 21.17, 21.19, 23.1C1, 24.2, 24.4, 26.2, 26.3, 26.5, 26.7, 27.1, 27.5, 27.12, 28.19, 28.20, 28.21, 28.22.

Laurie O'Keefe:
21.8, 21.9, 21.10, 21.12, text art, page 450

Rolin Graphics:
19.1A, 20.3A&B, 20.9A, text art, page 344; 21.11, 21.14, 21.15, 21.16A, 22.1, 22.5, 23.3, 23.5A, text art, page 377; 24.8, 24.9, 24.12, 24.16, 24.18A, 24.19A, text art, page 395; 25.4A, 25.5, 25.7, text art, page 405; 25.12A, text art, page 414; 27.2, 27.4B, 27.6B.

Rolin/Iverson:
25.3

Part 4

Kathleen Hagelston:
29.16A

Illustrious, Inc.:
29.A, text arts, page 493; Fig. A

Mark Lefkowitz:
31.2, 31.8A-F

Rolin Graphics:
29.1, 29.3A2, 29.3B2, 29.3C2, 29.6, 29.14B, 29.15, 29.17A-B, 29.19A&C, 29.20, 30.1, 30.2, 30.3A-B, 30.5B, text art, page 498; 30.6B&C, text art, page 500; 30.9, 30.11, 30.12, text art, pages 504-505; 31.1, text art, pages 509, 521; 32.1, 32.2A-C, 32.3, 32.8, text art, page 529; 32.12A&B, 32.13.

Rolin/Iverson:
31.5

Part 5

Molly Babich:
34.2, 38.6B

Chris Creek:
35.3, 38.1A

Anne Greene:
405, text art, page 663; 44.10A-E

Kathleen Hagelston:
43.12, 43.13

Hagelston/Leggitt:
36.13

Hagelston/O'Keefe:
35.5

Illustrious, Inc.:
text art, page 539; 33.1, 33.2A, text art, page 545, 558; 34.11B, 35.A, text art, page 596; 44.2, text arts, page 717; Figs. A&B

Carlyn Iverson:
33.8, 33.12, 34.2, 34.3, 34.11A, 35.6, 36.7, 38.11, 39.14A-E, 40.10, 40.11A-D, 41.9A, 43.7B, 44.11, 44.12B

Laurie O'Keefe:
33.5, 35.1

Mark Lefkowitz:
34.5, 34.6, 36.2, 36.6, text art, page 604; 37.4, 37.5, 37.7, 37.8, text art, page 618; 38.5, 38.6A, 38.7, 38.8, 38.9, 39.1, 39.6, 39.7, 39.8, 39.10, 39.12, 40.1, 42.2

Ron McClean:
43.8A

Steve Moon:
33.9, 43.11

Diane Nelson:
34.B.B, 44.13

Rolin Graphics:
33.6A.2, 33.6B.2, 33.6C.2, 33.11, 34.7, 34.8,

34.12A, 34.13, 34.14, 35.9A-B, 35.10, text art, pages 591, 593, 594, 595; 36.12, text art, page 605; 37.10B&C, text art, pages 614, 616; 38.1B, 38.2A, 38.3, 38.10, 38.12, text art, page 631; 39.3, 39.4, 39.5, 40.2A-D, 40.6C, 40.7A-D, 40.8, 41.2, 41.3, 41.4, 41.5, 41.6, text art, pages 671, 672, 675; 42.3A&B; 42.4, 42.5, 42.7A-C, 42.8A-C, 42.12, 42.13, text art, page 693; 43.1A, 43.6, 43.9, 44.5, 44.9

Mike Schenk:
42.11, 43.5C

Tom Waldrop:
text art, page 550; 39.2A, 39.9, 41.11B&C, 42.A, 43.3, 43.4A&B, 43.7A, 43.10, text art, page 710

Part 6

Illustrious, Inc.:
45.A, 45.12, 46.3A, 46.8, 50.6

Carlyn Iverson:
49.5

Mark Lefkowitz:
49.10, 49.12

Rolin Graphics:
45.2A-C, 45.3, 45.4, 45.5B, 45.7A&B, 45.8, 45.9, 46.1, 46.2, 46.3B, 46.4B, 46.5, 46.6A-C, text art, page 753; 47.1, 47.2, 47.3, 47.4B, 47.5B, 47.A.A-D, 48.1, 48.2, 48.4, 48.5, 48.6B, 48.8, 48.11A-C, 49.1A, 49.2A&B, 49.3, 49.4, 50.8, 50.9B, 50.11

How Do You Do That? Boxes

p. 292, Holly Ahern; p. 345, Steve Miller; p. 487, Dr. Kingsley Stern; p. 581, Dr. Andy Anderson; p. 737, Lee Drickamer

INDEX

Chick, 311, 712
Chiggers, 438
Chihuahua, 306
Childhood, 722
Chilopoda, 438, 439
Chimpanzee, 312, 342, 461, 463, 645, 735, 740
Chinese cabbage, 307
Chironomus, 270, 295–96
Chironomus pallidivittatus, 295–96
Chironomus tentans, 295–96
Chiroptera, 459
Chitin, 45–46, 47, 437, 665
Chlamydia, 706, 707
Chlamydomonas, 384, 385, 386
Chlorella, 130
Chlorides, 627
Chlorine, 26, 30
Chlorofluorocarbons (CFCs), 801
Chlorophyll, 124, 125, 126, 371, 384
Chlorophyta, 384
Chloroplasts, 68, 69, 74–75, 81, 123–25, 476, 489
CHNOPS, 26, 42
Cholecystokinin-pancreozymin (CCK-PZ), 596, 680
Cholesterol, 50, 57, 87, 223, 566–67, 599–600
Cholesterol blood level, 566–67
Chondrichthyes, 448
Chordae tendineae, 557
Chordates, 4, 417, 429, 431, 443–44, 448, 665
Chorion, 455, 718
Chorionic villi sampling, 226, 227
Choroid, 655, 657
Chromatids, 157, 159, 160, 208
Chromatin, 67, 68, 69, 70, 157, 269–70
Chromosome mapping, 206–8
Chromosome mutations, 208–11, 316–17
Chromosome number
    changes in, 208–9
    and organism types, 157
Chromosome puffs, 269, 295
Chromosomes, 66, 68, 69, 157
    and autosomal chromosome abnormalities, 216–18, 222–24
    and chromosome mapping, 206–8
    and chromosome number, 208–9
    and chromosome puffs, 269, 295
    comparison of meiosis and mitosis, 178, 179
    contents of, 158, 159, 160
    duplicated, 159, 160
    eukaryotic, 157–59
    and genes, 196–214
    and genetic mapping, 288–92
    and human chromosomes, 216–19
    independent assortment of, 176
    and life cycles of algae, 385
    and meiosis, 171–76
    and mutations, 208–11
    and pangenesis, 177
    and polytene chromosomes, 269
    prokaryotic, 156–57
    and sex chromosomes, 204, 218–19, 220, 222, 228–31
    structure changes, 209–11
Chromosome sex determination, 204
Chromosome theory of inheritance, 202
Chrysanthemum, 471
Chrysophyta, 388–89
Chthamalus stellatus, 755–56

Chyme, 593
Cicada, 665
Cigarettes, 615
Cilia, 69, 78, 79, 81
Ciliary body, 657
Ciliates, 381–82
Ciliophora, 381
Cilium, 67
Circadian rhythms, 530, 731
Circulatory system, 546, 552–71
    closed circulatory system, 448, 552, 553–54
    invertebrates, 553–54
    and open circulatory system, 431, 553–54
    vertebrates, 555–57
Cirrhosis, 597
Citric acid cycle, 144
Cladistics, 342–43
Cladogenesis, 340, 347–48
Cladogram, 342, 343
Clams, 417, 431, 433, 590
Class, 4, 340
Classical conditioning, 735
Classifications
    of bacteria, 374–76
    and five kingdoms system of classification, 360–62
    hierarchy of, 340
    of leaves, 490
    of organisms, 4–5, A-1–A-2
    of primates, 459
    of skeletons, 665
Clathrin, 98
Clavicle, 666
Cleavage, 713, 716, 718
Climax community, 769
Clitellum, 435
Clitoris, 700, 701
Cloaco, 621
Clonal selection theory, 577, 578
Cloning, 279–81, 286–87, 288, 577, 578
Closed circulatory system, 448, 552, 553–54
Clostridium botulinum, 373
Clotting, 284, 563–65
Clotting factor, 284
Clover, 409
Clownfishes, 761
Club fungi, 392
Club mosses, 405, 410–11
Cnidarians, 417, 419–21, 426, 553, 665
Coastal communities, 766–68
Cobamide coenzymes, 112
Cocaine, and abuse, 648–49
Coccolithophore, 359
Coccus, 371, 372
Coccyx, 667
Cochlea, 659, 660
Code, and genes, 252–54
Codominance, 197
Codon, 257, 259
Coelocanth, 451, 452
Coelom, 424, 434, 443, 665, 714
Coelomates, 431
Coenzyme A, 143, 145
Coenzymes, 57, 112–14, 143, 145
Coevolution, of plants and pollinators, 518–19
Cofactor, 112
Cohesion, 34, 35, 497–98
Cohesion-tension model of xylem transport, 497–98

Coitus interruptus, 706
Cole, K. S., 637
Coleoptile, 517, 523, 525
Collagen, 57, 540, 724
Collagen fibers, 540
Collar cells, 417
Collecting duct, 625, 626
Collenchyma cells, 476, 477
Colon, 597
Colonial green algae, 385
Color, of skin, 224, 225
Color blindness, 228–29
Color vision, 658
Colostrum, 705
Columbia University, 204
Columnar epithelium, 539
Commelinales, 4
Commensalism, 373, 374, 759, 760–61
Commensalistic bacteria, 373, 374
Common descent, 308–10
Communal groups, 741
Communication, and behavior, 738–40
Community, 3, 744, 754–64, 766–69, 773, 781
Companion cells, 478
Comparative anatomy, 310–12
Comparative biochemistry, 311–12
Comparative embryology, 311
Compartmentalization, 66
Competition, 755–56
Competitive exclusion principle, 756
Competitive inhibition, 111–12
Complementary base pairing, 55, 56, 239, 240
Complementary DNA (cDNA), 280, 289, 368, 369
Complement system, 576–77
Complete gut, 590–91
Complete linkage, 207
Complete proteins, 598
Complete ventilation, 610
Complex carbohydrates, 598
Compound eye, 437, 440, 654, 655
Compound fruits, 514
Compounds, and molecules, 29–33
Concentration, and substrates, 111
Concentration gradient, 89
Concepts and critical thinking, answers to, A-7–A-18
Conclusions, 14
Condensation, and hydrolysis, 44
Conditioning, classical and operant, 734–35
Condom, 706
Condor, 412
Conduction, 639
Cones, 397, 406, 655, 656, 657, 658
Configurations, 29
Conidia, 391
Coniferophyta, 410–11
Conifers, 397, 405–7, 410–11
Conjugation, 282, 283, 371, 382, 385–86
Connective tissue, 539, 540–42
Connell, Joseph, 755
Conscious brain, 646
Conservation, 412, 776–77
Constant regions, 578
Consumers, 8, 781, 783
Contact inhibition, 271
Continental drift, 344
Continuous feeders, 589
Contour farming, 786
Contour feathers, 454

Index

Trichonympha collaris, 383
Triglycerides, 48, 57
Triiodothyronine, 680
Trilobite, 438
Trinity College, 340
Triplet code, 253, 311–12
Trisomy, 209, 210, 217, 218
tRNA. *See* Transfer RNA (tRNA)
Trophic levels, 782
Trophoblast, 718
Tropical conservation, 776–77
Tropical Forest Resources Assessment Project, 777
Tropical rain forests, 777–78, 788, 793–95
Tropism, 529
Trout, 451
Trp operon, 268
True-breeding plants, 183
True coelom, 424
Trypanosome infection, 383
Trypsin, 110, 111, 594
Tryptophan, 253, 265, 268
TSH. *See* Thyroid-stimulating hormone (TSH)
T-tubules, 671
Tubal ligation, 706
Tube cell, 510
Tubers, 486
Tube-within-a-tube body plan, 421, 423, 424, 435, 443, 448
Tubeworms, 373, 435
Tubular secretion, 625–26
Tubules, 623–24, 625, 626
Tumor angiogenesis factor, 690
Tumor necrosis factor, 284
Tumors, 272, 273
Tumor supressor genes, 273, 274
Tundra, 765, 773–75
Tune, 451
Tunicate, 443
Turbellaria, 421
Turgor pressure, 92, 93, 498
Turitella, 302
Turner syndrome, 217, 218, 219
Turnips, 481
Turtles, 307, 453
T virus, 234, 236
Twins, 225
Tympanic membrane, 659, 660
Tympanum, 441
Tyrosine, 51, 253, 274
Tyrosine phosphatases, 274

U

Ulcers, 584
Ulna, 667
Ultimate causes, 731
Ultrasound, 226, 227
Ultraviolet (UV) radiation, 260, 267, 273, 544
Ulva, 386, 388, 398
Unconscious brain, 644–46
United Nations (UN), 777
United States Bureau of Land Management, 793
United States Department of Agriculture, 531, 598, 786, 793
United States Department of Health and Human Services, 598
United States Environmental Protection Agency (EPA), 802

United States National Museum Natural History, 776
Unit membrane model of membrane structure, 86
Units of length, A-4
Units of volume, A-5
Units of weight, A-5
Unity of life, 11
Universal solvent, 33
University of Arizona, 487
University of California, Berkeley, 467
University of Colorado, 356
University of Halle, 637
University of Tennessee Noise Laboratory, 661
University of Toronto, 687
University of Vienna, 183
Unsaturated fatty acids, 47
Unwinding, 241
Uptake of water, 495–96
Uracil, 251, 253
Uranium, 345
Urban sprawl, 793
Urchins, 417, 442
Urea, 108, 621, 627
Urease, 108
Ureter, 624
Urethra, 624, 697
Urethritis, NGU. *See* Nongonoccal urethritis (NGU)
Uric acid, 621, 627
Uridine, 55
Urinary bladder, 624
Urine, 624, 625–30
Urochordata, 443
Uropods, 440
Uterine cycle, 703
Uterus, 700, 701
Utricle, 659, 660

V

Vaccines, 283–84, 580, 708
Vacuoles, 41, 67, 68, 69, 74, 81, 381, 553
Vagina, 700, 701
Vaginal infection, 393
Vagus nerves, 640
Valine, 51, 251
Valve, cross section of, 562
Van Helmont, Jean-Baptiste, 126
Variability of bases, 238, 240
Variable regions, 578
Variables, experimental and dependent, 15
Variation, 305, 324, 327–28
Varicose veins, 561
Variety, in cells, 61–66, 67
Vascular bundles, 483–84
Vascular cambium, 479, 484
Vascular cylinder, 481
Vascularization, 272
Vascular pathways, 560–61
Vascular plants, 399–404
Vascular system, 441
Vascular tissue, 475, 477–78, 481
Vas deferentia, 697
Vasectomy, 706
Vasopressin, 680
Vectors, 279, 289
Vegetarian birds, 454
Vegetation, 10
Vegetative organs and major tissues of plants, 492

Vegetative propagation, 515–19
Veins, of body, 555
Veins, of leaf, 489, 491
Venae cavae, 557, 560
Ventilation, complete and incomplete, 610
Ventral solid nerve cord, 437
Ventricles, 642
Venus's-flytrap, 490, 491
Vertebrae, 666, 667
Vertebral column, 448, 664, 666, 667
Vertebrates, 416
  brains among, 645
  circulatory system, 555–57
  and diversity, 447–71
  endocrine system, 678–90
  evolution of, 448–58
  phylogenetic tree of, 449
  sexual reproduction, 696
Vesicles, 67, 69, 71
Vespula arenaria, 758
Vessel elements, 477–78
Vestigial structures, 311, 667
Vicia fava, 499
Victoria, Queen, 229, 230
Vidarabine, 370
Vietnam War, 524, 793
Villi, 594, 595
Viral DNA, 279–80, 282–83, 289
Viral encephalitis, 370
Viral hepatitis, 597
Virchow, Rudolf, 61
Viroids, 368–70
Viruses, 13, 89–90, 234, 236, 237, 366–70
  and animal viruses, 368
  and cancer, 271–75
  and cloning, 279–80
  HIV. *See* Human immunodeficiency virus (HIV)
  HPV. *See* Human papilloma virus (HPV)
  and life cycles, 367–70
  and Monera, 365–78
  and retroviruses as vectors, 289
  and STD, 706
  and vectors, 279, 280, 282–83
  and in vivo therapy, 288
Visceral mass, 431, 432, 434
Visible light, 122, 123
Vision, 655–58
Visual communication, 738
Vitamins, 112, 113, 600–603, 658
Vitis vinifera, 524
Vitreous humor, 655
Voice box, 611, 612
Voltage, 637
Volume, units of, A-5
Volvox, 380, 385, 387
Von Baer, Karl E., 714
von Seysenegg, Erich Tschermak, 202
Vulva, 700, 701

W

Wading birds, 454
Wallace, Alfred Russel, 305, 308
Wallace's Line, 308
Wang, Andrew, 243
Warblers, 756
Warm chemistry, 115
Warming, global, 800